Theory of Fusion Plasmas, 13th Joint Varenna-Lausanne International Workshop 2012

Journal of Physics: Conference Series Volume 401

Varenna, Italy
27-31 August 2012

ISBN: 978-1-62276-867-7
ISSN: 1742-6588

Printed from e-media with permission by:

Curran Associates, Inc.
57 Morehouse Lane
Red Hook, NY 12571

Some format issues inherent in the e-media version may also appear in this print version.

Printed by Curran Associates, Inc. (2013)

For permission requests, please contact the Institute of Physics
at the address below.

Institute of Physics
Dirac House, Temple Back
Bristol BS1 6BE UK

Phone: 44 1 17 929 7481
Fax: 44 1 17 920 0979

techtracking@iop.org

Additional copies of this publication are available from:

Curran Associates, Inc.
57 Morehouse Lane
Red Hook, NY 12571 USA
Phone: 845-758-0400
Fax: 845-758-2634
Email: curran@proceedings.com
Web: www.proceedings.com

TABLE OF CONTENTS

PREFACE

The 2012 joint Varenna-Lausanne international workshop on the theory of fusion plasmas has been very fruitful. A broad variety of topics were addressed, as usual covering turbulence, MHD, edge physic, RF wave heating and a taste of astrophysics. Moreover the scope of the meeting was extended this year to include the physics of materials and diagnostics for burning plasmas. This evolution reflects the complexity of problems at hand in fusion, in particular in the context of ITER construction. Long-standing problems without immediate consequences have sometimes become an urgent matter in that context. One may quote for instance the choice of plasma facing components or the design of control systems.

Another characteristic of the meeting is the interplay between various domains of plasma physics. For instance MHD modes are now currently investigated with gyrokinetic codes, kinetic effects are more and more included in MHD stability analysis, and turbulence is now accounted for in wave propagation problems. This is the proof of cross-fertilization and it is certainly a healthy sign in our community.

Finally introducing some novelty in the programme does not prevent us from respecting the traditions of the meeting. As usual a good deal of the presentations were dedicated to numerical simulations. Combining advanced numerical techniques with elaborated analytical theory is certainly a trademark of the Varenna-Lausanne conference, which was respected again this year.

The quality and size of the scientific production is illustrated by the 26 papers which appear in the present volume of *Journal of Physics: Conference Series*, all refereed. We would also like to mention another set of 20 papers to be published in *Plasma Physics and Controlled Fusion*. We hope the readers will enjoy this special issue of JPCS and the one to come in PPCF.

Xavier Garbet and Olivier Sauter
October 26, 2012

Peer review statement

All papers published in this volume of *Journal of Physics: Conference Series* have been peer reviewed through processes administered by the proceedings Editors. Reviews were conducted by expert referees to the professional and scientific standards expected of a proceedings journal published by IOP Publishing.

Implementing zero-banana-width quasilinear operator for fast ICRF simulations

R Bilato, M Brambilla

Max Planck Institute for Plasma Physics, EURATOM Association, Boltzmannstr. 2, 85748 Garching, Germany

Z Jiang

State Key Laboratory of Advanced Electromagnetic Engineering and Technology, Huazhong University of Science and Technology, Wuhan, 430074, China

E-mail: `roberto(dot)bilato(at)ipp(dot)mpg(dot)de`

Abstract. Numerical simulations of plasma heating with waves in the ion cyclotron range of frequencies (ICRF) require to iteratively couple a solver for wave propagation in plasmas with a solver of the quasilinear kinetic equation. Among the codes developed for this purpose, the `TORIC-SSFPQL` package is characterized by its high execution speed. The kinetic code `SSFPQL`, however, was based on a somewhat simplified physical model, in which some important effects of the toroidal geometry were omitted. We have recently improved this model by taking into account in the zero–banana–width limit the influence of toroidal trapping on the ICRF quasilinear operator. To make the extended model compatible with the representation of the ion distribution functions as a truncated series of Legendre polynomials in the velocity pitch-angle adopted in `SSFPQL`, a special approach based on the multiprecision arithmetic had to be developed. We describe these new developments, and present first results obtained with the improved model.

1. Introduction

Plasma heating in the ion cyclotron range of frequencies (ICRF) has been extensively used in the past and present fusion devices, and should be available from the very beginning of the ITER operation [1]. This has motivated a considerable effort in simulating ICRF wave propagation and absorption. Since the wavelengths at these frequencies are comparable with the characteristic lengths of the plasma equilibrium profiles, asymptotic methods, such as geometrical optics, are not applicable, and the wave equations have to be solved numerically in full toroidal geometry, implying rather high execution costs. A brief review of the handful of full-wave codes developed and used for this purpose in the last few years is given by Budny and co-authors in [2].

During ICRF heating the distribution functions of the resonant species deviate appreciably from Maxwellians because of the production of important populations of energetic ions. In turn, suprathermal ions influence the propagation and absorption of the waves. As a consequence, a fully consistent simulation of ICRF heating requires the simultaneous solution of the full–wave equations and of the quasilinear kinetic equation for the heated ion species [3–6], thereby further increasing the computational costs. The kinetic code is also needed to predict the collisional redistribution of the absorbed high-frequency (hf) power among different species.

The numerical complexity of the problem can be considerably reduced by omitting in the kinetic equation some details of the physics which are important to determine the fate of the most energetic ions produced by resonant ion-cyclotron (IC) interactions, but have only little influence on wave propagation and absorption. The kinetic code `SSFPQL` [7] which has been coupled to the full-wave code `TORIC` [8] looks for the steady-state solution of the surface-averaged kinetic equation, which amounts to neglect the final radial width of the guiding-center orbits of the heated ions. More importantly, in its first implementation the evaluation of the quasilinear (QL) diffusion coefficient was simplified by neglecting the effect of toroidicity on the parallel ion motion, thereby assuming that all ions cross IC resonances with constant velocity. In reality, the mirror force due to the inhomogeneity of the confining tokamak magnetic field has the consequence that some ions revert their parallel velocity near or even before reaching the resonance. The effect of toroidal trapping on wave absorption, although not dominant, is not really negligible.

In this note we discuss the control of round-off and truncation errors when solving the kinetic equation by expanding the solution in Legendre polynomials, and the extension of the numerical scheme used by the `SSFPQL` code in order to take into account toroidal trapping. In Section (2) we briefly recall the `SSFPQL` model and its implementation. In order to include the effects of toroidal trapping in the numerical scheme of this code it is necessary to avoid the accumulation of round-off errors in the numerical evaluation of certain recursive relations needed to build the QL diffusion coefficient in the Legendre representation. This has been achieved by using the multiprecision arithmetic facility offered by the `GMP` library [9]. In Section (3) the sources of truncation are discussed, and two empirical criteria to control them are proposed. In Section (4) we present the implementation of the effects of toroidal trapping in the numerical scheme of `SSFPQL`. An example of quasilinear distribution function in which these effects are taken into account is presented there. In the last Section we draw some conclusions and summarize what has yet to be done to complete this work.

2. The surface–averaged kinetic equation solved by `SSFPQL`

In the `TORIC-SSFPQL` package, the code `TORIC` solves the wave equation in axisymmetric toroidal plasmas [8], and the code `SSFPQL` evaluates the steady-state solution of the surface-averaged Fokker-Planck (FP) equation [7], written as balance between collisions, radio-frequency heating, and any other power source or sink S_j, including, if present, NBI heating [10]:

$$0 = \left(\frac{\partial F^{(i)}}{\partial t}\right)_{\rm coll} + \left(\frac{\partial F^{(i)}}{\partial t}\right)_{\rm rf} + S^{(i)}\,. \tag{1}$$

The linearized collisional operator describing test ions of species i colliding with a background Maxwellian plasma can be written as the sum of two differential operators acting separately on the speed v and on the pitch angle $\mu = v_{\|}/v$:

$$\left(\frac{\partial F^{(i)}}{\partial t}\right)_{\rm coll} = \mathcal{C}_v(F^{(i)}) + \mathcal{C}_\mu(F^{(i)})\,.$$

The operator $\mathcal{C}_\mu(F^{(i)})$ describing pitch-angle scattering is proportional to the Legendre operator L_μ,

$$\mathcal{C}_\mu(F^{(i)}) = \frac{\Theta_c(v)}{2v^3}\, L_\mu(F^{(i)}) = \frac{\Theta_c(v)}{2v^3}\, \frac{\partial}{\partial\mu}\left[(1-\mu^2)\,\frac{\partial}{\partial\mu}\right] F^{(i)}(v,\mu)\,, \tag{2}$$

whose eigenfunctions are the Legendre polynomials $P_n(\mu)$. This peculiarity of the operator $\mathcal{C}_\mu$ suggests to represent the solution of Equation (1) as a superimposition of Legendre polynomials,

$$F^{(i)}(v,\mu;\psi) = \sum_{i=0} \tilde{F}_n^{(i)}(v;\psi)\ P_n(\mu)\ . \tag{3}$$

The two dimensional partial differential equation (1) is thereby reduced to a set of coupled ordinary-differential equations (ODEs) for the coefficients $\tilde{F}_n^{(i)}$ of the expansion.

The QL operator describing IC heating can be approximated as a purely perpendicular diffusion operator in velocity space,

$$\left(\frac{\partial F^{(i)}}{\partial t}\right)_{\rm rf} = \frac{1}{v_\perp}\frac{\partial}{\partial v_\perp}\left(v_\perp\ D_p(v_\perp)\ \frac{\partial F^{(i)}}{\partial v_\perp}\right)\ , \tag{4}$$

with the QL coefficient $D_p(v_\perp)$ for the pth harmonic heating given by

$$D_p(v_\perp) = D_0\ \left|J_{p-1}(\xi_\perp v_\perp) + \lambda_p\ J_{p+1}(\xi_\perp v_\perp)\right|^2\ . \tag{5}$$

Here $J_p(x)$ are the Bessel functions of the first kind, $v_\perp = v\sqrt{1-\mu^2}$, $\xi_\perp = k_\perp v_{\rm thi}/\Omega_{\rm ci}$, with $\Omega_{\rm ci} = Z_i eB/m_i c$ the cyclotron angular frequency, and $\lambda_p = (E_-/E_+)_{\rm res}$ is the wave polarization at the resonance, with E_- and E_+ the right and left circular components of the wave electric field. Transforming the QL operator to spherical coordinates and projecting on $P_n(\mu)$, the hf contribution to the equation for $\tilde{F}_n^{(i)}(v;\psi)$ becomes

$$\left(\frac{\partial \tilde{F}_n^{(i)}}{\partial t}\right)_{\rm rf} = \sum_m \frac{1}{v^2}\left\{\frac{\partial}{\partial v}\left[v\left(\tilde{D}_p^{(0,0)}(n,m,v)\ v\frac{\partial}{\partial v} - \tilde{D}_p^{(0,1)}(n,m,v)\right)\right]\right.$$
$$\left.\left(\tilde{D}_p^{(1,0)}(n,m,v)\ v\frac{\partial}{\partial v} - \tilde{D}_p^{(1,1)}(n,m,v)\right)\right\}\tilde{F}_m^{(i)}(v) \tag{6}$$

with

$$\tilde{D}_p^{(j,k)}(n,m,v) = \int_{-1}^{+1}\left(\mu^j\frac{{\rm d}^j}{{\rm d}\mu^j}P_n\right)\ (1-\mu^2)\ D_p(v,\mu)\ \left(\mu^k\frac{{\rm d}^k}{{\rm d}\mu^k}P_m\right){\rm d}\mu \tag{7}$$

and $D_p(v,\mu) = D_p(\xi_\perp v\sqrt{1-\mu^2})$. The evaluations in closed form of the integrals (7) involving the Bessel functions (5) is made possible by using the multiplication theorem [11],

$$J_q(\xi_\perp v\sqrt{1-\mu^2}) = (1-\mu^2)^{q/2}\sum_{k=0}^{\infty}\frac{\mu^{2k}}{k!}\left(\frac{\xi_\perp v}{2}\right)^k\ J_{q+k}(\xi_\perp v)\,. \tag{8}$$

Using (8) in (7) we have

$$\tilde{D}_p^{(j,k)}(n,m,v) = D_0\sum_{s=0}^{\infty}\left[\tilde{Q}_{p-1}^{(j,k)}(n,m,s)\mathcal{J}_{p-1,p-1}^{s}(\xi_\perp v) + 2{\rm Re}(\lambda_p)\ \tilde{Q}_p^{(j,k)}(n,m,s)\mathcal{J}_{p-1,p}^{s}(\xi_\perp v)\ +\right.$$
$$\left.|\lambda_p|^2\ \tilde{Q}_{p+1}^{(j,k)}(n,m,s)\mathcal{J}_{p,p}^{s}(\xi_\perp v)\right]\ , \tag{9}$$

where

$$\tilde{Q}_p^{(j,k)}(n,m,s) = \int_{-1}^{+1}\left(\mu^j\frac{{\rm d}^j}{{\rm d}\mu^j}P_n\right)\ (1-\mu^2)^{1+p}\ \mu^{2s}\ \left(\mu^k\frac{{\rm d}^kP_m}{{\rm d}\mu^k}\right){\rm d}\mu \tag{10}$$

and

$$\mathcal{J}^{s}_{p,q}(\xi_\perp v) = \left(\frac{\xi_\perp v}{2}\right)^s \sum_{r=0}^{s} \frac{1}{r!\,(s-r)!} J_{p+r}(\xi_\perp v)\, J_{q+s-r}(\xi_\perp v)\,. \tag{11}$$

Once the coefficients of the kinetic equations are calculated the set of ODEs in the variable v are solved with cubic–Hermite finite-element technique, the same used in `TORIC` code.

The critical numerical issue here is the evaluation of the integrals $\tilde{Q}^{(j,k)}_p(n,m,s)$. More precisely, we have to calculate the following three sets of integrals

$$X_p(n,m,k) = \int_{-1}^{+1} P_n(\mu)\,(1-\mu^2)^{1+p}\,\mu^k\,P_m(\mu)\mathrm{d}\mu \tag{12}$$

$$Y_p(n,m,k) = \int_{-1}^{+1} P_n(\mu)\,(1-\mu^2)^{1+p}\,\mu^k\left(\mu\frac{\partial P_m}{\partial\mu}\right)\mathrm{d}\mu \tag{13}$$

$$Z_p(n,m,k) = \int_{-1}^{+1} \left(\mu\frac{\partial P_n}{\partial\mu}\right)(1-\mu^2)^{1+p}\,\mu^k\left(\mu\frac{\partial P_m}{\partial\mu}\right)\mathrm{d}\mu \tag{14}$$

Since Legendre polynomials are characterized by oscillations, whose period decreases when approaching the domain boundaries ± 1 and with increasing order of the polynomial, it is unfeasible to evaluate these integrals numerically with sufficient accuracy. They satisfy, however, a set of two-dimensional, two-way recurrence relations. Unfortunately these relations are not stable against round-off error in the forward direction [12]. While totally negligible for Legendre polynomials up to a certain order, the round-off errors rapidly spoil any further evaluation when this order is exceeded. Working in IEEE 64-bit floating-point arithmetic typical of double-precision Fortran implementations, the maximum order of the polynomials which can be included is thereby limited to between 20 and 25. While this number might seem rather large, it turns out to be insufficient to guarantee convergence of the solution to a positive definite distribution function in a sufficiently large energy domain as soon as the QL operator is dominant over the collisional one. Empirically, this occurs when the local power deposited in the ions by IC resonances exceeds a fraction of Watt/cm^3, the precise limit depending on the collisionality of the plasma. The limitation is even more severe, if the simultaneous presence of NBI drives the distribution functions further away from Maxwellians already in an intermediate range of energies.

An important property of the recurrence relations (12,13,14) is that only rational numbers are involved both in the coefficients of the relation themselves and in the starting values [4]. Therefore, the use of the arithmetic of rational numbers offers a convenient way of circumventing the round-off error problem: the recurrence relations are evaluated using the rational representation of the numbers, and only one final division is needed to convert the result to a floating-point number. This makes the round-off errors equal to machine precision. One faces the problem, however, that the recurrence relations involve fractions of very quickly increasingly large integers, which cannot be represented with the integer arithmetic of standard compilers. This obstacle is removed by using the long–integer arithmetic of the multi-precision `GMP` library [9]. A first implementation of long–integer arithmetic for this purpose was made using `Mathematica` [13]. This, however, implied preparing a very large table of data to be read in the initialization phase of `SSFPQL`; the order of the Legendre expansion, moreover, cannot exceed the order available in this table. In addition, for tables extending to polynomials of order sixty or more, it is appreciably faster to calculate the integrals with the `GMP` library rather than to read them from the external file created with `Mathematica`.

As an example, Figure (1) shows the execution time needed to evaluate $X_p(n,m,k)$ in the range $n,m \le N_L$, $k < 2(N_L+100)+3$, and $0 \le p \le 3$, varying the maximum order N_L of the Legendre polynomials included. Compared are the evaluation in Fortran using double and quadruple floating-point precision, and with the GMP library. In all three cases the computational time increases quadratically with N_L, but the GMP library is almost an order of magnitude slower than Fortran in quadruple precision. Nevertheless, the increased overhead for SSFPQL is still very tiny, and is more than compensated by the gain in accuracy. In Figure (1) the results of the Fortran evaluation are plotted with dashed line when the accumulated error exceeds 100%. In practice, use of the GMP library is the only solution for $N_L > 20$.

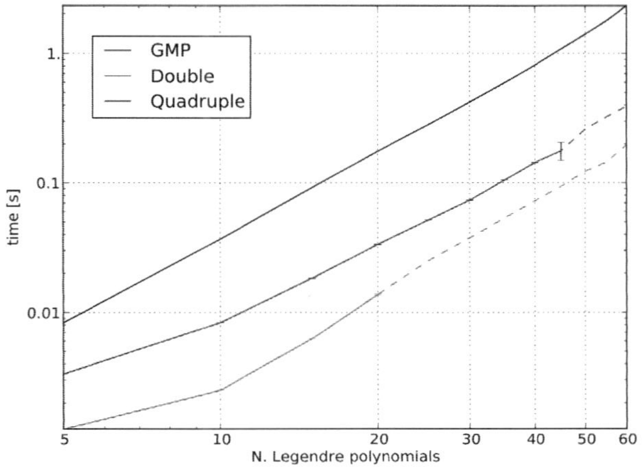

Figure 1. Computational time for double (`selected_real_kind(15, 307)`), quadruple (`selected_real_kind(33, 4931)`) floating-point arithmetic, and GMP multiprecision, required to calculate $X_p(n, m, k)$ as function of the maximum number of Legendre polynomials. Here $0 \le p \le 3$ and $0 \le k \le 2(N_L + 100) + 3$. The solid lines changes to dashed when the error becomes larger than 100%; in the GMP case the accuracy is always equal to the machine precision.

3. Controlling the truncation errors

In addition to the accumulation of round-off errors in the recurrence relations, the precision of SSFPQL solutions depends also on the errors introduced by truncating the sum (8) and, thus, the expansion of the QL diffusion coefficient in Bessel polynomials. In turn, the minimum number of polynomials which have to be included to ensure a good convergence of the Legendre representation of the QL operator determines how far the recursive relations discussed in the previous section have to be pushed.

While the r.h.s. of Equation (8) reduces to a single term for $\mu = 0$, convergence is slowest at the extrema $\mu = \pm 1$ of the velocity pitch-angle variable (ions with purely parallel velocity). In particular, at these points the slowest convergence of the series is when $q = 0$, where (8) becomes

$$\sum_{k=0}^{K=\infty} \frac{1}{k!} \left(\frac{x}{2}\right)^k J_k(x) = 1 \,. \tag{15}$$

The requirement that (15) is satisfied within a given numerical accuracy selects the minimum number of terms to be retained in the numerical evaluation of the series (8). Figure (2.left) shows the difference between the truncated l.h.s. of (8) (evaluated in double precision) and unity as function of the argument x of the Bessel functions, and the number K of the terms

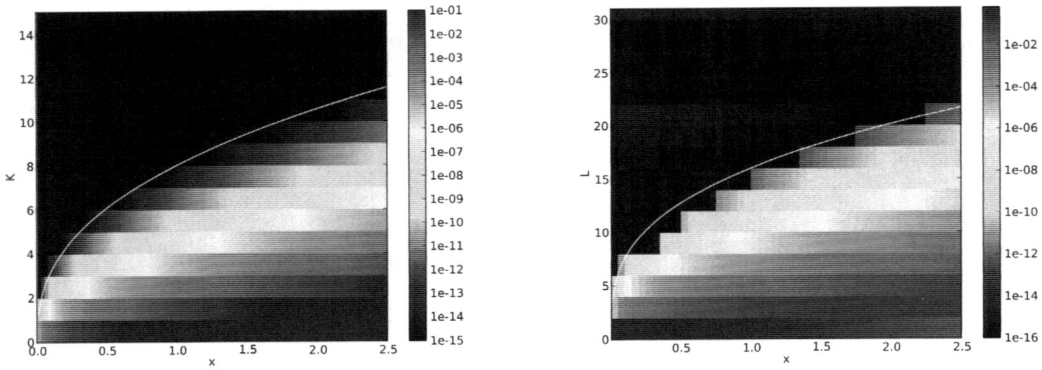

Figure 2. Left: Difference between the l.h.s. of (15) and unity as function of x and K. Right: l.h.s. of (18) as function of x and L. The white lines show the criteria (16) and (19), respectively.

kept. From this figure we can deduce the heuristic criterion

$$K \gtrsim 8 \cdot x^{2/5} . \tag{16}$$

In a similar way, we can derive a heuristic criterion for the minimum number of Legendre polynomials necessary to represent the QL diffusion coefficient. For this purpose, we note that due to the use of spherical velocity coordinates the integrand of Equation (7) contains the factor $(1-\mu^2)$, and thus vanishes at $\mu \pm 1$. However, since $P_n(\pm 1) = (\pm 1)^n$ for all n, convergence of the Legendre series to zero can be critically slow at these points (this is analogous to the Gibbs phenomenon of Fourier series). The worse situation is for $p=0$, in which the expansion

$$(1-\mu^2)\, J_0^2(x\sqrt{1-\mu^2}) = \sum_{n=0}^{L=\infty} \left(n+\frac{1}{2}\right) P_n(\mu) \sum_{k=0}^{K=\infty} \mathcal{J}_{0,0}^k(x)\, \tilde{Q}_0^{(0,0)}(n,0,k) \tag{17}$$

becomes for $\mu = \pm 1$

$$\sum_{n=0}^{L=\infty} (\pm 1)^n \left(n+\frac{1}{2}\right) \sum_{k=0}^{K=\infty} \mathcal{J}_{0,0}^k(x)\, \tilde{Q}_0^{(0,0)}(n,0,k) = 0 . \tag{18}$$

From the identity (18) we can derive a criterion for the minimum number L of Legendre polynomials to be used in order to approximate the QL diffusion coefficient with a given numerical accuracy. Figure (2.right) shows the l.h.s. of Equation(18) as function of x and L; the number K of terms in the inner sum is chosen so as to satisfy the criterion (16). A rough rule of thumb to achieve double precision numerical accuracy is

$$L \gtrsim 16 \cdot x^{1/3} . \tag{19}$$

To gain a feeling of what the two criteria (16) and (19) imply for ICRF simulations, we can assume $\omega \approx p\, \Omega_{ci}$ and $k_\perp \approx \omega / v_A$, with v_A the Alfvén speed: then $\xi_\perp \approx p\ 0.063\ \sqrt{n_i\, T_i}/B_0$, with n_i, B_0 and T_i respectively in 10^{19} m^{-3}, Tesla and keV. For typical plasma parameters, $\xi_\perp \approx p\, 0.1$. We conclude that to obtain an accurate Legendre representation of the quasilinear operator up to values of $v/v_{\rm thi}$ of a few tens, the number of Legendre polynomials and of the terms in (8) has to be of the order of a few tens. This is only a necessary condition to obtain

Journal of Physics: Conference Series **401** (2012) 012001 doi:10.1088/1742-6596/401/1/012001

a well converged and positive-definite solution of the FP equation in the same velocity range. Experience shows that as soon as the heating rate is faster than the collisional relaxation the number of polynomials to be included exceeds the values foreseen by (19), and increases rapidly with increasing hf power per ion.

4. Toroidal effects in the zeroth–banana–width limit

In a tokamak the combination of the (nearly) horizontal gradient of the confining magnetic field B with the adiabatic invariance of the magnetic moment $\mu_b = mv_\perp^2/2B$ of the charged particles causes a toroidal modulation of their parallel velocity and the well-known trapping of a class of particles in the outer part of the device. Sufficiently deeply trapped ions have their turning point to the low-field side of the surface $\omega = p\Omega_c$, and do not see the resonance until diffused by collisions on a less deeply trapped orbit. To describe the effect of trapping in `SSFPQL`, the QL diffusion coefficient $D_p(v_\perp)$ has to be multiplied by a function describing its dependency on the velocity pitch-angle:

$$D_p(v_\perp;\psi) = D_0\; g\,(\mu;\psi)\,|J_{p-1}(\xi_\perp v_\perp) + \lambda_P\; J_{p+1}(\xi_\perp v_\perp)|^2 \qquad (20)$$

(in reality g depends also weakly on the ion energy; to avoid an excessive slowing down of the execution time, however, in `SSFPQL` an average value has to be used). The function g approaches unity for well passing particles ($\mu \to \pm 1$), has a maximum for the value of μ characterizing ions which turn close to resonance, and thus spend more time interacting with the waves, and decreases rapidly to zero for smaller values of μ, i.e. for ions reflected before reaching the resonance. An example is shown in Figure (3.left). To include the factor g in the numerical

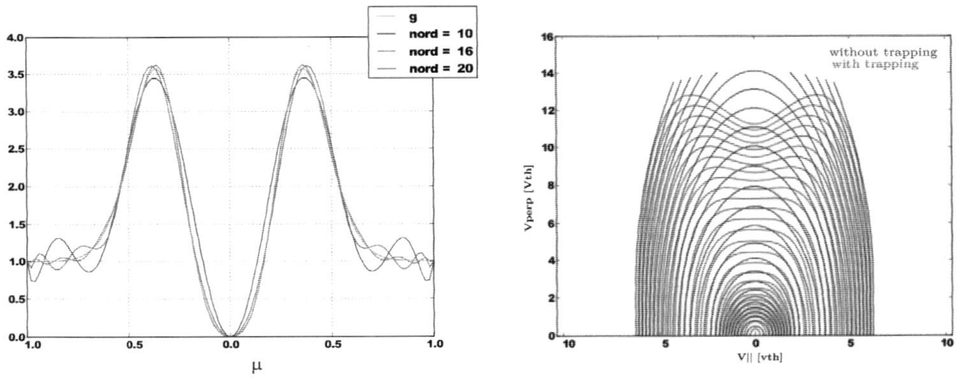

Figure 3. Left: Factor modeling trapping effects, and Lagrange-Chebyshev interpolants corresponding to three values of N_g. Right: Distribution function of hf heated species with and without trapping effects.

scheme of `SSFPQL`, we note that if g is approximated with a polynomial G_{N_g},

$$g(\mu;\psi) \simeq G_{N_g}(\mu) = \sum_{q=0}^{N_g} g_q(\psi)\;\mu^q\,, \qquad (21)$$

the integrals X_p in Equation (12) become, for example,

$$X_p^{(g)}(n,m,k;\psi) = \sum_{q=0}^{N_g} g_q(\psi)\; X_p(n,m,k+q) \qquad (22)$$

and similar expressions hold for Y_p and Z_p. It is sufficient, therefore, to evaluate the tables of these integrals without the factor g in the initialization phase of the run, and to use the convolution (21) with the appropriate function $g(\mu;\psi)$ to build the values needed on each magnetic surface ψ.

If the 'interpolant' polynomial G_{N_g} is looked for in the standard form (21), however, it is difficult to achieve a satisfactory accuracy. Instead, we use the Lagrange interpolant polynomial based on the set of $N_g + 1$ zeros of the Chebyshev polynomial $T_{N+1}(\mu)$ [12]. This minimize the Runge phenomenon, i.e. the tendency of the interpolant to oscillate between the base point with increasing amplitude towards the extremes of the interval of interpolation. Starting from the construction rules of the Lagrange interpolant, moreover, we have derived an efficient recurrence relation for the coefficients of G_{N_g} in the standard representation. Figure (3.left) shows three interpolants for three different values of N_g. The impact of the trapping effects on the solution of the FP equation is illustrated in Fig. (3.right). It is responsible for the characteristic rabbit-ears in the contour plot of the distribution functions, and for a reduction of number of fast ions produced by IC heating, an important issue for off-axis heating schemes.

5. Conclusion

We have discussed the control of truncation and round-off errors in the solution of the quasilinear kinetic equation based on the expansion in Legendre polynomials, and the extension of this technique in `SSFPQL` in order to take into account toroidal trapping effects in the limit of the zero–banana–width approximation. The extension of the interface between `TORIC` and `SSFPQL` to accept distribution functions modified by toroidal trapping is in progress.

Acknowledgment

We are grateful to Stephan Brunner for his careful reading of the manuscript and the useful discussions that followed. Work by Z.J. was supported by a grant of the National Nature Science Foundation of China under Contract No 115450 and No 10990214.

References

[1] Gormezano C, Sips A, Luce T, Ide S, Becoulet A, Litaudon X, Isayama A, Hobirk J, Wade M, Oikawa T, Prater R, Zvonkov A, Lloyd B, Suzuki T, Barbato E, Bonoli P, Phillips C, Vdovin V, Joffrin E, Casper T, Ferron J, Mazon D, Moreau D, Bundy R, Kessel C, Fukuyama A, Hayashi N, Imbeaux F, Murakami M, Polevoi A and John H S 2007 *Nuclear Fusion* **47** S285–S336

[2] Budny R, Berry L, Bilato R, Bonoli P, Brambilla M, Dumont R, Fukuyama A, Harvey R, Jaeger E, Indireshkumar K, Lerche E, McCune D, Phillips C, Vdovin V, Wright J and members of the ITPA-IOS 2012 *Nuclear Fusion* **52** 023023

[3] Jaeger E, Harvey R, Berry L, Myra J, Dumont R, Phillips C, Smithe D, Barrett R, Batchelor D, Bonoli P, Carter M, D'azevedo E, D'ippolito D, Moore R and Wright J 2006 *Nuclear Fusion* **46** S397–S408

[4] Brambilla M and Bilato R 2009 *Nuclear Fusion* **49** 085004

[5] Green D L, Jaeger E F, Berry L A and the rf SciDac team 2009 *Journal of Physics: Conference Series* **180** 012058

[6] Jucker M, Cooper W and Graves J 2012 *Nuclear Fusion* **52** 013015

[7] Brambilla M 1994 *Nuclear Fusion* **34** 1121–43

[8] Brambilla M 1999 *Plasma Physics and Controlled Fusion* **41** 1–34

[9] GNU multiple precision arithmetic library `http://gmplib.org/`

[10] Bilato R, Brambilla M, Maj O, Horton L, Maggi C and Stober J 2011 *Nuclear Fusion* **51** 103034

[11] Truesell C 1950 *Proc. N.A.S.* **36** 752

[12] Gautschi W 2012 *Numerical Analysis* (Birkhäuser Boston)

[13] Wolfram S Mathematica `http://www.wolfram.com/mathematica/`

Journal of Physics: Conference Series **401** (2012) 012002 doi:10.1088/1742-6596/401/1/012002

Magnetic island evolution under the action of electron cyclotron current drive

D Borgogno[1], L Comisso[1,2], D Grasso[1,2], E Lazzaro[3]

[1] Dipartimento Energia, Politecnico di Torino, Italy
[2] Istituto Sistemi Complessi - CNR, Roma, Italy
[3] Istituto di Fisica del Plasma "P.Caldirola", Associazione Euratom-ENEA-CNR, Milano, Italy

Abstract. The magnetic island evolution under the action of an externally current generated by electron cyclotron (ECCD) wave beams is studied using a reduced resistive magnetohydrodynamics (RRMHD) plasma model. We found interesting and somewhat unexpected features of the actual nonlinear 2-D evolution of the magnetic perturbation depending on the injection time of the radio frequency control. In particular in the linear phase of the magnetic island growth we observe that the complete annihilation of the island width is followed by a spatial phase shift of the island, referred as "flip"instability. On the other hand, a current drive deposition in the Rutherford regime can be accompained by the occurrence of a Kelvin-Helmholtz type shear flow instability, responsible for the onset of a plasma turbulent behavior.

1. Introduction

In plasma confinement devices the magnetic reconnection process causes a local topological transition of the magnetic surfaces, accompanied by a transformation of magnetic energy into plasma kinetic energy and heat in a short time. In tokamak devices, these tearing modes are a serious cause of degradation of plasma confinement and an important open issue consists of finding the appropriate means of control of this perturbation. Various methods for the mitigation and control of tearing modes have been implemented in existing experiments. One of the most promising methods presently considered to counteract robustly the tearing instabilities in a tokamak is based on the localized injection of an external control current inside the magnetic island [1, 2]. Due to its localized deposition [3], the electron cyclotron current drive (ECCD) is very appropriate for this stabilization purpose, enabling the driven currents to both locally modify the equilibrium [4] and compete with the perturbed island [5]. Experimental results have successfully demonstrated stabilization on several devices [6, 7, 8].

The conventional approach on which practical control systems are being designed is based on the 0-D model of the (generalized) Rutherford equation [9] that describes the time evolution of a nominal width of the island. Although largely applied, in this approach the fundamental topological aspects of the problem are neglected.

In this paper we present numerical results corcerning the effect of ECCD injection on the dynamical behavior of magnetic islands associated with resistive tearing modes, in a fluid plasma. In order to analyze the effect on different size islands, the current drive is turned on at linear, Rutherford and saturated phases of a highly unstable reconnection process. A continuously driven control current peaked at the island O-point has been adopted, whose amplitude is con-

Journal of Physics: Conference Series **401** (2012) 012002 doi:10.1088/1742-6596/401/1/012002

sistent with the magnetic field dynamics.

2. Model and island evolution without current drive

The analysis has been performed by adopting a two-dimensional fluid plasma description based on the standard Reduced Resistive Magneto-Hydro-Dynamics Model (RRMHD). In order to consider the ECCD contribution, an extra term simulating an externally imposed current density source J_{ec} has been added to the plasma Ohm's law. The poloidal magnetic flux function ψ and the vorticity U evolution equations are [10]:

$$\frac{\partial \psi}{\partial t} + \mathbf{v}_\perp \cdot \nabla \psi = -\eta(J - J^{(0)} + J_{ec}), \tag{1}$$

$$\frac{\partial U}{\partial t} + \mathbf{v}_\perp \cdot \nabla U - \mathbf{B}_\perp \cdot \nabla J = 0 \tag{2}$$

The total magnetic field is $\mathbf{B} = B_0\mathbf{e}_z + \mathbf{B}_\perp$, the in-plane magnetic field is $\mathbf{B}_\perp = \nabla\psi \times \mathbf{e}_z$, the in-plane velocity is $\mathbf{v}_\perp = -\nabla\varphi \times \mathbf{e}_z$ and $U = \mathbf{e}_z \cdot (\nabla \times \mathbf{v}_\perp) = \nabla_\perp^2 \varphi$, where φ is the stream function. The current density is $J = \mathbf{e}_z \cdot (\nabla \times \mathbf{B}_\perp) = -\nabla_\perp^2 \psi$ and $J^{(0)}$ is its equilibrium component. The parameter η represents the plasma resistivity, which corresponds to the physical effect responsible for the decoupling between the magnetic surfaces and the plasma motion and, as a consequence, for the triggering of the magnetic reconnection instability. All lengths are scaled to the macroscopic scale length L, while the time is normalized by the Alfvén time $\tau_A = L/B_y^{(0)}$, with $B_y^{(0)}$ the characteristic value of the equilibrium magnetic field in the perpendicular plane and L its shear length.

The control density current is assumed to have the following spatial gaussian form

$$J_{ec}(x, y, t) = J_m(t) exp(-(\psi(x, y, t) - \psi_O(t))^2/\delta^2), \tag{3}$$

where ψ_O is the magnetic flux at the O-point of the magnetic island. This allows a uniform distribution of the control current on the magnetic surfaces $\psi = const.$ The aim of an auxiliary driven control current like eq.(3) is to restore the ideal frozen flux condition in eq.(1), by reducing the perturbed current density $J - J^{(0)}$. It is expected to counterbalance the unstable reconnection process and eventually to stop the magnetic island growth [11, 12, 13].

The maximum height J_m has a step function profile like

$$J_m(t) = \begin{cases} 0 & t < t_1 \ V \ t > t_2 \\ A & t_1 < t < t_2 \end{cases} \tag{4}$$

where t_1 and t_2 are the switching on and switching off time of the current control, respectively. The amplitude A is a constant proportional to the difference between the maximum and the minimum current density value along the rational surface $x = x_s$ at $t = t_1$, $A = a \cdot (J_{max}(t_1) - J_{min}(t_1))$. Finally, δ is a constant, representing the EC beam width in ψ space, and taken proportional to the difference between the magnetic flux at the X- and the O-point of the magnetic island when $t = t_1$, i. e. $\delta = b \cdot (\psi_X(t_1) - \psi_O(t_1))$.

The equations (1)-(2) are integrated numerically by splitting all the fields in two parts: an equilibrium, independent on time, and an evolving perturbation. The perturbed component is advanced in time by an explicit, fourth order Adam-Bashforth scheme. The equations are solved in a $L_x \times L_y$ slab, with periodic boundary conditions along the y direction. Dirichlet conditions are applied at the edges of the x axis, imposing that all the perturbed fields go to zero. A compact finite difference algorithm [14], suitable for non-equispaced grid, is used for the spatial operations along the x direction, while pseudo-spectral methods are adopted for the periodic direction. Numerical filters are introduced in order to control the numerical error propagation. As described in [14], these filters smooth out the small spatial scales below a chosen cutoff, while leaving unchanged the large scale dynamics on all the time evolution of the process. In order to address the proper spatial resolution to have a reasonably good scale separation in the simulations, the code has been parallelized adopting the MPI libraries.

We set up a numerical experiment of spontaneous magnetic reconnection process in a static, "Harris pinch" equilibrium configuration with

$$B_{\perp y}^{(0)} = B_y^{(0)} \tanh(x/L) \qquad \text{and} \qquad \mathbf{v}_\perp^{(0)} = 0, \tag{5}$$

where $B_y^{(0)} = 1$ and $L = 1$. The linear theory shows that this equilibrium becomes tearing mode unstable if the instability parameter $\Delta' = 2 \cdot (1/k_y - k_y) > 0$, where $k_y = 2\pi m/L_y$ is the instability wave number along y, m being the corresponding mode number [15]. In this analysis we considered $L_y = 8\pi$, which corresponds to a rather large value of $\Delta' = 7.5$ for $m = 1$. It is accepted that in the regimes $\Delta' \gg 1$ magnetic reconnection can develop on a relatively fast time scale both in collisional [16] and collisionless [17, 18, 19] regimes. Moreover there is experimental evidence for $\Delta' \gg 1$ during the sawtooth instability (see, e.g., [20] and references therein). The box extention along the x direction is $L_x = 22.64$, which avoids any influence of boundary conditions on the reconnection dynamics. We destabilized the equilibrium (5) by introducing the following current density perturbation

$$\delta J(x, y) = \hat{J}(x) cos(k_y y), \tag{6}$$

where $k_y = 1/4$ and $\hat{J}(x)$ is a function localized within a width of order $(\eta/k_y)^{1/3}$ around the rational surface $x_s = 0$, whose amplitude is 10^{-4}. We assumed a plasma resistivity $\eta = 5 \cdot 10^{-4}$. In order to properly treat the small scale structures we expect in the large Δ' regime [16], a mesh of $nx = ny = 1024$ grid points has been adopted.

The time evolution of the reconnection process we considered is shown in fig.1. The figure describes the variation of the magnetic island width when the current control is switched off ($J_{ec} = 0$), through the logarithmic plot of the magnetic flux function at the X-point versus the time.

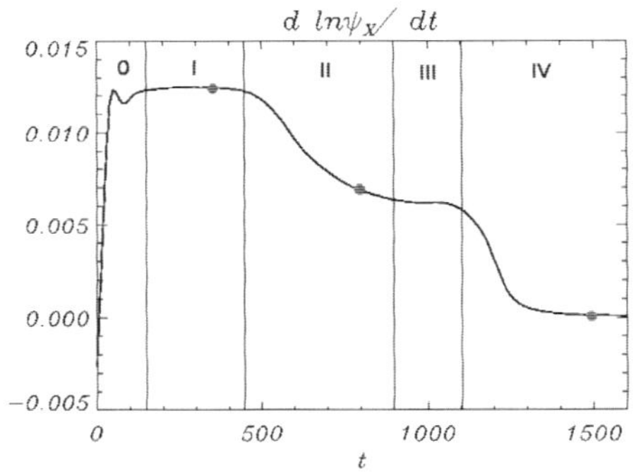

Figure 1. Effective growth rate at the X-point vs time for the free evolving system. After the initial transient (0), four main stages can be identified: the linear exponential growth (I), the Rutherford regime (II), the second exponential phase (III) and finally the saturation (IV). The red dots indicate the instants at which the current drive has been switched on.

After a short transient, four distinct stages can be identified. In the time interval $150 < t < 450$ the single helicity magnetic perturbation exhibits a linear exponential growth in time [21]. Then at $t = 500$ the Rutherford regime starts, lasting till $t = 900$, during which the magnetic island growth is algebraic in time [22]. Note that in the case we present here, the Rutherford stage does not lead to the saturation regime, as predicted by the resistive MHD for $\Delta' \ll 1$. Indeed, a second exponential growth starts around $t = 900$ and it lasts till $t = 1100$. Contrary to the simulations with large Δ' reported in [16, 23], no collapse of the cross shape current configuration, centered at the X-point of the island, develops at this stage as well as current sheet formation. The disagreement depends on the characteristic Δ' value of the initial equilibrium configuration we considered. Even though we analyzed a strongly driven tearing mode, the value of Δ' is not large enough to satisfy the analytical Waelbroeck's transition criterion [24], which predicts the current sheet collapse when the magnetic island width W is larger than the threshold value W_c, with $W_c = 25/\Delta'$. Indeed, in our case the magnetic island width at the $t = 1000$, when the collapse is expected, is $W \approx 3.23$ which is slightly lower than $W_c \approx 3.33$. Finally, at $t = 1100$ the magnetic island growth rate falls quickly down and at $t = 1300$ the island width saturates.

3. Effect of current drive with gaussian profile on the island width

In order to analyze the influence of an externally driven current control on the magnetic island, we applyed the current drive J_{ec} in the linear, Rutherford and saturated phase of the reconnection process described in the previous section. In all the cases we explored the J_{ec} parameters lay in the intervals $1 < J_m/(J_X - J_O)|_{t_1} < 10$ and $0.1 < \delta/(\psi_X - \psi_O)|_{t_1} < 1$, where J_X and J_O represent the current density at the X and at the O-point of the magnetic island, respectively. The basic response of the tearing mode perturbation to the injection of the electron cyclotron (EC) driven current is monitored by the time history of the effective area $A_{isl}(t)$ enclosed by the separatrix of the magnetic island. The island contour is found identifying numerically the cordinates of the hyperbolic X-point, (x_X, y_X), where $\partial_x\psi = \partial_y\psi = 0$, and tracking accurately the surface level $\psi(x, y, t) = \psi(x_X, y_X, t)$, whose area A_{isl} is calculated.

Fig.2 shows the evolution of the area of a linear, small size magnetic island, after the injection of the current density J_{ec} at $t_1 = 350$. The left frame displays several $A_{isl}(t)$ traces corresponding to different values of the shine-on spot δ, for a fixed value of the amplitude $J_m = 4\cdot(J_X - J_O)|_{t_1}$, J_X and J_O being the current density values at the hyperbolic and elliptic point of the magnetic island respectively. It is apparent that the dependence of $A_{isl}(t)$ on δ is strongly nonlinear. The optimum control corresponds to the absorption depth $\delta = 0.5\cdot(\psi_X - \psi_O)|_{t_1}$. In this case, after a short transient in which the island slightly grows, its area falls rapidly to zero, in the limit of the numerical precision we assumed, in a time interval comparable to the linear growth time of the free system, $\gamma_{lin}^{-1} \sim 1/0.012$. Lower values of δ allow a very poor control on the magnetic island growth mainly because they are too small compared to the original island size. On the other hand, when $\delta = (\psi_X - \psi_O)|_{t_1}$ we numerically verified that almost the 20% of the externally imposed ECCD is deposited ouside the magnetic island at $t = 350$ and, as a consequence, does not contribute to the island reduction. This is responsible for the lower rate of quench compared to the case with $\delta = 0.5\cdot(\psi_X - \psi_O)|_{t_1}$. By varying the current drive amplitude we observed that no significant control can be reached for $J_m < 4\cdot(J_X - J_O)|_{t_1}$ for any $0.1 < \delta/(\psi_X - \psi_O)|_{t_1} < 1$. On the contrary, when $J_m > 4\cdot(J_X - J_O)|_{t_1}$ the auxiliary control current is suitable to counteract significantly the current pertubation growth due to the tearing mode instability. In this case the magnetic island annihilation is reached on a time scale that decreases monotonically with the inverse of the current drive amplitude.

As shown in the right frame of fig.2, a remarkable fact occurs for a critical value $\delta = 0.5\cdot(\psi_X - \psi_O)|_{t_1}$ at later times. After the suppression at $t = 550$, the area of the island bounces back with a growth rate equal to the previous rate of quench. In other words the effect of the applied current meant to restore stability in the linear stage (so called "early" control action) may lead on the contrary to another unstable state. This is a state bifurcation known as *flip* instability [25, 26] because the value of the reconnected flux ψ at the resonant surface

Journal of Physics: Conference Series **401** (2012) 012002 doi:10.1088/1742-6596/401/1/012002

changes sign, which is equivalent to a shift of $L_y/2$ of the equilibrium position of the elliptic O-point of the tearing perturbation. It is noteworthy that this behaviour, already studied in the context of classical reconnection driven by perturbation of boundary conditions reappears, with universal chacteristics, in the frame of the so-called non-inductive current drive effects where it has been always ignored. The consequences of this effect on the strategy of control of tearing instabilities can be better appreciated inspecting also the 2D structure of the islands before, during and after the ECCD injection.

The effect of the localized current injection on a macroscopic magnetic island in the Rutherford

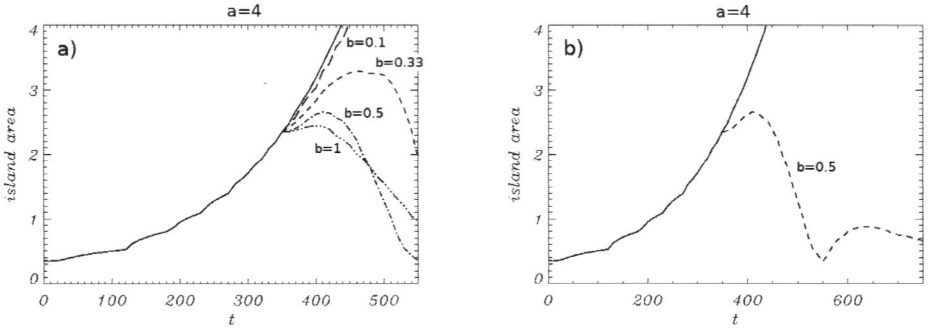

Figure 2. (a) Time evolution of the magnetic island area for four different values of the normalized current drive width $b = \delta/(\psi_X - \psi_O)|t_1$. The control current source has been switched on at $t = 350$ with a maximum height of $J_m = 4 \cdot (J_X - J_O)|t_1$. (b) In the case $b = \delta/(\psi_X - \psi_O)|_{t_1} = 0.5$, once the magnetic island reaches a sufficiently small size the island reverts back to instability. The actual cause of the abrupt change is the result of a phase shifted island being driven by the radio frequency current. We refer to this behavior as *flip* instability.

regime has been investigated by turning on J_{ec} at $t_1 = 800$. A characteristic feature of the system in this phase, as was already shown in recent research works [13, 28], is that the island suppression process does not proceed on a reversible path through a gradual shrinking of the island size and a restoration of a state similar to the initial one. An alternative equilibrium is reached with two straight, thin and parallel to the y axis current sheets symmetrically distributed on both sides of the resonant surface $x_s = 0$. The current layers are localized around the maximum width of the magnetic island at $t = 800$, independently from the strength and the width of the current drive. In [27] an interpretation of this behavior is found by analogy with the classical Hahm-Kulsrud-Taylor problem [29, 30]. An exemple of the plasma current distribution, in response to the EC driven current pulse with parameters $J_m = 4 \cdot (J_X - J_O)|t_1$, $\delta = 0.5 \cdot (\psi_X - \psi_O)|_{t_1}$, exhibiting the formation of a twin current sheet structure is shown in fig.3.

Numerical simulations have shown that the complete suppression of the magnetic island in the Rutherford regime is more difficult than in the linear stage. It occurs just for a rather narrow range of combinations of the J_{ec} parameters, on much longer time scales than the ones observed in the linear island control case. As an example, fig.4 shows the evolution of magnetic island area for different values of δ and a fixed $J_m = (J_X - J_O)|_{t_1}$.

The new nonlinear equilibrium has shown to be prone to secondary instabilities. The current sheet formation is coupled to highly localized, bar shaped, vorticity structures giving rise to strong shears that lead to the onset of fluid like, Kelvin-Helmholtz instabilities. The vorticity layers are aligned to the y axis and picked around the current sheet positions, where $\mathbf{B}_\perp \sim B_y$ and $\partial_y J \sim 0$. As a consequence the contribution of the nonlinear term $\mathbf{B}_\perp \cdot \nabla J$ in the equa-

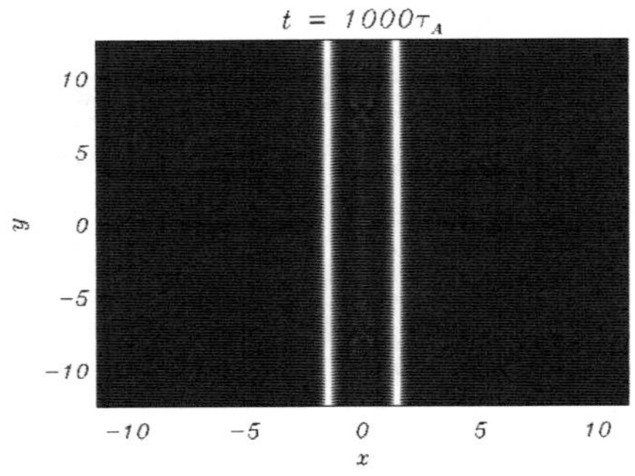

Figure 3. Contour plot of the current density J at $t = 1000$ in response to the ECCD pulse turned on in the Rutherford regime ($t_1 = 800$).

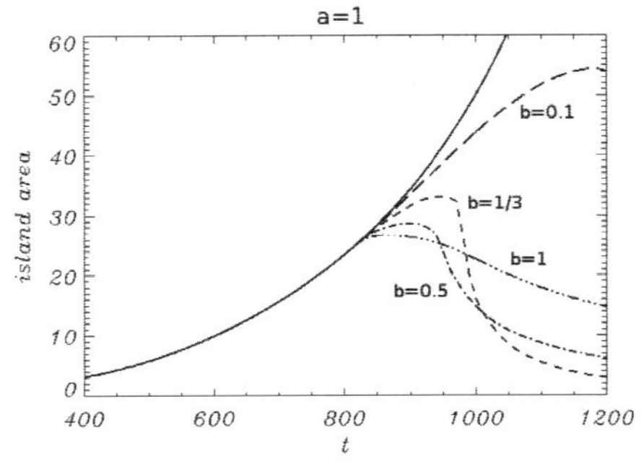

Figure 4. Time evolution of the magnetic island area for control current injections at $t_1 = 800$, with various normalized EC beam widths $b = \delta/(\psi_X - \psi_O)|t_1$ and a fixed maximum height $J_m = (J_X - J_O)|_{t_1}$

tion (2) is almost negligible, wich makes the vorticity evolution almost completely fluid. The Kelvin-Helmholtz like secondary instability is responsible for the turbulent redistribution of the patterns in the spatial region confined by the current layers. It affects not only the fluid field, but also the magnetic field, whose surfaces are advected by the turbulent motion towards the resonant surface $x = 0$, forcing them to reconnect. According to the recent results that highlighted the role of the turbulence on the enhancement of the reconnection rate [31, 32], the generation of the Kelvin-Helmholtz instability could represent a strong limitation for the efficiency of the control startegy. Fig.5 shows the evolution of the vorticity structures in presence of the Kelvin-Helmholtz instability at four different stages of a numerical simulation with external control of parameters $J_m = 3 \cdot (J_X - J_O)_{t_1}$, $\delta = 0.1 \cdot (\psi_X - \psi_O)|_{t_1}$.

In order to analyze the effect of an ECCD power deposition on a tearing mode at the nonlinear saturation stage we carried out a campaign of numerical simulations where the external current control has been injected at the instant $t = 1500$ of the free system evolution. Compared to the control in the linear and in the Rutherford phase, in this case the magnetic island never shrinks to zero. A new equilibrium is reached on a time scale comparable to the time the free system takes to reach the magnetic island saturation, with two current sheets on each side of the singular $x = 0$ surface and a macroscopic deformation of the magnetic topology.

The ECCD control models of tearing modes based on the conventional 1-D Rutherford equation predict very stringent requirements of the focusing of the EC wave beam on the magnetic island. The present study which includes finite 2-D nonlinear effects shows that a successful control action requires broader current deposition, less demanding from the control point of view. However a broader ECCD deposition causes a rapid fall of the control efficiency, due to the fact that a relevant fraction of the injected current falls outside the separatrices, even in presence of small magnetic island reduction.

4. Conclusion

The results of the detailed 2-D nonlinear dynamics of tearing modes under the action of a finite width EC wave beam allow an improved understanding of the role of effects occurring on different space scales, such as the beam width, the island width and the shielding current sheets. When the ECCD is applied to a large size island, in particular, these effects strongly change the

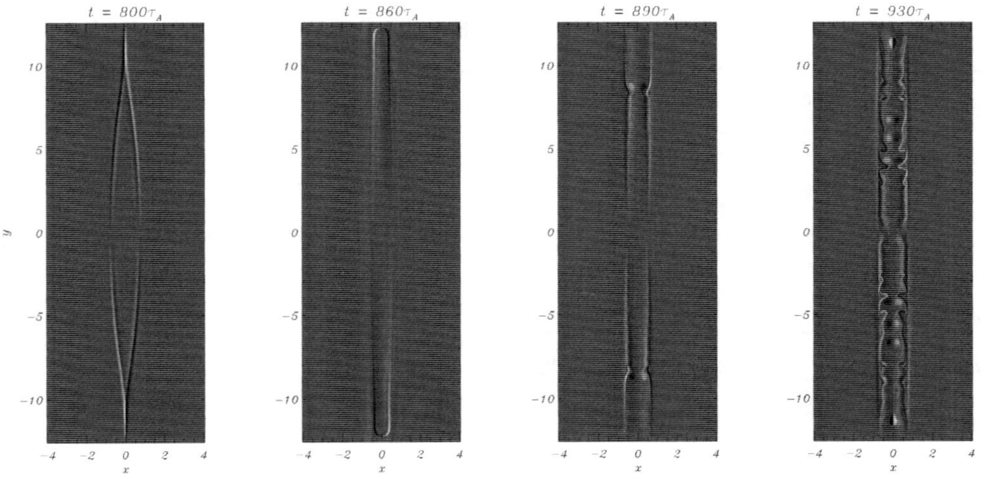

Figure 5. Nonlinear evolution of the vorticity patterns in response to the radio frequency power current drive injection at $t_1 = 800$.

perturbation spectrum, with the apperance of secondary harmonics and the modification of the original magnetic equilibrium, which reduce the control action, designed specifically for decreasing the primary, most unstable armonic. These effects are different from those of conventional Rutherford models and should be considered carefully in practical applications.

5. Acknowledgments

This work was partly supported by the Euratom Communities under the contract of Association between EURATOM/ENEA. The views and opinions expressed herein do not necessarily reflect those of the European Commission.

References

[1] Lazzaro E and Nowak S, *Plasma Phys. Control. Fusion*, **51**, 035005, (2009).
[2] Zohm H, *Phys. Plasmas*, **4**, 3433 (1997).
[3] Fisch J, *Rev. Mod. Phys.*, **59**, 175 (1987).
[4] Pletzer A and Perkins F W, *Phys. Plasmas*, **6**, 1589 (1999).
[5] Hegna C and Callen J D, *Phys. Plasmas*, **4**, 2940 (1997).
[6] Gantenbein G et al., *Phys. Rev. Lett.*, **85**, 1242 (2000).
[7] Maraschek M et al., *Phys. Rev. Lett.*, **98**, 025005 (2007).
[8] Goodman T P et al., *Phys. Rev. Lett.*, **106**, 245002 (2011).
[9] La Haye R J, *Phys. Plasmas*, **13**, 055501 (2006).
[10] Strauss H R, *Phys. Fluids*, **19**, 134 (1976).
[11] Prater R, *Phys. Plasmas*, **11**, 2349 (2003).
[12] Woodby J, Schuster E, Bateman G, Kritz A H, *Phys. Plasmas*, **15**, 092504 (2008).
[13] Comisso L, Lazzaro E, *Nucl. Fusion*, **50**, 125002 (2010).
[14] Lele S K, *J. Comp. Phys.* **103**, 16 (1992).
[15] Biskamp D, *Magnetic Reconnection in Plasmas*, Cambridge University Press 2000.
[16] Loureiro N F, Cowley S C, Dorland W D, Haines M G and Schekochihin A A, *Phys. Rev. Lett.*, **95**, 235003 (2005).
[17] Aydemir A Y, *Phys. Fluids B*, **4**, 3469 (1992).
[18] Ottaviani M and Porcelli F, *Phys. Rev. Lett.*, **71**, 382 (1993).

[19] Kleva R G, Drake J F and Waelbroeck F L, *Phys. Plasmas* **2**, 23 (1995); Wang X and Bhattacharjee A, *Phys. Rev. Lett.* **70**, 1627 (1993).
[20] Angioni C et al., *Plasma Phys. Control. Fusion*, **44**, 205 (2002).
[21] Furth H P, Killen J, Rosenbluth M N, *Phys. Fluids*, **6**, 459 (1963).
[22] Rutherford P H, *Phys. Fluids*, **16**, 1903 (1973).
[23] Jemella B D, Shay M A, Drake J F, Rogers B N, *Phys. Rev. Lett.*, **91**, 125002 (2003).
[24] Waelbroeck F L, *Phys. Rev. Lett.*, **70**, 3259 (1993).
[25] Monticello D A, White R B, Rosenbluth M N, Plasma Physics and Controlled Nuclear Fusion Research 1978, *Proc. 7th IAEA Int, Conf. Tokyo*, 1978, paper IAEA-CN- 37/K-3, Vol. 1, Vienna, p. 605.
[26] Lazzaro E, Coelho R, *Eur. Phys. J. D* **19**, 97 (2002).
[27] Lazzaro E, Comisso L, *Plasma Phys. Control. Fusion*, **53**, 054012 (2011).
[28] Lazzaro E, Comisso L, Valdettaro L, *Phys. Plasmas*, **17**, 052509 (2010).
[29] Kulsrud R M and Hahm T S, *Phys. Scr.*, **T2B**, 525 (1982).
[30] Hahm T S and Kulsrud R M, *Phys. Fluids*, **28**, 2412 (1985).
[31] Loureiro N F, Uzdensky D A, Schekochihin A A, Cowley S C and Yousef T A, *Mon. Not. R. Astron. Soc.*, **399**, L146 (2009).
[32] Lapenta G, *Phys. Rev. Lett.*, **100**, 235001 (2008).

Joint Varenna-Lausanne International Workshop 2012 IOP Publishing
Journal of Physics: Conference Series **401** (2012) 012003 doi:10.1088/1742-6596/401/1/012003

MHD properties in the core of ITER-like hybrid scenarios

D Brunetti, W A Cooper, J P Graves, F Halpern
Centre de Recherches en Physique Des Plasmas, Association EURATOM-Confederation Suisse, EPFL, 1015 Lausanne, Switzerland

C Wahlberg
Department of Astronomy and Space Physics, EURATOM/VR Fusion Association, P.O. Box 515, Uppsala University, SE-751 20 Uppsala, Sweden

H Lütjens, J F Luciani
Centre de Physique Thorique, Ecole Polytechnique, CNRS, F-91128 Palaiseau Cedex, France

E-mail: daniele.brunetti@epfl.ch

Abstract. MHD stability is studied in the frame of ITER like hybrid scenarios, characterized by weak reversed shear, with particular attention on ideal internal kink modes and infernal mode stability. Numerical simulations for ITER like equilibria, with hollow q profiles with an off axis minimum close to unity, were carried out using the 3-D equilibrium code ANIMEC, which showed the presence of a 3-D helical core [1] with the characteristics of a saturated internal kink mode. The internal kink perturbation has been investigated non-linearly using the XTOR code, in the ideal frame. A scan in the current was performed and it has been found that when the minimum of q is above the unity (low currents), the helical distorsion shows good agreement with the results provided by the 3-D ANIMEC simulations while for q_{min} below 1 (high currents), XTOR gives a residual distorsion in contrast with the ANIMEC results. Moreover infernal mode stability in hybrid scenarios has been studied analytically extending the quasi-interchange model with the inclusion of resistive and both electron and ion diamagnetic effects. This enables us to investigate kinetic effects on plasma scenarios susceptible to saturated kink structures exhibited into ANIMEC simulations.

1. Introduction

Standard tokamak operation scenarios are characterized by monotonically increasing safety factors q , with usually $q_0 < 1$ (q_0 is the value of the safety factor on the magnetic axis), full inductive plasma current and the standard high-performance operation. Another mode of operation is the advanced scenario which has weak reversed shear, produced by bib-inductive current drive and bootstrap current. A mode of operation is the hybrid scenario, which is an intermediate step between the H-mode standard scenario and an advanced scenario characterized by an extended low shear region where the safety factor $q \approx 1$. In recent hybrid experiments, good confinement and high beta limit have been achieved [3] with stationary density profiles enclosing a large plasma volume with low magnetic shear with q_0 close to 1 (hybrid scenario).

As pointed out in [2, 4] these kind of discharges can be characterized by a soft magnetohydrodynamic (MHD) activity, whose most important modes can be divided as follows:

- Long lived modes - typically MAST operation regime (ideal saturated MHD instabilities infernal modes, $q_{min} > 1$ [5])
- Global oscillations - TCV (sawteeth cycles, $q_{min} > 1$)
- Snakes - JET (ideal infernal modes)

Internal kink instabilities are accessible in the regime of high β and q profiles which are close to the unity (either monotonic with $q_0 < 1$, or with low or weak reversed shear having $q_{min} \approx 1$ and $q > 1$ everywhere). Under these conditions the internal kink instability develops and 3-D helical structures appear in the plasma core, whose saturation gives rise to the above mentioned long lived modes; these saturated states have been found experimentally in the spherical tokamak MAST discharges [5] and also in *"standard"* tokamaks like TCV operating in the above mentioned hybrid scenario (viz. weak reversed shear) and it is conjectured that this new plasma state can be regarded as a new kind of equilibrium.

Moreover these scenarios could be affected by *infernal mode* activity either ideal or resistive: in particular coupling the main kink harmonic ($m = n = 1$) to its resistive *"tearing"* sidebands, namely the $m' = m \pm 1$ harmonics, give rise to the *infernal mode* instability [6, 7], which is closely related to the *quasi interchange mode* [8], giving an enhanced growth rate for the internal kink mode; The resistive infernal mode has both MHD characteristics in the core region and tearing characteristics at $q = (m + 1)/n$ surface: this suggests that diamagnetic effects should be important since the tearing mode is required to rotate at the electron diamagnetic frequency at that surface [7]. Therefore an extension the MHD model has been studied in order to include higher order corrections, viz. resistivity and plasma diamagnetism and other advanced effects including viscosity and sheared $\boldsymbol{E} \times \boldsymbol{B}$ toroidal rotation.

2. Ideal kink instabilities in ITER-like scenarios

Since the simplest long lived mode, in the hybrid scenarios, is the saturated internal kink, the first part of this work was devoted to the study of such MHD activity investigating the relation which occurs between magnetohydrodynamical modes and the shape of q profile (i.e. density current profile). Numerical simulations for ITER like equilibria, with hollow q, were carried out using the 3-D equilibrium code ANIMEC [1] and the nonlinear MHD stability code XTOR [9] in the ideal frame: ANIMEC predicts the presence of 3-D helical cores in these particular configurations [1] with the characteristics of a saturated internal kink mode. XTOR simulations were performed starting from a given 2-D (i.e. axisymmetric) ITER like equilibrium ($\beta \approx 3\%, 0.96 < q_{min} < 1.03$), with the same profiles which give rise to the 3-D helical cores [1] prepared with 2-D equilibrium code CHEASE [10]. A scan in the current, ranging from 13.2 to 13.8 MA, was performed and the associated saturated state is shown in figure 1. Both ANIMEC and XTOR results predict the presence of kink-like 3-D structures, however it has been found that when the minimum of q is above the unity (low currents), the helical distorsion δ_h shows good agreement with the results provided by the 3-D equilibrium simulations performed using ANIMEC and also with the analytical predictions for the nonlinearly saturated kink given by

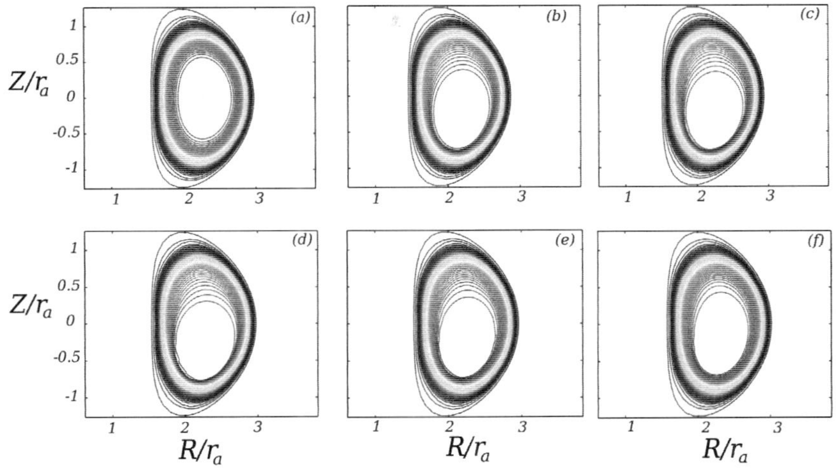

Figure 1. Contour of constant pressure surfaces at fixed toroidal angle $\phi = 3/2\pi$ for saturated internal kink mode evaluated by the ideal stability code XTOR starting from an ANIMEC equilibrium, with toroidal plasma current (a) 13.0 MA, (b) 13.3 MA,(c) 13.4 MA, (d) 13.5 MA, (e) 13.7 MA, (f) 13.8 MA; r_a denotes the plasma minor radius.

[11, 12] (here $' \equiv \partial/\partial_r$):

$$\xi_h^2 q'' = \frac{8}{71}\left(\frac{8\pi}{3}\right)^3 \Delta q \left[\left\{\frac{1}{2}\left(\frac{\gamma^2}{\omega_A \Delta q}+1\right)\left[\left(\frac{\gamma^2}{\omega_A \Delta q}+1\right)^{1/2}+1\right]\right\}^{1/2}-1\right]_{r=r_1} \tag{1}$$

$$\xi_h q' = \left.\frac{13}{\pi}\frac{\gamma}{\omega_A}\right|_{r=r_1} \tag{2}$$

where ξ_h is the radial displacement, $\omega_A = v_A/(\sqrt{3}R_0)$ with v_A the Alfvén velocity, R_0 the major radius, γ is the linear growth rate of the internal kink mode, $\Delta q = |q_{min-1}|$ and r_1 is the position of the resonant $q = 1$ surface. Eq. (1) derives from the calculation of the nonlinear $m = n = 1$ saturated kink mode for nonmonotonic q profiles with reversed shear having $q_{min} \approx 1$ at $r = r_1$ [11] while equation (2) is valid for the $m = 1$ mode for monotonic safety factor profiles with the $q = 1$ surface at $r = r_1$ [12]. However, when $q_{min} < 1$ we found that the analytical predictions almost agree with the nonlinear results, but there is a significant deviation between the 3-D equilibrium calculations and the nonlinear simulations: while the helical distorsion predicted by ANIMEC decreases to 0 when $q_{min} \approx 0.96$, XTOR gives a residual distorsion in this region (as shown in figure 2) accordingly with the linear stability calculations performed with TERPSICHORE code [13]. Indeed this is to be expected because the analytic linear internal kink mode of Bussac [14] is unstable. A convenient value for the viscosity was chosen in order to avoid numerical instabilities in the core region; however, even if the growth rate for the $n = m = 1$ mode is not so sensitive to variations in the viscosity ν [15], we should take care of this parameter, since its value can affect significantly the behaviour of $n \neq m$ modes as shown in figure 3, damping potentially unstable modes.

3. Infernal mode stability

The nonlinear code XTOR-2F [9] solves an extended set of magnetohydrodynamic equations in toroidal tokamak geometry in which effects like viscosity and plasma diamagnetism are also

Journal of Physics: Conference Series **401** (2012) 012003 doi:10.1088/1742-6596/401/1/012003

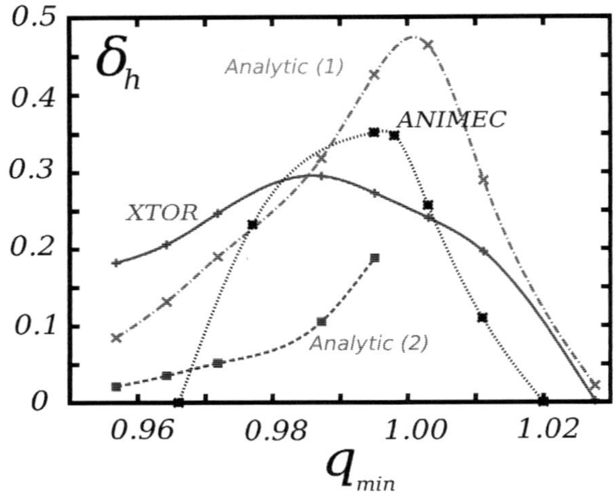

Figure 2. Helical distorsion of the magnetic axis for an ITER like scenario as a function of the minimum of the safety factor q. The black line shows the ANIMEC results ($r_a\delta_h = \sqrt{R_{01}^2 + Z_{01}^2}$ where $R_{01}(Z_{01})$ is the $(m = 0, n = 1)$ Fourier component of $R(Z)$ at the magnetic axis [1]), while the blue curve refers to the nonlinear XTOR simulations ($r_a\delta_h = \sqrt{(R - R_0)^2 + (Z - Z_0)^2}$, with (R, Z) being the position of the magnetic axis calculated at $\phi = 3/2\pi$ (ϕ is the toroidal angle) and (R_0, Z_0) being the position of the magnetic axis for the axisymmetric equilibrium); The red and green lines correspond to the analytical predictions given respectively by (1) and (2).

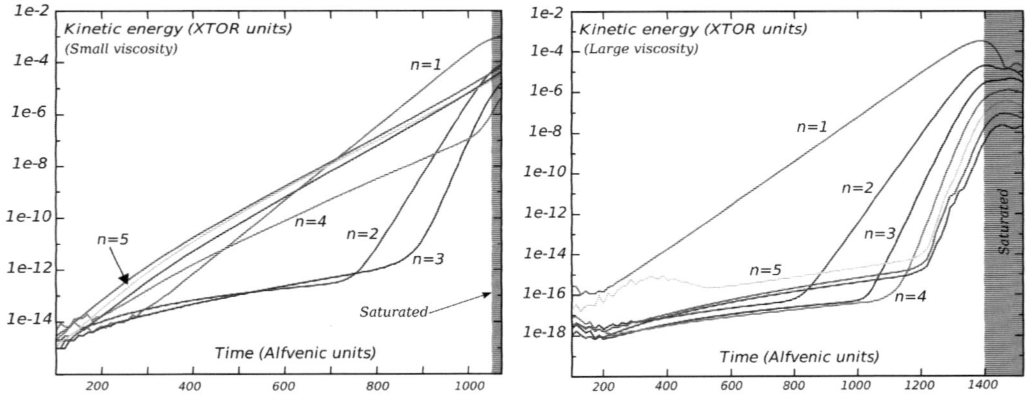

Figure 3. Behaviour of the kinetic energy for $n = (1, ..., 9)$ modes for two different values of the viscosity ($\nu = 3 \times 10^{-6}$(XTOR units) in left picture and $\nu = 5 \times 10^{-5}$(XTOR units) in the right one); the other equilibrium parameters were kept the same for both simulations. We can clearly see that the $n = (4, 5)$ is strongly reduced in the linear phase. We notice also that (around 1000 alfvenic times in the picture on the right) the growth rate γ for $n > 1$ modes is given by $\gamma_{n>1} = n\gamma_{n=1}$ [16] as expected by toroidal couplings.

included. Therefore the quasi-interchange model has been analytically extended in order to retain these higher order effects. The model assumes a parabolic pressure profile with a constant subsonic toroidal rotation in the low-shear region. The analysis is split in three regions: the low-shear region, the sheared region and the resonant layer of the $m+1$ sideband. The equation describing the toroidal MHD modes are derived from the full set of MHD equations [17] rather than performing a minimization of the energy integral [18].

Perturbed quantities are assumed flute like having an $exp(-i\omega t)$ time dependence. The inverse safety factor profile is given by $\mu = n/m_0 - \hat{S}(r/r^*)^\lambda$. Using standard techniques [17, 19, 20, 21] the following dispersion relation in the regime of $constant - \psi$ approximation including diamagnetic effects is derived:

$$a\Delta' \sim S^{3/4}\left(\frac{\widehat{\omega}_s(\widehat{\omega}-\omega_i^*)_s}{\omega_A^2}\left(\frac{(\widehat{\omega}-\omega_e^*-\omega_e^T)_s}{\omega_A}\right)^3\right)^{1/4} \tag{3}$$

where $\widehat{\omega}$ is the Doppler shifted frequency due to toroidal rotation; the subscript s indicates that the quantity is evaluated at the resonant $m+1$ surface. At leading order in the shear-free region the equation for the main m_0 harmonic reads [18]:

$$\frac{d}{dr}\left[r^3 Q\frac{dX_{m_0}}{dr}\right] + r\left[(1-m_0^2)Q - \frac{\alpha r}{2R_0}\left(\frac{1}{q^2}-1\right)\right]X_{m_0}$$
$$+\mathbf{\Lambda}\alpha r^{m_0+1}\int_0^{r^*}\alpha r^{1+m_0}X_{m_0}dr = 0$$

where $Q = \delta q^2/q^4 - \widehat{\omega}(\widehat{\omega}-\omega_i^*)(1+2q^2)/(m_0^2\omega_A^2)$, $\delta q = q - m_0/n$, $\mathbf{\Lambda} = q^2(r^*)^{-2-2m_0}\mathbf{\Lambda}^{\omega}_{m_0,n}$, $\mathbf{\Lambda}^{\omega}_{m_0,n} = [(m+1)/2](2+m+\boldsymbol{C})/(m-\boldsymbol{C})$, $\boldsymbol{C} = r^*X'_+(r^*_+)/X_+(r^*_+)$ and $\alpha = 2R_0p'_0/B_0^2$ being the usual ballooning parameter. The factor $\boldsymbol{C}$ is evaluated by means of (3) solving the sideband equations in the sheared region with the model safety factor written before. Hence the following dispersion relation for the infernal mode instability for an almost resonant $q \approx 1$ safety factor is derived:

$$\frac{\widehat{\omega}(\widehat{\omega}-\omega_i^*)}{\omega_A^2} = \frac{n^2}{1+2q^2}\left[\left(\frac{\delta q}{q}\right)^2 - \frac{2\varepsilon_*^2\mathbf{\Lambda}^{\omega}_{m,n}\beta_p^2}{(m+1)^2(m+2)}\right]$$

where $\varepsilon_* = r^*/R_0$ and $\beta_p = \bar{p}q^2/(B_0^2\varepsilon^2)$, $\bar{p}$ denoting the plasma pressure on the magnetic axis. For a fixed δq_c the ideal MHD stability boundary is identified by $\beta_p = \beta_{crit}$; however, small positive deviations in δq are allowed, namely $\delta q^2 = \delta q_c^2 + \Delta^2$. Defining $\gamma = -i\omega$, neglecting toroidal flow and diamagnetic corrections, close to marginal MHD stability the behaviour $\gamma^{5/4}(\gamma^2+\Delta^2)$ is recovered [7], giving the $S^{-3/13}$ scaling for $\Delta^2 \ll \gamma^2$ and the $S^{-3/5}$ behaviour for $\Delta^2 \gg \gamma^2$ as shown in figure (4). Using standard techniques diamagnetic corrections are introduced in the expression for Δ' which is used to evaluate the jump of the m_0+1 solution at its own rational surface giving the following dispersion relation

$$\omega\left(\omega-\omega_i^*\right) \sim 1/r_s\Delta'\left(\omega,\omega_{i,e}^*\right)$$

which has to be solved numerically.

Neglecting diamagnetic effects but taking into account toroidal flows the following expression is derived:

$$\left(\omega-\Omega_{E\times B}^{(s)}\right)^{5/4}\left[\left(\omega-\Omega_{E\times B}^{(l-s)}\right)^2-\Delta\right] \sim \beta_p^2 S^{-3/4}$$

where the symbols (s) and $(l-s)$ indicate that the quantity is evaluated at the rational surface or in the low shear region respectively. The solution procedure is similar to the one presented

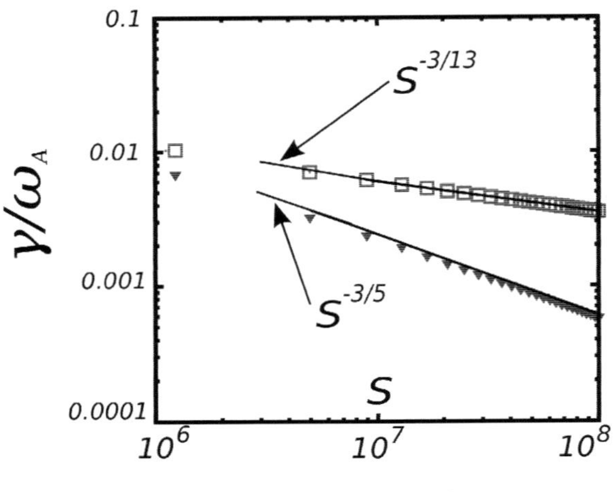

Figure 4. Scaling of the growth rates for $\Delta = 0$ ($S^{-3/13}$) and $\Delta = 0.001$ ($S^{-3/5}$) for the $m_0 = n = 7$ mode with $\delta q = 0.003$, $\varepsilon_* = 0.18$, $\lambda = 6$ and $\hat{S} = 0.03$.

before considering three different cases with $\Delta = 0$: $\omega \ll \Omega^{(s)}_{E\times B} \ll \Omega^{(l-s)}_{E\times B}$, $\Omega^{(s)}_{E\times B} \ll \omega \ll \Omega^{(l-s)}_{E\times B}$, $\Omega^{(s)}_{E\times B} \ll \omega \simeq \Omega^{(l-s)}_{E\times B}$. The nonlinear stability code XTOR will be used in order to find the unstable roots found analytically for the different cases presented before.

4. Conclusions

A comparison between numerical simulations of three dimensional equilibria (ANIMEC) and nonlinear saturated instabilities (XTOR) in the frame of ideal MHD has been performed. When q_{min} is above unity good agreement between ANIMEC and XTOR results was found in accordance with analytical predictions for saturated internal kinks [11, 14]. However nonlinear results predict a residual distorsion when $q_{min} < 1$ while in three dimensional simulations this distorsion disappears for $q_{min} \lesssim 0.96$.

Note that ANIMEC equilibrium calculations, analytic results and the nonlinear simulation carried out with XTOR neglect resistivity and this in principle allows current sheets to develop and the way in which they are handled may explain the difference observed in figure 2. However, since the $q_{min} > 1$ case is the most physically relevant, when the role of resistivity is localized at the resonant $m_0 + 1$ surface, it has been decided to drop resistive corrections in the low shear region. The quasi-interchange mode (and infernal modes) model has been extended analytically by including combined resistivity and diamagnetic effects. Moreover, recent TCV experiments [22] show that above a certain value of the plasma elongation κ sawtooth activity is suppressed and it is replaced by continuous oscillations with frequencies around 10kHz in the direction of the electron diamagnetic drift. Thus we plan to model these experiments through examination of infernal mode stability with the extended quasi-interchange model. It is envisaged that this work will guide the non-linear numerical work to follow.

Acknowledgments

This work was supported in part by the Swiss National Science Foundation and the Swedish Research Council (VR). The authors are very grateful to S P Hirshman for his invaluable contributions to the development of the 3D equilibrium solvers . The views and the opinions expressed herein do not necessarily reflect those of the European Commission.

References

[1] Cooper W A et al. 2011 *Plasma Phys. Control. Fusion* **53** 024002
[2] Gormezano C et al. 2007 *Nucl Fusion* **47** S285
[3] Hobirk J et al. 2009 *36th EPS Conference on Plasma Phys.*, ECA vol.**33E**, O-5.057 (2009)
[4] Gormezano C et al. 2004 *Plasma Phys. Control. Fusion* **46** B435
[5] Chapman I T et al. 2010 *Nucl. Fusion* **50** 045007
[6] Manickam J et al. 1987 *Nucl Fusion* **27** 1461
[7] Charlton L A et al. 1989 *Phys. Fluids* B **1** 798
[8] Wesson J A 1986 *Plasma Phys. Controlled Fusion Research* **28** 243
[9] Lütjens H and Luciani J F 2010 *J. Sci. Comput* **229** 8130
[10] Lütjens H et al. 1996 *Comput. Phys. Commun* **97** 219
[11] Avinash K 1990 *Phys. Fluids* B **2** 2373
[12] Rosenbluth M et al. 1973 *Phys. Fluids* **16** 1894
[13] Cooper W A et al. 2011 *Plasma Phys. Control. Fusion* **53** 124005
[14] Bussac M N et al. 1975 *Phys. Rev. Lett.* **35** 1638
[15] Porcelli F 1987 *Phys. Fluids* **30** 1734
[16] Lütjens H and Luciani J F 1997 *Phys. Plasmas* **4** 4192
[17] Mikhailovskii A B 1998 *Instabilities In A Confined Plasma* (IoP Publishing, Bristol, UK, 1998)
[18] Hastie R J and Hender T C 1988 *Nucl. Fusion* **28** 585
[19] Ara G et al. 1978 *Ann. Phys (N.Y.)* **112** 443
[20] Migliuolo S 1993 *Nucl. Fusion* **33** 1721
[21] Porcelli F 1987 *Phys. Fluids* **30** 1734
[22] Reimerdes H et al. 2006 *Plasma Phys. Control. Fusion* **48** 1621

Joint Varenna-Lausanne International Workshop 2012 IOP Publishing
Journal of Physics: Conference Series **401** (2012) 012004 doi:10.1088/1742-6596/401/1/012004

Conditions for Lower Hybrid Current Drive in ITER

R Cesario[1], L Amicucci[2], A Cardinali[1], C Castaldo, S Ceccuzzi[2], F Napoli[2], A A Tuccillo[1], A Galli[3], G Schettini[3]
[1]Associazione EURATOM/ENEA sulla Fusione, Centro Ricerche Frascati, Italy
[2]Università Roma Tre, Rome, Italy
[3]Università La Sapienza, Rome, Italy

Abstract
To control the plasma current profile represents one of the most important problems of the research of nuclear fusion energy based on the tokamak concept, as in the plasma column the necessary conditions of stability and confinement should be satisfied. This problem can be solved by using the lower hybrid current drive (LHCD) effect, which was demonstrated to occur also at reactor grade high plasma densities provided that a proper method should be utilised, as assessed on FTU (Frascati Tokamak Upgrade). This method, based on theoretical predictions confirmed by experiment, produces relatively high electron temperature at the plasma periphery and scrape-off layer (SOL), consequently reducing the broadening of the spectrum launched by the antenna produced by parasitic wave physics of the edge, namely parametric instability (PI). The new results presented here show that, for kinetic profiles now foreseen for the SOL of ITER, PI is expected to hugely broaden the antenna spectrum and prevent any penetration in the core of the coupled LH power. However, considering the FTU method and assuming higher electron temperature at the edge (which would be however reasonable for ITER) the PI-produced spectral broadening would be mitigated, and enable the penetration of the coupled LH power in the main plasma. By successful LHCD effect, the control of the plasma current profile at normalised minor radius of about 0.8 would be possible, with much higher efficiency than that obtainable by other tools. A very useful reinforce of bootstrap current effects would be thus possible by LHCD in ITER.

Introduction

Experiments of JET (Joint European Torus), utilising plasma configuration at high triangularity (d□0.4) and main heating power up to 30 MW applied at the end of the plasma current ramp-up (q_{95}□5, I_P=1.5MA B_T=2.3T), were performed for developing fully non-inductive steady-state regimes relevant for ITER [1]. High values of normalised b (b_N) and confinement were obtained (β_N□2.8, H_{98}□1.05) during an ELMy H-mode phase of discharges with radially broad transport barriers (TBs). By the relatively high main heating power applied in these experiments (□30 MW), a radially broad and steep pressure gradient occurred, which consequently produced a relatively high bootstrap current density (j_{BS}) at the plasma periphery [2,3]. However, this current was insufficient for freezing significantly the current relaxation. As a consequence, the improved confinement lasted only several confinement times due to the onset of strong low-*n* MHD activity.
Thus, for a successful development of reactor-graded plasma operations, it would be important to have different and flexible tools capable of producing and controlling steady-

state profiles with a high fraction of non-inductive plasma current driven in the outer half of the plasma column. The solution of this problem could be simplified by means of the lower hybrid current drive (LHCD) effect [4]. By coupling radiofrequency power in the range of a few GHz to tokamak plasmas, non-inductive current was driven with high efficiency in many experiments [5–7]. Unfortunately, the LHCD effect was routinely observed only operating at plasma densities markedly lower that these required for ITER, and the extrapolation to reactor grade higher densities resulted to be problematic. Indeed, in LHCD experiments on FTU at high densities, operating in standard regime, LHCD signatures disappear for high line-averaged plasma densities (n_{e_av} □1.1×10^{20} m^3) [8]. In Alcator C-Mod, which operates at the frequency f_0=4.6 GHz (the closest used frequency to that, 5 GHz, envisaged for ITER), the LHCD signatures disappear for high line-averaged plasma densities (n_{e_av} □1.5×10^{20} m^3) and decreased much more markedly than expected at medium densities, n_{e_av} □$0.8–1.0\times10^{20}$ m^3, depending on the operating parameters [9]. The effect of collisional absorption [10] was invoked for interpreting the observed anomalous LH power damping at the edge. However, these plasmas have relatively low electron plasma temperature in the core, which prevents the full deposition of the coupled LH power at the first radial pass. The effect of collisional absorption was ruled out by analysis performed considering the available data of experiments of JET operating in condition of first-pass LH regime, and presenting the same effect of LH power damped at the very edge, unexpected by quasi-linear wave theory [13,9]. In the latter experiments, the lack of penetration of the coupled LH power in the core, was observed to occur at relatively high density even at the periphery ($n_{e0} \sim n_{e\_0.8} \sim 0.4\times10^{20}\times10^{20}$ m^{-3}, where $n_{e\_0.8}$ represents the density at normalised minor radius r/r_{Sep} ~0.8, where r_{Sep} is the plasma minor radius).

Considering the available data of these experiments of JET, a modelling work showed that the anomalous LH damping at the plasma edge can be a consequence of the effect of a huge spectral broadening due to parametric instability (PI), which is expected to occur in the cold radial region of the plasma periphery [12,13].

The problem of how extrapolating the LHCD effect at high densities, which resulted unsuccessful for decades, was recently solved by experiments on FTU [9]. These experiments were performed following the guidelines of previously formulated theoretical prediction on the PI-produced spectral broadening effect [12,13]. This prediction consists in relatively high electron temperature of the plasma SOL and periphery, having the effect of diminishing the local parasitic LH power damping due to the PI-produced spectral broadening. Consequently, the LH power penetration in a high density plasma core is enabled.

The phenomenon of the LH spectral broadening was monitored in the experiments at high density on FTU. A simple tool was utilised consisting in a RF (radio-frequency) loop antenna probe located outside the machine and connected to a spectrum analyser, centred at the LH operating frequency [14,15]. This diagnostic gives indication of effects of wave physics of the edge that are decisive for determining the LH power propagation and damping in tokamaks.

The RF probe showed that, in standard operating conditions of FTU high density plasmas, in which no LHCD effect was observed to occur as in other experiments, the operating line frequency was hugely broadened by LH sideband waves (up to about 15 MHz) [8].

Considering the successful interpretation of the LH experiments performed at high plasma densities on FTU [8] and JET [11], we have extended the model approach to the case of a possible LH current drive experiment on ITER. In the next section, some details of the model approach are shown, and in Sec. 3 preliminary results are presented.

The modelling approach

The LH^{star} code [16,12,13,8], developed at the ENEA Laboratory of Frascati, is a unique available tool for LH modelling in tokamak plasmas that considers the effect of non-linear physics of the edge produced by parametric instability. The code describes the wave propagation and damping and produces the radial density profiles of LH absorbed power and driven current. The *initial* $n_{//}$ spectrum modified by the PI phenomenon at the plasma edge is first calculated and then is inputted in the ray-tracing and Fokker-Planck modules, together with the kinetic profiles and the magnetic reconstructed equilibrium data. As further special feature of the LH^{star} code, the ray-tracing and Fokker-Planck analyses are consistently coupled as shown hereafter. The ray trajectory is provided to the Fokker-Planck code that computes the absorption on the modified distribution function of electrons.

We consider here the case of a possible LHCD experiment on ITER, considering the magnetic equilibrium data and the kinetic profiles foreseen for steady-state scenario. An operating frequency of 5 GHz has been taken into account.
The kinetic profiles utilised for the modelling [17] are shown in Figure 1.

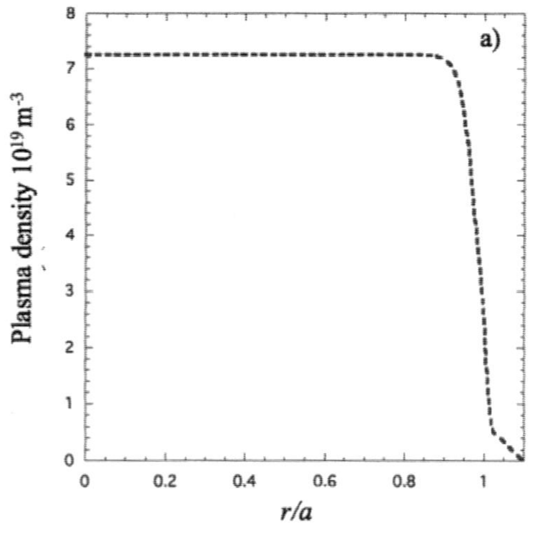

Figure 1a. Electron plasma density radial profile foreseen for the steady-state scenario of ITER

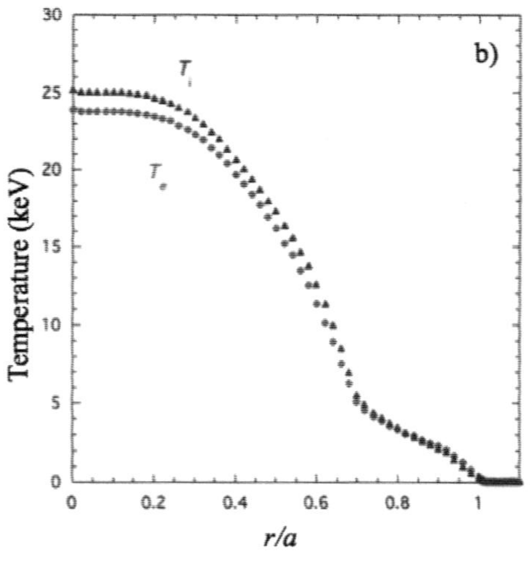

Figure 1b. Electron (red curve) and ion (blue curve) plasma temperature radial profiles foreseen for the steady-state scenario of ITER [17].

The radial profile of LH-driven current density has been modelled considering the effects of LH wave propagation in toroidal geometry. These effects can broaden and up-shift the $n_{//}$ initial spectrum inputted in the ray-tracing and Fokker-Planck modules of the LH^{star} code. In the code, the inputted LH $n_{//}$ spectrum (initial spectrum) does not coincide with the nominal antenna spectrum, as usually done in other modelling tools, but is produced as output of a dedicated module (LHPI: lower hybrid parametric instability). This module calculates the spectral broadening of PI produced at the plasma edge, utilising a slab geometry as described in Refs. 12,13.

In summary the runs of the code are performed in the following steps.

i) The non-linear physics of plasma edge produced by the PI phenomenon is taken into account for calculating the broadening of the spectrum launched by the antenna. This result allows calculating the amplification factor that considers, for any LH sideband spectral $n_{//}$ component, the convective losses of PI that occur in the realistic conditions of experiment. These conditions consist in considering the plasma inhomogeneous, and the finite spatial extent of the region illuminated by the antenna. Significant convective losses are produced indeed by the plasma inhomogeneity and finite extent of the pump wave region imposed by antenna geometry. Consequently, considering the kinetic profiles of the SOL (scrape-off layer) and plasma periphery (typically for $r/r_{\rm Ant}$ □r/a □0.8, where $r_{\rm Ant}$ is the radial layer of the antenna mouth), the initial $n_{//}$ spectrum is obtained and inputted to the ray-tracing and Fokker-Planck modules.
The availability of such initial spectrum is essential for determining, via quasi-linear effects, the coupled LH power propagation and damping in the plasma. To neglect the effect of spectral broadening due to PI should be allowed only provided that PI effects are negligible. However, this circumstance does not ever occur for typical tokamak plasma edge parameters [13]. A more complete evaluation of the spectral broadening (not done in the LH^{star} code) should be obtained by including also the effect of linear scattering by density fluctuations, as done in a recent work [18]..

ii) The initial spectrum is inputted in ray-tracing Fokker-Planck modules, which calculate the radial profiles of LH deposition by retaining the effects of ray propagation in toroidal geometry [16]. The effect of toroidicity can produce a further change of the spectrum launched by the antenna, consisting in an $n_{//}$ broadening and up-shift. The ray-tracing and Fokker-Planck modules are properly coupled: in a code run, starting from the edge layer, the ray tracing and Fokker-Planck analyses are performed step by step together and consistently for each ray. Consequently, the spatial evolution of the LH power distributed in the $n_{//}$ spectrum is calculated consistently with the quasi-linear effects, and the LH deposition profile is accurately calculated.
In case a ray reaches a LH wave cut-off layer, located at the plasma edge, the geometric optic limit fails, since as the wavelength becomes locally too large. Consequently, the ray-tracing technique cannot be further used. Therefore, in these cases, the LH^{star} code does not provide a complete output of LH deposition profiles.

In summary, the original features of the LH^{star} code consist in considering the PI-produced spectral broadening and in producing consistent quasi-linear and ray-tracing analyses. The broadened initial spectrum provides an essential contribution for filling the gap in phase

Journal of Physics: Conference Series **401** (2012) 012004 doi:10.1088/1742-6596/401/1/012004

velocities for LHCD, and for taking into account the parasitic LH power deposition at the edge at high densities. The calculations are performed only in the framework of validity of the WKB limit [12,13]. Consequently, the LH^{star} code is asked to not treat completely cases in which rays reach the LH cut-offs at the plasma edge.

Conversely (and incidentally), other widely utilised LH modelling tools, e.g., in Refs. 21,22, are however enabled continuing the ray-tracing runs also in case LH cut-offs are reached. Moreover, differently from the LH^{star} code, these tools calculate the toroidicity effects by ray-tracing independently of Fokker-Planck analysis. In these tools, the ray tracing runs are generally performed before initiating the Fokker-Planck analysis of quasi-linear damping. In this way, the toroidicity effect should underestimate the effect of initial spectral broadening. This problem can be solved by coupling together the ray-tracing and Fokker-Planck analysis as done in the LH^{star} code.

Results

We show here preliminary results of the modelling producing the radial profiles of LH current density driven expected for ITER. A coupled power density of 10 MW/m^2 at frequency of 5 GHz has been considered. The nominal antenna spectrum has a Gaussian distribution at around $n_{//}$=1.95 and width of 0.5. The modelling has been performed taking into account the effect of spectral broadening produced at the edge by PI, which is expected to occur with the following two cases of different electron temperatures in the SOL. *i*) The same SOL profiles foreseen for ITER are considered (T_e□20 eV for r/a ~1.0 – 1.05); *ii*) relatively higher electron temperature are assumed in most of the SOL, as further reasonable guess for ITER (T_e□700 eV for r/a ~1.0 – 1.01, T_e□500 eV for r/a ~1.01 – 1.04, T_e□20 eV for r/a ~1.04 – 1.05).

Higher electron temperature might be obtainable with operations aimed at reducing the particle recycling from the vessel walls. A local heating produced by coupling radiofrequency power at the electron cyclotron frequency should be also considered.

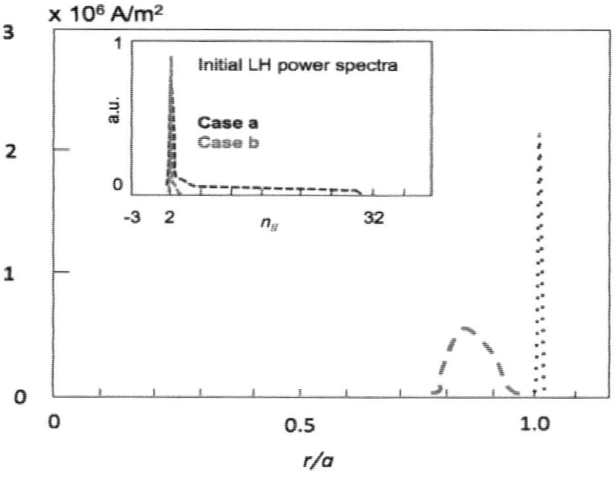

Figure 2. LH-driven current density profiles expected for ITER considering the same plasma parameters but different electron temperatures in the SOL. Consequently, a different broadening in the initial $n_{//}$ spectrum is produced by parametric instability. Case a (blue lines): profile foreseen for ITER. Case b (red lines): higher electron temperature has been assumed, see the text.

As consequence of the modelling performed with the LH^{star} code, the LH-driven current density profiles expected for ITER for the aforementioned cases are shown in Figure 2. For the case of electron temperature profile foreseen for ITER, a huge spectral broadening and a consequent LH power deposition at the very edge are expected to occur. Conversely, assuming a warmer edge, the coupled LH power is enabled to penetrate into the plasma bulk thank to the consequently diminished PI-produced spectral broadening.
Such result is supported by observation during the LH experiment on FTU performed in standard regime at ITER-relevant high plasma density. In these experiments, no effect of the coupled LH power occurred in the core when operating with an electron temperature profile in the periphery and SOL of FTU very similar to that of ITER [9]. Moreover, the FTU experiment result was optimistic for extrapolating LHCD scenarios to ITER, as weaker PI spectral broadening effect are expected to occur for an operating frequency (8 GHz) higher than the maximum foreseen for ITER (5 GHz).

Conclusions

The parametric instability is a phenomenon that broadens the LH spectrum launched by the antenna. Both a huge spectral broadening and the failure of the LH power penetration in the core occur in presence of plasma with SOL and periphery relatively cold, as typically occurs in LH experiments at high plasma densities. In this condition, also the collisional damping effect acts, preventing the LH power penetration in the core. However, this effect resulted small in experiments of JET at high density and in single pass LH regime, in which the same failure of LH power penetration was observed to occur, as in other LH experiments at high densities.
The PI is generally expected to occur at the antenna-plasma interface, where the relatively high densities necessary for coupling ($w_{pe} > w_0$) are generally accompanied by relatively low temperature of plasma edge. Low electron temperature favours indeed the spectral broadening, with consequent increase of the power deposition at the periphery. A case with higher temperature in the SOL has been considered (to account also for the strong uncertainty at this moment of the design), and inputted in the LH^{star} code for modelling the LHCD in ITER.
Operations producing a relatively high electron temperature in the SOL are necessary in ITER for reducing the risk of interaction of the coupled LH power with parametric instability at the plasma edge, which is detrimental for LH power penetration in the core.

References

1. Mailloux J et al. 36th EPS Conference on Plasma Phys. Sofia, June 29 - July 3, 2009 ECA Vol.**33E**, P-5.164 (2009)
2. Cesario R, Castaldo C, Fonseca A, De Angelis R et al., Conference on Radio-Frequency Power in Plasma, Clearwater, USA 2007
3. Cesario R et al., "Low recycling operations and improved core confinement in advanced scenarios of JET (Joint European Torus)" submitted to Plasma Phys. and Contr. Fus.
4. Fisch N J, "Theory of current drive in plasmas" *Rev. Mod. Physics,* 59, (1987) 175-234
5. Bernabei S, Daughney C, Efthimion P, Hooke W, Hosea J, Jobes F et al., "Lower Hybrid Current Drive in the PLT Tokamak", *Phys. Rev. Letters* **49**, (1982) 1255-1258

6. Porkolab M, Schuss J J, Lloyd B, Takase Y, Texter S, Bonoli P et al., "Observation of Lower Hybrid Current Drive at High Densities in the Alcator C Tokamak", *Phys. Rev. Letters,* **53** (*1984*) 450
7. De Marco F in Course and Workshop on Applications of RF waves to Tokamak Plasmas, Varenna (Italy) 1985, Editors, S Bernabei, U. Gasparino and E. Sindoni, International School of Plasma Physics (1985) 316
8. Cesario R, Amicucci L, Cardinali A, Castaldo C, Marinucci M et al., Nature comms., 1 (5) 55, 2010
9. Wallace G M, Hubbard A E, Bonoli P T, Faust I C, Harvey R W, Hughes J W et al., Nucl. Fusion 51 (2011) 083032
10. Brambilla M, *Kinetic Theory of Plasma Waves*, Oxford Press, 1999
11. Cesario R, Amicucci L, Castaldo C, Kempenaars M, Jachmich S et al., Plasma Physics and Controlled Fusion, Volume 53, Issue 8, pp. 085011 (2011)
12. Cesario R, Cardinali A, Castaldo C, Paoletti F, Mazon D and the EFDA-JET Contributors, "Modelling of Lower Hybrid Current Drive in Tokamak Plasmas by including the Spectral Broadening induced by Parametric Instability", *Phys. Rev. Letters* **92** 17 (2004) 175002
13. Cesario R, Cardinali A, Castaldo C, Paoletti F, Fundamenski W, Hacquin S and the JET-EFDA workprogramme Contributor·, "Spectral broadening of lower hybrid waves produced by parametric instability in current drive experiments on tokamak plasmas", *Nucl. Fusion* **46** (2006) 462-476.
14. Takase Y et al. Phys. Fluids **26**, 2992 (1985)
15. Cesario R and Cardinali A, Nucl. Fusion **29**, 1709 (1989)
16. Cardinali A, 2000 Recent Research Developments in Plasmas, Vol. 1, 185.
17. Colas L, Milanesio D, Faudot E, Goniche M and Loarte A, *Journ. of Nucl. Materials* 390–391 (2009) 959–962
18. Bertelli N Submitted to Plasma Phys. and Controlled Fusion
19. Bonoli P T and Englade R C, Phys. Fluids, **29**, 2937 (1986)
20. Giruzzi G et al., Nucl. Fusion **37**(5), (1997) 673

Journal of Physics: Conference Series **401** (2012) 012005 doi:10.1088/1742-6596/401/1/012005

Fast gyrofluid reconnection in high temperature plasmas

L Comisso[1], F L Waelbroeck[2] and D Grasso[1]

[1] Dipartimento di Energia, Politecnico di Torino, Corso Duca degli Abruzzi 24, 10129, Torino, and Istituto dei Sistemi Complessi - CNR, Via dei Taurini 19, 00185, Roma, Italy
[2] Institute for Fusion Studies, The University of Texas at Austin, Austin, TX 78712-1060, USA

E-mail: luca.comisso@polito.it

Abstract. The nonlinear evolution of collisionless magnetic reconnection in the presence of a strong guide field is analyzed on the basis of a gyrofluid model for compressible plasmas. It is found that, in a certain regime of plasma parameters, ion gyration contributes to generate two distinctive nonlinear acceleration phases of the growth rate. Furthermore, in the advanced nonlinear phase, finite values of the ion Larmor radius are identified to be responsible for a splitting of the narrow layer structures of ion guiding-center parallel velocity and density perturbations around the magnetic equilibrium null line.

1. Introduction

Magnetic reconnection is a process that changes the topology of the magnetic field lines and results in the conversion of magnetic energy into kinetic energy, thermal energy, and particle acceleration. It is observed to occur both in laboratory and in space plasmas, where it plays a key role in a number of events, classical examples of which are sawtooth relaxations in fusion devices, substorms in the Earth's magnetosphere, and solar and stellar flares [1, 2]. A general feature of these phenomena are the short time scales involved, which are much shorter than those associated with classical collisional dissipation processes. In the absence of collisions, the finite mass of electrons limit their response to a magnetic-field-aligned electric field and allows magnetic field lines to reconnect. Another common characteristic of these events is that the reconnection rate is not only fast, but also exhibits a sudden increase in its time derivative (see, e.g., [3, 4]). Explanation of the observed growth rates constitutes one of the main challenges of magnetic reconnection theory (see, e.g., [5]).

In this work we consider the regime of high temperature plasmas in the presence of a strong guide field, that is relevant for tokamaks as well as the solar corona. In this regime the sonic and ion Larmor radius, ρ_s and ρ_i, exceed the width of the electron diffusion region [6]. Therefore, sonic and ion finite Larmor radius effects cannot be neglected and indeed are known to produce a speed up of the reconnection rate [7 - 14]. Here we report new results of gyrofluid simulations that enable us to examine the effects of ion gyration on the growth rate and the field structures of spontaneous collisionless magnetic reconnection. This paper is organized as follows. In Section 2 we present the model equations that include the nonlocal response of the plasma resulting from ion gyration. In Section 3 these equations are used to simulate spontaneous magnetic reconnection in a current sheet. Finally, conclusions are drawn in Section 4.

2. The electromagnetic gyrofluid model

We consider a Hamiltonian gyrofluid model [15] obtained by taking the two lowest velocity space moments of the five-dimensional gyrokinetic equations for electrons and ions [16], assuming constant temperatures and neglecting collisions and the electron gyroradius. Electron inertia terms, on the other hand, are retained in order to break the frozen-in condition and allow for magnetic reconnection phenomena. The system of equations is completed using Ampère's law and the quasi-neutrality condition. As in Ref. [17], the magnetic curvature effects are neglected and the dynamics is assumed two-dimensional, with an ignorable coordinate along the strong, out-of-plane, magnetic field $B_0\hat{\mathbf{z}}$ (the guide field, that is taken to be constant). Furthermore, a plasma with single ion species and charge number $Z = 1$ is assumed.

We adopt a normalization scheme such that all the lengths are normalized to a characteristic magnetic equilibrium scale length L, and all times to the Alfvén time $\tau_A = L/v_A$, where $v_A = B_0/(\mu_0 n_0 m_i)^{1/2}$, with n_0 a constant background density, μ_0 the permeability of free space, and m_i the ion mass. Thus, dependent variables are normalized in the following way:

$$\left(\hat{n}_{i,e}, \hat{u}_{i,e}, \hat{\psi}, \hat{\phi}\right) = \left(\frac{L}{d_i}\frac{n_{i,e}}{n_0}, \frac{L}{d_i}\frac{u_{i,e}}{v_A}, \frac{\psi}{B_0 L}, \frac{\phi}{B_0 L v_A}\right), \tag{1}$$

where dimensionless quantities appear on the left hand side. Hereafter the carets denoting normalized quantities will be omitted for simplicity of notation. The fields n_i and $u_i = \hat{\mathbf{z}} \cdot \bar{\mathbf{v}}_i$ represent the perturbed density and the parallel (out-of-plane) velocity of the ion *guiding centers*, whereas n_e and $u_e = \hat{\mathbf{z}}\cdot\mathbf{v}_e$ are the perturbed density and the parallel velocity of the electrons. We indicate with ψ the magnetic flux function, related to the in-plane magnetic field by $\mathbf{B}_\perp = \nabla\psi\times\hat{\mathbf{z}}$, and with ϕ the electrostatic potential.

In a right-handed Cartesian coordinate system (x, y, z), the evolution equations of the model are:

$$\frac{\partial n_i}{\partial t} + [\Phi, n_i] + [u_i, \Psi] = 0, \tag{2}$$

$$\frac{\partial}{\partial t}\left(\Psi + d_i^2 u_i\right) + [\Phi, \Psi + d_i^2 u_i] + \rho_i^2[n_i, \Psi] = 0, \tag{3}$$

$$\frac{\partial n_e}{\partial t} + [\phi, n_e] + [u_e, \psi] = 0, \tag{4}$$

$$\frac{\partial}{\partial t}\left(\psi - d_e^2 u_e\right) + [\phi, \psi - d_e^2 u_e] - \rho_s^2[n_e, \psi] = 0, \tag{5}$$

where $[f, g] = \partial_x f \partial_y g - \partial_y f \partial_x g$ is the canonical Poisson bracket between two generic fields f and g. The four dimensionless parameters appearing in the above equations are the (normalized) electron and ion skin depth, $d_e = (c/\omega_{pe})/L$ and $d_i = (c/\omega_{pi})/L$ respectively, and the (normalized) sonic and ion Larmor radius, $\rho_s = (c_s/\omega_{ci})/L$ and $\rho_i = (v_{ti}/\omega_{ci})/L$ respectively. Here, $c_s = (T_e/m_i)^{1/2}$ is the sound speed based on the electron temperature, $v_{ti} = (T_i/m_i)^{1/2}$ is the ion thermal speed, and the other symbols have their usual meaning. Furthermore, $\Phi = \Gamma_0^{1/2}\phi$ and $\Psi = \Gamma_0^{1/2}\psi$ are the gyro-averaged electrostatic and parallel magnetic potentials, where the symbol $\Gamma_0^{1/2}$ refers to the gyro-averaged operator introduced by Dorland and Hammet [18], that we adopt in its lowest-order Padé approximant form $\Gamma_0^{1/2} = \left(1 - \rho_i^2\nabla_\perp^2/2\right)^{-1}$. Eqs. (2)-(5) are closed by the parallel component of Ampère's law

$$\nabla_\perp^2\psi = u_e - \Gamma_0^{1/2}u_i, \tag{6}$$

and by imposing quasi-neutrality on the particle density (not the guiding-center density)

$$n_e = \Gamma_0^{1/2}n_i + \left(\frac{\Gamma_0 - 1}{\rho_i^2}\right)\phi, \tag{7}$$

where the last term of Eq. (7) represents the polarization density due to the variation of the electric field around a gyro-orbit.

The system of equations (2)-(7) describes both the inertial ($\beta_e \ll 2m_e/m_i$) and the kinetic ($\beta_e \gg 2m_e/m_i$) Alfvén wave regimes, whereas for $\beta_e \sim 2m_e/m_i$ the model equations need to be extended to account for the electron Landau damping occuring in this regime $v_{te} \sim v_A$ [19]. Here, the parameter that distinguishes between these regimes is the electron beta, defined as $\beta_e \equiv 2\mu_0 n_0 T_e/B_0^2$, which represents the ratio of the electron thermal pressure to the magnetic pressure.

3. Reconnection rate and field structures of ion guiding centers

In order to investigate the evolution of spontaneous magnetic reconnection instabilities, the system of equations (2)-(7) is solved numerically considering the following static equilibrium:

$$\psi_{eq}(x) = \sum_{n=-11}^{11} a_n e^{inx}, \qquad n_{i_{eq}}(x) = n_{e_{eq}}(x) = n_{eq}, \qquad u_{i_{eq}}(x) = 0, \tag{8}$$

where a_n are the Fourier coefficients of the function $f(x) = 0.1/\cosh^2 x$, and n_{eq} represent a uniform, nondrifting background density. The fields in the model equations are decomposed in a time independent equilibrium and an evolving perturbation that is advanced in time according to a third order Adams-Bashforth algorithm. Double periodic boundary conditions are imposed and a pseudospectral method is used for the space discretization on a domain $\{(x, y) : -\pi \leq x < \pi, -2\pi \leq y < 2\pi\}$, with a resolution up to 2048×256 grid points. The reconnection instability is initiated by perturbing the equilibrium with a small disturbance of the ion guiding-center density defined as $\tilde{n}_i = 10^{-6}\left(\cos(x + y/2) - \cos(x - y/2)\right)$.

The equilibrium (8) is tearing unstable with linear stability parameter $\Delta' = 59.9$ for the longest wavelength mode in the system $k_y = 2\pi/L_y = 1/2$, with L_y the length of the domain along the y direction. This choice allows us to explore magnetic reconnection in the large-Δ' regime, relevant to the sawtooth oscillations in tokamaks [20], for example, or the Geospace Environmental Modeling (GEM) Challenge [21]. The value of the ion skin depth d_i is taken to be equal to the half width of the magnetic equilibrium scale length L. Considering the challenge to clearly separate all relevant spatial scales of the problem, the electron to ion mass ratio, m_e/m_i, is chosen to be 1/100, making $d_e = (m_e/m_i)^{1/2} d_i = 5 \times 10^{-2}$. This mass ratio is larger than the real one, but it has been shown by Ricci *et al.* [22] that most features of collisionless reconnection with a guide field are not sensitive to the mass ratio, except for the magnitude of the electron velocities in the electron diffusion region. Finally, in order to distinguish ion gyration effects from those of electron pressure gradient and electron inertia, ρ_i and ρ_s are varied independently.

As a first step, we consider a range of parameters such that $\rho_i \ll \rho_s < d_e \ll d_i \lesssim L$. In this regime, with $\rho_s = 2.5 \times 10^{-2}$ and $\rho_i = 10^{-5}$, the instantaneous growth rate of the reconnection instability is shown in Fig. 1(a). The growth rate is defined as $\gamma = d(\ln \delta\psi_X)/dt$, where $\delta\psi_X$ is the reconnected flux at the X-point. After an initial transient ($\gamma_L t \lesssim 1.5$), the reconnection instability evolve through three different stages: the linear phase, scaling as $e^{\gamma_L t}$ (constant growth rate $\gamma_L = 2.9 \times 10^{-3}$) ($1.5 \lesssim \gamma_L t \lesssim 4.5$), followed by the faster-than-exponential phase, during which the growth rate increases up to a peak value $\gamma = 1.3 \times 10^{-2}$ ($4.5 \lesssim \gamma_L t \lesssim 7.7$), and finally the saturation period in which the growth rate slows down ($\gamma_L t \gtrsim 7.7$).

A similar evolution of the reconnection instability is found for $\rho_i \ll d_e < \rho_s \ll d_i \lesssim L$, but with higher linear ($\gamma_L = 4.9 \times 10^{-3}$) and nonlinear growth rates (peak value $\gamma = 1.83 \times 10^{-2}$). Fig. 1(b) shows the instantaneous growth rate for $\rho_s = 0.1$ and $\rho_i = 10^{-5}$. The speed up of the reconnection can be linked to the different dispersive character of the kinetic Alfvén wave respect to the shear Alfvén wave. In fact, the term proportional to ρ_s^2 in Eq. (5) introduces

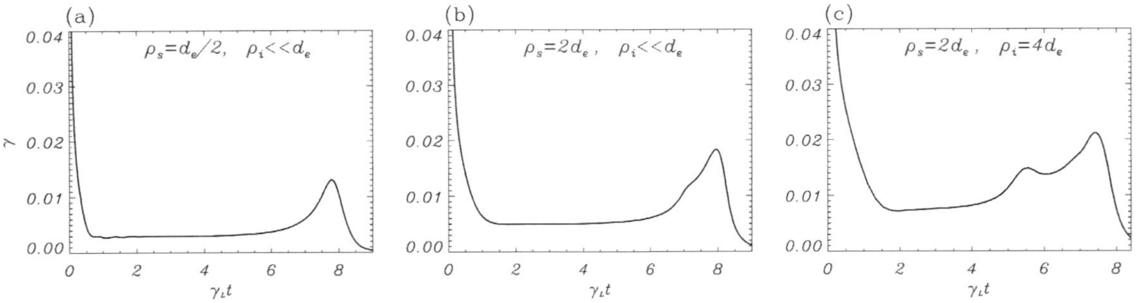

Figure 1. Growth rate of the reconnection instability, $\gamma = d(\ln \delta\psi_X)/dt$, as a function of $\gamma_L t$, for (a) $\rho_s = 2.5\times10^{-2}$, $\rho_i = 10^{-5}$, (b) $\rho_s = 0.1$, $\rho_i = 10^{-5}$, and (c) $\rho_s = 0.1$, $\rho_i = 0.2$. The linear growth rate γ_L is determined numerically: (a) $\gamma_L = 2.9\times10^{-3}$, (b) $\gamma_L = 4.9\times10^{-3}$, and (c) $\gamma_L = 7.3\times10^{-3}$. The values of the electron and ion skin depth are $d_e = 5\times10^{-2}$ and $d_i = 0.5$.

the dynamics of the kinetic Alfvén wave, which has a dispersive character, i.e. $\omega \propto k_\perp^2$, where ω is the wave frequency and $k_\perp$ the wave number. For $\rho_s > d_e$ (equivalent to $\beta_e > 2m_e/m_i$), the structure of the electron inertial region is controlled by the dynamics of the kinetic Alfvén wave rather than the Alfvén dynamics because this region has an intrinsic scale length which is well below the ion inertial length. Indeed, the quadratic dispersion character of the kinetic Alfvén wave leads to an increase in the phase speed ($v_\phi \propto k_\perp$) with decreasing spatial scales and therefore to an increase in the velocity at which the electrons can be ejected from the X-point as the spatial scales of the electron diffusion region decrease [23]. In the absence of a guide field (strictly antiparallel field merging), the counterpart of the electron pressure term is the Hall term in the generalized Ohm's law, which also introduces a high frequency dispersive wave, the whistler, which plays a role analogous to that of the kinetic Alfvén wave [24 - 26].

For plasma parameters of interest to magnetic confinement fusion experiments, ion Larmor radius effects and sonic Larmor radius effects should be considered simultaneously, since $\rho_i/\rho_s = (T_i/T_e)^{1/2} \sim 1$ and $\rho_i/d_e = (\beta_i m_i/2m_e)^{1/2} > 1$, with $\beta_i \equiv 2\mu_0 n_0 T_i/B_0^2$. For this reason and to highlight the contribution of the ion gyration, we consider a regime such that $d_e < \rho_s \lesssim \rho_i \ll d_i \lesssim L$. In particular, the instantaneous growth rate for $\rho_s = 0.1$ and $\rho_i = 0.2$ is shown in Fig. 1(c). In comparison to the cold ion limit, the evolution of the reconnection instability exhibits a novel behaviour: nonlinearly, the growth rate is characterized by two distinct phases of strong increase. After an initial transient, which lasts until $\gamma_L t \lesssim 2$, reconnection evolves through a linear phase from $\gamma_L t \approx 2$ to $\gamma_L t \approx 3$ with $\gamma_L = 7.3\times10^{-3}$, and then develops into a first growth rate acceleration phase that begins at an island width $w \sim d_e$, and persists until $\gamma_L t \approx 5.5$, when $\gamma = 1.49\times10^{-2}$. The first acceleration stage is followed by a phase in which the growth rate decreases until $\gamma_L t \approx 6$, when a second remarkable increase takes place. This second acceleration phase is accompanied by an opening up of the electron parallel velocity layer into an "X"-geometry similar to the island separatrix. Moreover, from $\gamma_L t \gtrsim 7$ also the parallel velocity of the ion guiding centers splits into an "X"-geometry configuration centered at the X-point, but with lower opening compared to the electron velocity. The formation of an "X"-geometry in the current distribution was linked to an increase of the reconnection rate in a number of works [7 - 9, 23 - 28]. Here we observe a similar behaviour in the parallel velocity of the ion guiding centers, which enters the parallel electron momentum equation through Ampère's law. In fact, making use of Eq. (6) we can rewrite Eq. (5) as

$$\frac{\partial}{\partial t}\left(\psi - d_e^2\nabla_\perp^2\psi - d_e^2\Gamma_0^{1/2}u_i\right) + [\phi, \psi - d_e^2\nabla_\perp^2\psi - d_e^2\Gamma_0^{1/2}u_i] - \rho_s^2[n_e,\psi] = 0, \quad (9)$$

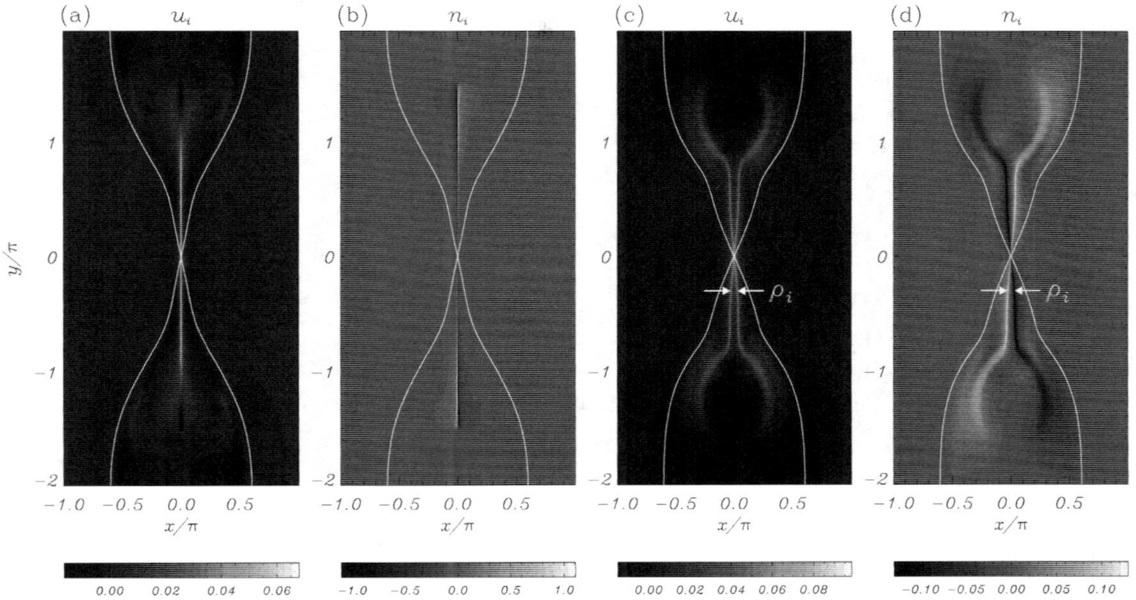

Figure 2. Contour plots of the ion guiding-center parallel velocity u_i and perturbed density n_i at an island width $w \approx 1.2\pi$ for (a)-(b) $\rho_s = 0.1$, $\rho_i = 10^{-5}$, and for (c)-(d) $\rho_s = 0.1$, $\rho_i = 0.2$. The values of the other plasma parameters are: $d_e = 5 \times 10^{-2}$, $d_i = 0.5$. The magnetic island separatrix at the corresponding time have been superimposed (white line). White arrows indicate the length scale ρ_i.

which puts in evidence how the frozen-in condition is violated also due to a term depending on u_i and often neglected in fluid models. However, for low-β regimes the influence of u_i on the rate of reconnection remains small in comparison to the role played by u_e. The increase of the growth rate during the second acceleration phase reaches a peak value $\gamma = 2.12 \times 10^{-2}$ at $\gamma_L t \approx 7.8$. After this time a saturation stage takes place.

The macroscopic structures that develop in the advanced nonlinear phase for the cold ion limit ($\rho_i^2/\rho_s^2 = T_i/T_e \ll 1$) and for the finite ion Larmor radius regime are compared in Fig. 2. Contour plots of u_i and n_i for $\rho_s = 0.1$, $\rho_i = 10^{-5}$ are shown in Figs. 2(a) and 2(b) for an island width $w \approx 1.2\pi$, corresponding at $\gamma_L t \approx 8.3$ in Fig. 1(b). Isocontours of the same fields for $\rho_s = 0.1$, $\rho_i = 0.2$ are shown in Figs. 2(c) and 2(d) for the same island width, corresponding at $\gamma_L t \approx 8$ in Fig. 1(c). In the cold ion limit narrow layer structures aligned with the magnetic equilibrium null line characterize the parallel velocity and perturbed density of the ion guiding centers. Finite values of the ion Larmor radius are identified to be responsible for a splitting of the regions of high ion parallel velocity. The same splitting happens in the regions of high density fluctuations. The result is that the quadrupolar density perturbation of the ion guiding centers, characterized by two regions of enhanced ion density and two regions of depleted ion density (cavities), instead of being aligned along the neutral line, opens on a spatial scale $\sim \rho_i$, as shown in Fig. 2(d). This behaviour is different from what was found for the antisymmetric signature of the quadrupolar electron density perturbation, where the regions of high and low electron density map the separatrices [8, 17, 22, 29 - 31]. The field structures of u_i and n_i were interpreted in terms of the change, due to ion Larmor radius effects, of the advection of two of the four Lagrangian invariants of the model (2)-(7), defined as $I_\pm \equiv \Psi + d_i^2 u_i \pm d_i \rho_i n_i$ [17]. We observe that the splitting of u_i and n_i for large values of ρ_i was not found in Ref. [17], where a regime of plasma parameters ($d_{e,i}$, $\rho_{s,i}$) four times larger than those of Figs. 1(c), 2(c), 2(d) was investigated (but with the same initial current sheet), in such a way that β_e and β_i remain the same. This is because the excursion of the ions in their Larmor gyration causes a smoothing of u_i and n_i for which ρ_i is the typical scale length [17, 32]. Ion gyration effects also influence the growth rate, that does not exhibit two distinct strong acceleration phases for the plasma parameters $d_e = 0.2$, $d_i = 2$, $\rho_s = 0.2$, $\rho_i = 0.8$, corresponding to the regime $d_e < \rho_s \lesssim \rho_i \sim L \lesssim d_i$.

4. Conclusions

In summary, we have shown that collisionless magnetic reconnection with a guide field undergoes a strong enhancement of the growth rate when a rarefied high temperature plasma is considered. Particularly noteworthy is the fact that the growth rate exhibits two distinct nonlinear acceleration phases when the ion Larmor radius exceeds the electron skin depth and is much less than the magnetic equilibrium shear length. We have also shown that finite values of the ion Larmor radius are responsible to produce an open configuration of the ion guiding-center parallel velocity, that is reminiscent of the cross-shaped configuration that characterize the parallel current density in magnetic reconnection mediated by the kinetic Alfvén wave. The same geometric pattern belongs to the quadrupolar ion guiding-center density perturbation.

Acknowledgments

The authors would like to acknowledge fruitful conversations with Dario Borgogno, Emanuele Tassi, Ahmet Aydemir and Vladimir Lakhin. One of us (L.C.) is grateful for the hospitality of the Institute for Fusion Studies at the University of Texas at Austin, where part of this work was done. This work was supported by the European Community under the contracts of Association between EURATOM and ENEA and by the U.S. Department of Energy under Contract No. DE-FG02-04ER-54742. The views and opinions expressed herein do not necessarily reflect those of the European Commission.

References

[1] Priest E R and Forbes T G 2000 *Magnetic Reconnection: MHD Theory and Applications* (Cambridge: Cambridge University Press)
[2] Biskamp D 2000 *Magnetic Reconnection in Plasmas* (Cambridge: Cambridge University Press)
[3] Hastie R J 1998 *Astrophys. Space Sci.* **256**, 177
[4] Bhattacharjee A 2004 *Annu. Rev. Astron. Astrophys.* **42**, 365
[5] Yamada M, Kulsrud R and Ji H 2010 *Rev. Mod. Phys.* **82**, 603
[6] Birn J and Priest E R (ed) 2007 *Reconnection of Magnetic Fields: Magnetohydrodynamics and Collisionless Theory and Observations* (Cambridge: Cambridge University Press)
[7] Aydemir A Y 1992 *Phys. Fluids B* **4**, 3469
[8] Kleva R G, Drake J F and Waelbroeck F L 1995 *Phys. Plasmas* **2**, 23
[9] Rogers B and Zakharov L 1996 *Phys. Plasmas* **3**, 2411
[10] Cafaro E, Grasso D, Pegoraro F, Porcelli F and Saluzzi A 1998 *Phys. Rev. Lett.* **80**, 4430
[11] Grasso D, Califano F, Pegoraro F and Porcelli F 2000 *Plasma Phys. Rep.* **26**, 512
[12] Porcelli F, Borgogno D, Califano F, Grasso D, Ottaviani M and Pegoraro F 2002 *Plasma Phys. Control. Fusion* **44**, 389
[13] Bhattacharjee A, Germaschewski K and Ng C S 2005 *Phys. Plasmas* **12**, 042305
[14] Biancalani A and Scott B D 2012 *Europhys. Lett.* **97**, 15005
[15] Waelbroeck F L and Tassi E 2012 *Commun. Nonlinear Sci. Numer. Simulat.* **17**, 2171
[16] Snyder P B and Hammett G W 2001 *Phys. Plasmas* **8**, 3199
[17] Comisso L, Grasso D, Tassi E and Waelbroeck F L 2012 *Phys. Plasmas* **19**, 042103
[18] Dorland W and Hammett G W 1993 *Phys. Fluids B* **5**, 812
[19] Hammett G W, Dorland W and Perkins F W 1992 *Phys. Fluids B* **4**, 2052
[20] Angioni C, Pochelon A, Gorelenkov N N, McClements K G, Sauter O, Budny R V, de Vries P C, Howell D F, Mantsinen M, Nave M F F, Sharapov S E and contributors to the EFDA-JET Workprogramme 2002 *Plasma Phys. Control. Fusion* **44**, 205

[21] Birn J, Drake J F, Shay M A, Rogers B N, Denton R E, Hesse M, Kuznetsova M, Ma Z W, Bhattacharjee A, Otto A and Pritchett P L 2001 *J. Geophys. Res.* **106**, 3715
[22] Ricci P, Brackbill J U, Daughton W and Lapenta G 2004 *Phys. Plasmas* **11**, 4102
[23] Rogers B N, Denton R E, Drake J F and Shay M A 2001 *Phys. Rev. Lett.* **87**, 195004
[24] Biskamp D, Schwarz E and Drake J F 1997 *Phys. Plasmas* **4**, 1002
[25] Shay M A and Drake J F 1998 *Geophys. Res. Lett.* **25**, 3759
[26] Drake J F, Shay M A and Swisdak M 2008 *Phys. Plasmas* **15**, 042306
[27] Lazzaro E, Ferrero M, Gianoli L and Valdettaro L 2000 *Phys. Scr.* **61**, 624
[28] Cassak P A, Shay M A and Drake J F 2005 *Phys. Rev. Lett.* **95**, 235002
[29] Pritchett P L 2005 *Phys. Plasmas* **12**, 062301
[30] Drake J F, Shay M A, Thongthai W and Swisdak M 2005 *Phys. Rev. Lett.* **94**, 095001
[31] Grasso D, Tassi E and Waelbroeck F L 2010 *Phys. Plasmas* **17**, 082312
[32] Del Sarto D, Marchetto C, Pegoraro F and Califano F 2011 *Plasma Phys. Control. Fusion* **53**, 035008

Journal of Physics: Conference Series **401** (2012) 012006 doi:10.1088/1742-6596/401/1/012006

Identifying the role of non-adiabatic passing electrons in ITG/TEM microturbulence by comparing fully kinetic and hybrid electron simulations

J Dominski[1], S Brunner[1], S K Aghdam[1], T Görler[2], F Jenko[2], and D Told[2]

[1] Centre de Recherches en Physique des Plasmas, Association EURATOM-Confédération Suisse, Ecole Polytechnique Fédérale de Lausanne, CH-1015 Lausanne, Switzerland.
[2] Max-Planck-Institut für Plasmaphysik, EURATOM Association, Boltzmannstrasse 2, D-85748 Garching, Germany.

E-mail: `julien.dominski@epfl.ch`

Abstract. The response of passing electrons in Ion Temperature Gradient (ITG) and Trapped Electron Mode (TEM) microinstability regimes is investigated in tokamak geometry making use of the flux-tube version of the gyrokinetic code GENE [Jenko *et al.* 2000 *Phys. Plasmas* **7** 1904]. Results are obtained with two different electron models: 1) fully kinetic, and 2) hybrid, in which passing particles are forced to respond adiabatically while trapped are handled kinetically. Comparing linear eigenmodes obtained with these two models enables to systematically isolate radially fine structures located at corresponding MRS's, clearly resulting from the non-adiabatic passing electron response. The analysis of preliminary non-linear simulations in the ITG regime shows that these fine structures on the non-axisymmetric modes survive in the turbulent phase. Furthermore, through non-linear coupling to axisymmetric modes, they induce modulations in the effective density and ion/electron temperature profiles: flattening at low order MRS's and steepening in between, as was already observed in Ref. [Waltz *et al.*, 2006 *Phys. Plasmas* **13** 052301].

1. Introduction

When studying Ion Temperature Gradient (ITG) and Trapped Electron Mode (TEM) microinstabilities and associated turbulence in magnetic fusion devices, it is often assumed that the frequency of the modes are sufficiently low such that the highly mobile *passing* electrons are responding adiabatically. The reduced adiabatic response for passing electrons has thus been extensively applied in gyrokinetic simulations of turbulent transport in the ITG/TEM regime. This approximation is indeed practical due to this important time scale separation between the ion and electron dynamics, such that carrying out simulations resolving the full kinetic evolution of all species remains a great numerical challenge.

The assumption of adiabatic response for the passing electrons is based on the argument that these particles are highly mobile in their motion along the magnetic field line, in such a way that the parallel phase velocity $\omega_r/k_\parallel$ of ITG/TEM microinstabilities is small compared to the electron thermal velocity $v_{th,e}$: $|\omega_r/k_\parallel| \ll v_{th,e}$, where ω_r is the (real) frequency of the mode and $k_\parallel$ its local wavevector component parallel to the magnetic field. However, it turns

Journal of Physics: Conference Series **401** (2012) 012006 doi:10.1088/1742-6596/401/1/012006

out that near low order Mode Rational Surfaces (MRS's) in a toroidal magnetized plasma, the adiabatic passing electron model is in fact not justified. Let us recall that a low order MRS is a surface at which the safety factor q_s is rational, $q_s = -m/n$, with m, n integer, and n a typical toroidal mode number of the considered instabilities (ITG/TEM in our case). At such a MRS, corresponding resonant Fourier modes, *i.e.* with poloidal-toroidal mode number pairs (m, n) such that $q_s = -m/n$, align with the magnetic field line and the associated parallel phase velocity $\omega/k_\parallel$ becomes larger than the electron thermal velocity $v_{th,e}$ within a radial width Δx around this surface. The condition for adiabatic response is thus violated within this interval.

Linear simulations of low frequency microinstabilities have shown that, as a consequence of the non-adiabatic response of passing electrons near MRS's, fine radial structures may form on the fields (e.g. electrostatic potential) of associated eigenmodes within the radial width Δx [1, 2]. The effect of this non-adiabatic electron response near MRS's in the non-linear turbulent regime and its repercussions on associated transport has however barely been addressed in the literature. To our knowledge, this subject has only been partly covered in Ref. [3].

Here, we present first results of a more systematic study of the non-adiabatic electron response in the vicinity of MRS's. This analysis is based on results obtained using the gyrokinetic GENE code [4] considering the simplest possible system. Despite the fact that GENE is a very comprehensive code, which in particular has been generalized more recently to enable global (finite $\rho^\star$) simulations [5], the flux-tube (*i.e.* local) version is considered for this study. Indeed, although all other profiles (temperature/density and their gradients, magnetic geometry coefficients) are assumed constant in the flux tube model, this local representation accounts for the (linearized) variation of the safety factor profile $q_s(x)$ and thus enables to address the particular dynamics in the vicinity of MRS's. Furthermore, a tokamak geometry with circular, concentric magnetic surfaces is chosen, described by the so-called analytic *ad hoc* model [6]. Let us finally point out that all simulations considered in this paper have been are carried out in the electrostatic limit.

The rest of this paper is organized as follows. In Sec. 2, general aspects of the GENE code are briefly described. Section 3 then presents the three different electron models implemented in GENE: 1) the fully kinetic, 2) the (fully) adiabatic, and 3) a hybrid electron model in which the trapped particles are described kinetically, while the passing ones are forced to respond adiabatically. Comparing simulation results from this later model with corresponding fully kinetic electron results provides a systematic way for identifying non-adiabatic passing electron effects. In Sec. 4, a simple analytical estimate for the radial width Δx of the non-adiabatic region around MRS's is derived. In Sec. 5, this theoretical estimate is then compared to the width of fine structures appearing in linear GENE simulations of ITG and TEM eigenmodes. To what extent these fine structures survive in the non-linear regime is then addressed in Sec. 6, where preliminary results of ITG microturbulence have been analyzed. The appearance of regular structures around MRS's on the flux-surface- and time-averaged fields, as a result of non-linear coupling with the unstable modes, is observed, in particular leading to flattening of density and temperature profiles. Conclusions are finally drawn in Sec. 7.

2. The GENE code

GENE evolves the distributions of an arbitrary number of different particle species by solving the associated gyrokinetic equations [7]. It makes use of an Eulerian-based representation and enables to carry out linear and non-linear simulations of microinstabilities and related turbulence in magnetic confinement devices. The system of gyrokinetic equations is closed with reduced Maxwell's equations for the electromagnetic potentials: the quasi-neutrality (or Poisson's) equation is sufficient if only electrostatic fluctuations are considered, while Ampère's law is invoked in addition if electromagnetic fluctuations are accounted for. More details on the system of gyrokinetic equations implemented in GENE can be found in Refs. [5, 8, 9].

In view of presenting the simulation results in Secs. 5 and 6, let us still briefly describe the configuration space coordinates considered in GENE: The code makes use of a field-aligned coordinate system (x, y, z), which is the natural choice for representing the fluctuating fields of microturbulence, themselves aligned with the equilibrium magnetic field $\vec{B}_0$ ($\vec{B}_0 \parallel \nabla x \times \nabla y$). For a tokamak configuration, in terms of the magnetic coordinates (ψ, χ, φ) = (poloidal magnetic flux, straight field line poloidal angle, toroidal angle), one defines the radial coordinate x as an appropriate function of ψ with units of length, the bi-normal coordinate $y = C_y[q_s(\psi)\chi - \varphi]$, and the "parallel" coordinate $z = \chi$. For flux-tube simulations, one chooses $C_y = x_0/q_s(0)$, where x_0 and $q_s(0)$ are respectively the minor radius and safety factor at the flux-tube center, so that y acquires units of length as well. In the coordinates (x, y, z), the flux-tube can then be expressed as a simple rectangular box of size $L_x \times L_y \times L_z$, where $L_z = 2\pi$.

3. Electron Models in GENE

We shall briefly describe here the implementation of the hybrid electron model, a new feature in GENE, which is a reduced description for the electron dynamics in which the trapped particles are described kinetically, while the passing particles are forced to respond adiabatically. This model is clearly only applicable for electrostatic fluctuations, such as ITG/TEM, as it is unable to account for fluctuations in the current carried by passing electrons, which are essential for describing electromagnetic modes. In the frame of the hybrid model, the Quasi-Neutrality Equation (QNE) is thus the only field equation that needs to be accounted for. Let us recall the form taken by the QNE in the fully kinetic and adiabatic description, before introducing its form in the hybrid approach.

3.1. Fully Kinetic

The starting form of the QNE simply reads:

$$\sum_{j=\text{ions, e}^-} q_j\, \rho_{1,j}(\vec{x}, t) = 0, \tag{1}$$

where $\rho_{1,j}(\vec{x}, t)$ is the fluctuating part of the density field of species j (here considering one ion species and electrons). Written in terms of the fluctuating part of the gyroaveraged particle distributions $f_{1,j}(\vec{X}, v_\parallel, \mu)$, expressed in gyrocenter variables $(\vec{X}, v_\parallel, \mu)$ [7], the QNE (1) becomes:

$$\sum_{j=\text{ions, e}^-} \left\{ \frac{2\pi q_j}{m_j} \int B^\star_{0,\parallel} \bar{f}_{1,j} dv_\parallel d\mu - \frac{q_j^2 n_{0,j}}{T_{0,j}} \left[\phi_1 - \frac{B_0}{T_{0,j}} \int \bar{\bar{\phi}}_1 \exp(-\frac{\mu B_0}{T_{0,j}}) d\mu \right] \right\} = 0, \tag{2}$$

where m_j and q_j are respectively the mass and electric charge of species j, B_0 the equilibrium magnetic field amplitude, and $B^\star_{0,\parallel}/m_j$ the Jacobian of the transformation from particle phase space variables $(\vec{x}, \vec{v})$ to gyrocenter variables $(\vec{X}, v_\parallel, \mu)$. The background density and temperature of each species are noted $n_{0,j}$ and $T_{0,j}$ respectively, while the fluctuating electrostatic potential is given by ϕ_1. As appearing in Eq. (2), the charge density of each species is decomposed into two contributions: the gyrodensity, explicitely expressed in terms of $\bar{f}_1$, and the polarization drift, expressed in terms of ϕ_1. The notation $\bar{\mathcal{A}}$ stands for the average of the field $\mathcal{A}(\vec{x})$ over the gyroangle θ (so-called gyroaverage):

$$\bar{\mathcal{A}}(\vec{X}, \mu) = \frac{1}{2\pi} \oint d\theta \mathcal{A}(\vec{X} + \vec{r}(\theta)),$$

where $\vec{r}(\vec{X}, \mu, \theta)$ is the gyroradius. The field $\bar{\bar{\phi}}_1$, appearing in the polarization drift terms, corresponds to the double gyroaverage of ϕ_1, which in (2) is furthermore weighted over the background distribution, assumed Maxwellian. Note that the integral $\int dv_{\parallel} d\mu$ for computing the gyrodensity term is carried out over the whole velocity space for both ions and electrons. Equation (2) thus corresponds to the fully kinetic electron model requiring the evolution of both the ion and electron distributions.

3.2. *Adiabatic*

In the (fully) adiabatic model, *all* electrons are forced to respond adiabatically. In this case the QNE reads:

$$\left\{ \frac{2\pi q_j}{m_j} \int B^{\star}_{0,\parallel} \bar{f}_{1,j} dv_{\parallel} d\mu - \frac{q_j^2 n_{0,j}}{T_{0,j}} \left[\phi_1 - \frac{B_0}{T_{0,j}} \int \bar{\bar{\phi}}_1 \exp(-\frac{\mu B_0}{T_{0,j}}) d\mu \right] \right\}_{j=\text{ions}} = \frac{e^2 n_{0,e}}{T_{0,e}} (\phi_1 - \langle \phi_1 \rangle_{\text{f.s.}}), \tag{3}$$

where the sum over species on the left hand side of Eq. (2) is now reduced to the sole ion contribution, while the charge density contribution from the adiabatic electron response appears on the right hand side of Eq. (3). The notation $\langle \phi_1 \rangle_{\text{f.s.}}$ stands for average over the magnetic flux surface. For this adiabatic model, there is clearly no need to evolve the gyrokinetic equation for the electron distribution.

3.3. *Hybrid*

Finally, in case of the hybrid model, the QNE is written:

$$\begin{aligned} &\left\{ \frac{2\pi q_j}{m_j} \int B^{\star}_{0,\parallel} \bar{f}_{1,j} dv_{\parallel} d\mu - \frac{q_j^2 n_{0,j}}{T_{0,j}} \left[\phi_1 - \frac{B_0}{T_{0,j}} \int \bar{\bar{\phi}}_1 \exp(-\frac{\mu B_0}{T_{0,j}}) d\mu \right] \right\}_{j=\text{ions}} \\ = &-\left\{ \frac{2\pi q_j}{m_j} \int_{\text{trap}} B^{\star}_{0,\parallel} \bar{f}_{1,j} dv_{\parallel} d\mu - \frac{q_j^2 n_{0,j}}{T_{0,j}} \left[\alpha_t \phi_1 - \frac{B_0}{T_{0,j}} \int_{\text{trap}} \bar{\bar{\phi}}_1 \exp(-\frac{\mu B_0}{T_{0,j}}) d\mu \right] \right\}_{j=\text{e}^-} \\ &+(1-\alpha_t) \frac{e^2 n_{0,e}}{T_{0,e}} (\phi_1 - \langle \phi_1 \rangle_{\text{f.s.}}), \end{aligned} \tag{4}$$

where the fluctuating electron charge density on the right hand side of Eq. (4) is formed of two contributions: The term $\{\ldots\}_{j=\text{e}^-}$ from the trapped particles, handled kinetically, and the contribution from the passing particles, handled adiabatically and weighted by the corresponding fraction $(1-\alpha_t)$. At a given position $\vec{x}$, the trapped electron contribution involves velocity integrals $\int_{\text{trap}} dv_{\parallel} d\mu$ over the trapping region, characterized by $m_e v_{\parallel}^2/2 + [B_0 - B_{0,\text{max}}]\mu < 0$, where $B_{0,\text{max}}(x)$ is the maximum magnetic field on the magnetic surface $x = \text{const.}$ on which $\vec{x}$ lies. The fraction of trapped particles can be estimated as $\alpha_t = [1 - B_0/B_{0,\text{max}}(x)]^{1/2}$.

In the hybrid model, the electron distribution of trapped particles is clearly required as source to the QNE. This is not the case however for the distribution of passing electrons, whose response is enforced to be adiabatic. It may thus appear that the trapped electron distribution can be neglected. This is in fact not true, as the effective particle trajectories, perturbed by the fluctuating fields associated to the microturbulence, can lead to passing particles becoming trapped and *vice versa*. Consequently, the complete electron distribution must nonetheless be evolved in the hybrid model, according to the same gyrokinetic equation considered in the fully kinetic model, except that the passing particle fraction in this case simply evolves passively, *i.e.* without contributing to the self-consistent field ϕ.

As an illustration, the linear spectra obtained with the GENE code using alternatively the fully kinetic, adiabatic, and hybrid electron model for the real frequency ω_r and growth rate γ of the most unstable mode at each ($\sim$ poloidal) wavenumber k_y are presented in Fig. 1. The physical parameters considered are for case ITG-1, summarized in Table 1. For small

For all cases: concentric circular geometry with $x_0/R = 0.16$, $q_s(0) = 1.4$, and $\hat{s} = 0.8$.						
Case	$\omega_n = R/L_n$	$\omega_{T_e} = R/L_{T_e}$	$\omega_{T_i} = R/L_{T_i}$	$k_{ymin}\rho_i$	$\tau = T_e/T_i$	$\mu = m_i/m_e$
ITG-1	2.22	6.96	6.960	0.3	1	1836
ITG-2	2.0	2.0	6.0	0.3 (0.07)	1 (1)	1836 (400)
TEM-2	3.0	6.0	0.0	0.3	1	1836

Table 1. Physical parameters for the different flux-tube GENE simulations. The local inverse aspect ratio, the safety factor and the magnetic shear at the center of the flux-tube are noted x_0/R, $q_s(0)$, and $\hat{s}$ respectively. The characteristic gradients $\omega_{n,T_e,T_i} = R/L_{n,T_e,T_i}$ for density, and electron/ion temperature are normalized wrt. the major radius R. Values in parenthesis are for non-linear runs if different from linear. Ionization degree $Z = 1$ considered for ions in all cases (Hydrogen plasma).

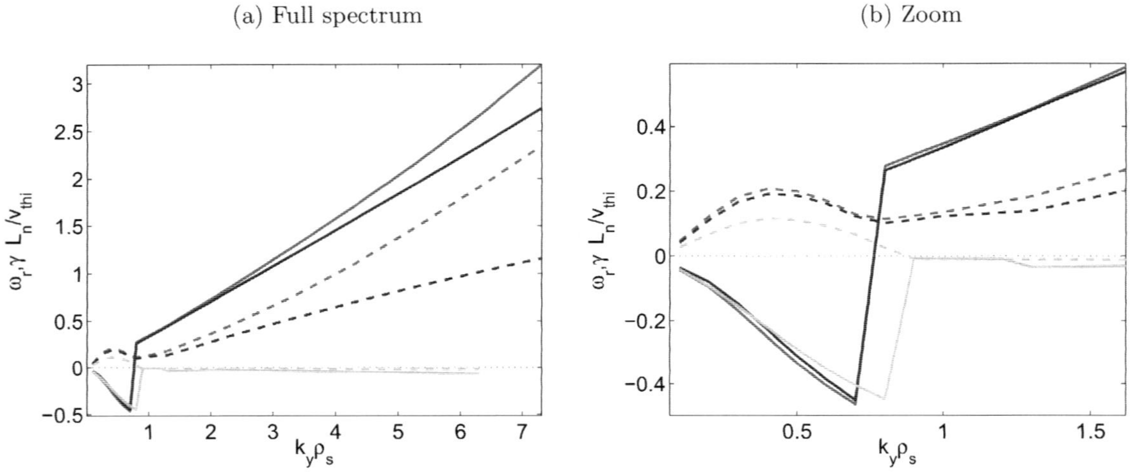

Figure 1. Linear results for Case ITG-1. Comparing the k_y spectra for the real frequency ω_r (full lines) and growth rate γ (dashed) of the most unstable microinstabilities, obtained with the three electron models: fully kinetic (red), adiabatic (green) and hybrid (blue).

wavenumbers, $0 < k_y\rho_i \lesssim 0.7$, the fully kinetic electron model provides an ITG instability ($\omega_r < 0$), well reproduced by the hybrid and approximately reproduced by the adiabatic model. There is indeed a $\sim 30\%$ error on γ with this latter model, resulting from the trapped electrons being forced to respond adiabatically, when in fact they are essentially passive for an ITG. At intermediate wavenumbers, $0.7 \lesssim k_y\rho_i \lesssim 2.0$, the fully kinetic model provides a TEM ($\omega_r > 0$), still well reproduced by the hybrid but absent in the adiabatic model. Finally, at high wavenumbers, $k_y\rho_i \gtrsim 2.0$, the fully kinetic spectra shows a progressive transition of the TEM mode towards an Electron Temperature Gradient (ETG) instability, at which point the hybrid representation starts to fail as well. This example thus clearly summarizes the different ranges of validity for the three different electron models.

4. Analytical Estimate for the Radial Width of Fine Structures

In this section, a simple analytical estimate is derived for the radial width Δx of the regions centered around MRS's, within which the passing electron response is expected to be non-adiabatic.

One considers a given MRS at the radial position $x = x_{m,n}$ such that $q_s(x_{m,n}) = -m/n$ and a corresponding resonant (poloidal, toroidal) Fourier mode (m, n), component of a microinstability with frequency ω_r. The local wavevector of this Fourier mode is given by $\vec{k} = m\nabla\chi + n\nabla\varphi$ and its component parallel to $\vec{B}_0$ reads:

$$k_\parallel = \vec{k} \cdot \frac{\vec{B}_0}{B_0} = \frac{\vec{B}_0 \cdot \nabla\chi}{B_0}[m + nq_s(x)] \simeq \frac{1}{Rq_s}[m + nq_s(x)],$$

having used $q_s = \vec{B}_0 \cdot \nabla\varphi / \vec{B}_0 \cdot \nabla\chi$, valid for the straight field line poloidal angle χ, and R standing for the major radius of the tokamak. Near the mode rational surface $x = x_{m,n}$, one obviously has $k_\parallel \to 0$ and the condition for adiabatic electron response, $|\omega_r/k_\parallel| \ll v_{th,e}$, breaks down. Taylor expanding around $x = x_{m,n}$ leads to $q_s(x = x_{m,n} + \delta x) \simeq q_s(x_{m,n})[1 + \hat{s}\,\delta x / x_{m,n}]$, as well as

$$k_\parallel(\delta x) = \frac{\hat{s}}{q_s}\frac{\delta x}{R} k_y,$$

where $\hat{s} = (x/q_s)\, dq_s/dx$ stands for the magnetic shear and $k_y = nq_s/x_{m,n}$ is the wavenumber with respect to the direction y (as $m \simeq nq_s$ for field-aligned modes, k_y in fact provides an estimate of the poloidal wavenumber). The region of non-adiabatic electron response is then estimated as $\Delta x = 2\delta x$ such that $\omega_r / k_\parallel(\delta x) = v_{th,e}$, finally providing:

$$\frac{\Delta x^{\rm th}}{\rho_i} = 2\,\frac{\omega_r}{v_{th,i}/R}\,\frac{q_s}{\hat{s}\,\sqrt{\tau\mu}\,k_y\rho_i}. \tag{5}$$

In this last relation, the width $\Delta x^{\rm th}$ and the wavenumber k_y have been normalized using the ion Larmor radius ρ_i, while the frequency ω_r is normalized with respect to the reference value $v_{th,i}/R$, where $v_{th,i}$ stands for the ion thermal velocity. Furthermore, $\tau = T_e/T_i$ and $\mu^{-1} = m_e/m_i$ respectively represent the electron/ion temperature and mass ratios. The dependence of $\Delta x^{\rm th}$ in Eq. (5) on μ obviously reflects the non-adiabatic nature of electrons which is accounted for in this context.

5. Linear Simulations

Linear flux-tube simulations of ITG and TEM eigenmodes have been computed making use of the spectral solver provided in the GENE code. In an axisymmetric system, such as in tokamaks considered here, linear eigenmodes have a fixed toroidal wavenumber n, which in GENE's field-aligned coordinate system (x, y, z) translates to a fixed wavenumber $k_y = n/C_y = nq_s/x_0$. Furthermore, as the unperturbed system is stationary, eigenmodes also have a fixed, complex frequency ω. In the GENE representation, an eigenmodes thus takes on the functional form (illustrated here on the electrostatic potential ϕ):

$$\phi(x, y, z; t) = \hat{\phi}(x, z) \exp[\imath(k_y y - \omega t)].$$

The eigenmode is therefore essentially characterized by the complex amplitude $\hat{\phi}(x, z)$, which represents the slow spatial variation, the fast phase variation being contained in the factor $\exp(\imath k_y y)$.

Figure 2 shows the typical eigenmode envelope $|\hat{\phi}(x, z)|$ for both an ITG and TEM mode, corresponding respectively to cases ITG-2 and TEM-2 of Table 1. The ITG and TEM results appear respectively in the left and right columns of sub-plots. In a linear, flux-tube simulation, due to the imposed periodic boundary conditions in x, the system size L_x along this direction must be taken as the distance between consecutive MRS's for the considered k_y mode number: $1 = n(dq_s/dx)L_x = k_y\hat{s}L_x$, leading to $L_x = 1/(k_y\hat{s})$. For $k_y\rho_i = 0.3$ and $\hat{s} = 0.8$ considered here, one thus obtains $L_x/\rho_i = 4.17$. Furthermore, the flux-tube is centered such that the single MRS

Journal of Physics: Conference Series **401** (2012) 012006 doi:10.1088/1742-6596/401/1/012006

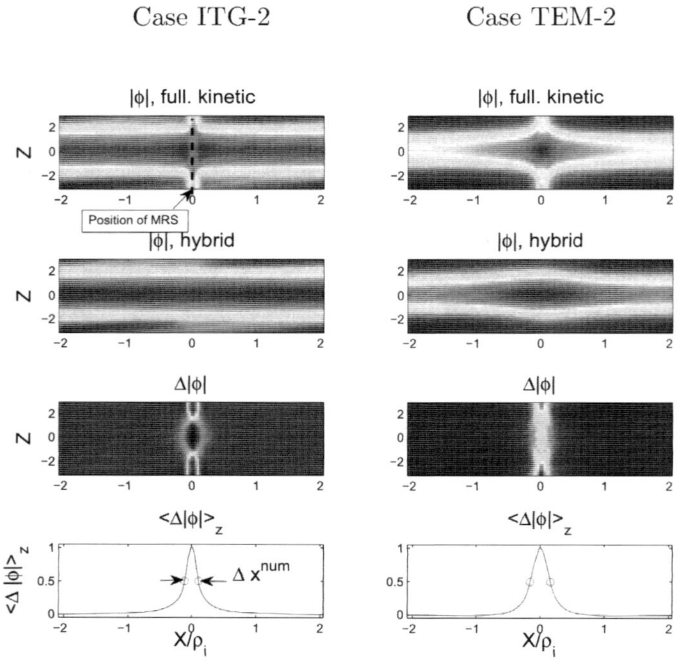

Figure 2. Envelopes $|\hat{\phi}|(x,z)$ of linear eigenmodes for ITG-2 and TEM-2 cases described in Table 1. The MRS for the considered $k_y\rho_i = 0.3$ is positioned at $x = 0$. The first and second row of sub-plots respectively present the results from the fully kinetic and hybrid electron models. The difference $\Delta|\hat{\phi}|$ is shown in the third row and its z-averaged profile in the fourth row, from which the numerical FWHM $\Delta x^{\rm num}$ estimate for the fine structure width can be obtained. Color coding: dark blue = zero, dark red = max. value.

contained in L_x is positioned at $x = 0$. The first row of sub-plots in Fig. 2 presents the results carried out with the fully kinetic model, while the second row shows the results obtained with the hybrid model. Note in all cases the modulation of the eigenmodes with respect to z: a large (resp. low) mode amplitude at $z = 0$ (resp. $z = \pm\pi$) corresponding to the outer (resp. inner) mid-plane of the tokamak, reflecting the characteristic ballooning structure of ITG/TEM modes. A notable difference however is the presence of a fine slab-like (non-ballooned) structure centered at $x = 0$ in the eigenmodes from the fully kinetic model, which is absent in the hybrid model results, thus clearly identified as a feature resulting from the kinetic passing electron dynamics and lying on the MRS's corresponding to the considered k_y. By subtracting the mode envelopes obtained from the two different electron models, $\Delta|\hat{\phi}|(x,z) = |\hat{\phi}^{\rm full.kin.}| - |\hat{\phi}^{\rm hybrid}|$ shown in the third row of sub-plots of Fig. 2, enables to cleanly isolate the structure. Furthermore z-averaging $\Delta|\hat{\phi}|(x,z)$ provides an average radial profile $\langle\Delta|\hat{\phi}|\rangle_z(x)$ shown in the last row of Fig. 2, from which a numerical estimate $\Delta x^{\rm num}$ of the structure width can be measured as the Full Width at Half Maximum (FWHM). The z-average is defined as $\langle.\rangle_z = \int .Jdz/\int Jdz$, where J is the Jacobian of the (x,y,z) coordinate system.

Estimates for $\Delta x^{\rm num}$ can be obtained while alternatively scanning the different physical parameters τ, μ, q_s, $\hat{s}$, and $k_y\rho_i$, all predicted to affect the fine structures according to the theoretical derivation for $\Delta x^{\rm th}$ leading to Eq. (5). Such scans have been carried out starting from the reference values given in Table 1 and considering both the ITG-2 and TEM-2 cases.

Journal of Physics: Conference Series **401** (2012) 012006 doi:10.1088/1742-6596/401/1/012006

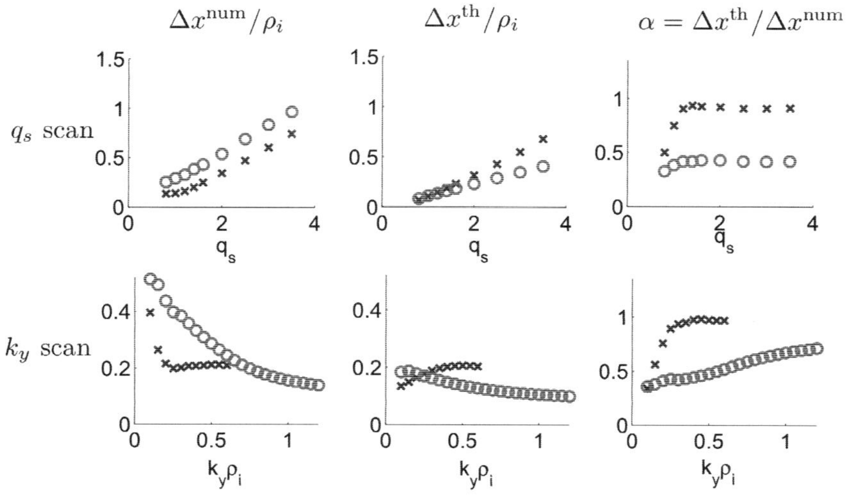

Figure 3. Comparison of the numerical and theoretical estimates, Δx^{num} and Δx^{th} respectively, for the width of fine structures for linear ITG/TEM modes along scans in the safety factor q_s and wavenumber $k_y\rho_i$. The factor $\alpha = \Delta x^{\mathrm{th}}/\Delta x^{\mathrm{num}}$ provides a measure of the agreement. Blue crosses: case ITG-2, red circles: TEM-2.

Corresponding results for scans in the safety factor q_s and wavenumber $k_y\rho_i$ are shown in Fig. 3. Note that, for evaluating the theoretical estimate (5) requires the knowledge of the real frequency ω_r of the eigenmode, which is taken from the simulation. Furthermore, the theoretical relation for Δx^{th} is based on the assumption that the boundary of the non-adiabatic passing electron response regions starts where the parallel phase velocity $\omega_r/k_\parallel$ equals *exactly* the electron thermal velocity $v_{th,e}$ (See Sec. 4). This is clearly arbitrary, as the transition from adiabatic to non-adiabatic is spatially continuous. In the same way, the numerical estimate Δx^{num}, defined as the FWHM of $\Delta|\tilde{\phi}|$, is ad hoc as well. One thus introduces the adjustment factor $\alpha = \Delta x^{\mathrm{th}}/\Delta x^{\mathrm{num}}$, which provides a good measure of the agreement between theoretical and numerical results. As shown in Fig. 3, the factor α is approximately constant over a significant range of the considered scans and furthermore takes on values ~ 1, which reflects that the theoretical relation (5) captures the essential parametric scalings and actually provides a good semi-quantitative estimate of the fine radial structures.

6. Non-Linear Simulations

Preliminary non-linear microturbulence results addressing the role of non-adiabatic passing electrons in the turbulent regime are presented here. One considers the case ITG-2, which leads to an ITG-dominated regime. Note that, contrary to the linear runs for which the physical mass ratio $\mu = 1836$ was chosen, the non-linear results were obtained with the reduced mass ratio $\mu = 400$ (see Table 1). Two simulations have again been performed, considering either the fully kinetic or hybrid electron model. These runs have been carried out over a simulation time of more than $150\,R/v_{th,i}$, ensuring good statistics (for reference, the corresponding maximum linear growth rates of the underlying ITG modes are of order $\gamma_{\max} \simeq 0.3\,v_{th,i}/R$). Contrary to linear calculations, non-linear runs obviously require to simultaneously account for the full spectrum of k_y modes. In a diagnostic process, one may nonetheless Fourier decompose the

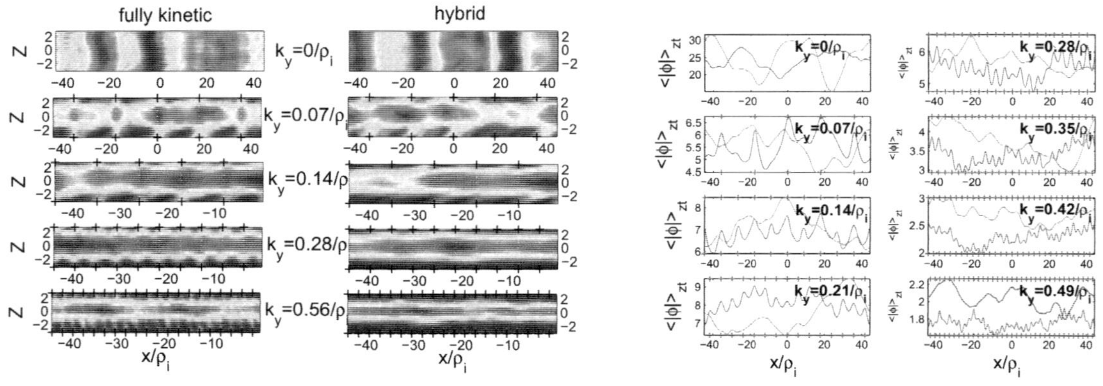

Figure 4. Non-linear ITG-2 case. Comparing results from both the fully kinetic and hybrid electron model. (left) Time-averaged Fourier amplitudes $\langle|\hat{\phi}_{k_y}|\rangle_t(x,z)$ of the electrostatic potential. Showing modes $k_y/k_{y,\min} = 0, 1, 2, 4 \,\&\, 8$, with $k_{y,\min}\rho_i = 0.07$. (right) Radial profiles of corresponding z-averages $\langle|\hat{\phi}_{k_y}|\rangle_{z,t}(x)$. Showing modes $k_y/k_{y,\min} = 0$ to 7. For each k_y, associated MRS's are pointed out with tick marks along the box edge of the corresponding figure. Blue lines: fully kinetic, red: hybrid electron model.

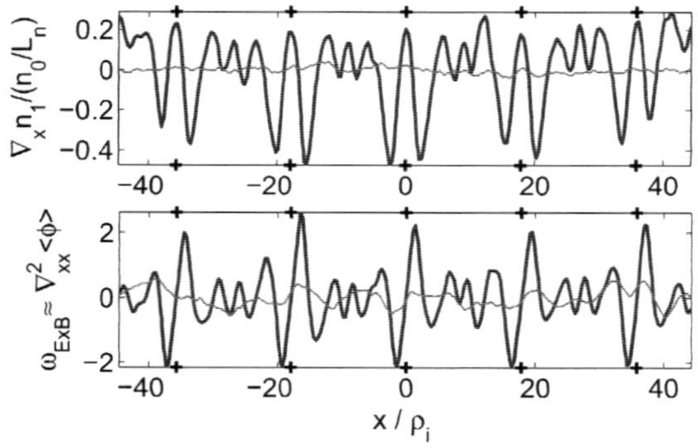

Figure 5. Non-linear ITG-2 case. (Top) Radial profiles of $(d/dx)\langle n_1\rangle_{\text{f.s.,t}}(x)$, where $\langle n_1\rangle_{\text{f.s.,t}}$ is the flux-surface- and time-averaged density fluctuation. (Bottom) Radial profile of the time-averaged $E \times B$ shearing rate $\langle\omega_{E\times B}\rangle_t(x)$ associated with the zonal flows.

turbulent fields and extract the different k_y modes, thus enabling to perform a similar analysis on the corresponding time-averaged mode amplitudes of the electrostatic field, $\langle|\hat{\phi}_{k_y}|\rangle_t(x,z)$, as done for the linear eigenmodes in Sec. 5 ($\langle.\rangle_t$ stands for the time average).

In the radial direction, the system size L_x considered for the non-linear flux-tube simulations must be chosen as an integer multiple of the distance $1/(k_{y,\min}\hat{s})$ separating lowest order MRS's, where the minimum k_y mode number $k_{y,\min}\rho_i = 0.07$ was set for the here considered simulations. The system length L_x must furthermore be significantly larger than the radial correlation length of the turbulence, which typically requires at least $L_x/\rho_i \sim 10^2$. One thus chose $L_x/\rho_i = 5/(k_{y,\min}\rho_i\,\hat{s}) = 89.3$, so that the system encompasses 5 lowest order MRS's. Concerning numerical resolution, $n_x = 256$, $n_y = 32$, and $n_z = 16$ grid points were considered in the x, y, and z-directions respectively. In particular, for the x-resolution one has $L_x/n_x = 0.35\,\rho_i$. This may appear marginal considering the widths of fine structures observed on the linear modes, but turns out to be appropriate for resolving the characteristic structures appearing in the nonlinear simulation as shown in the following. Concerning velocity space resolution, $n_{v_\parallel} = 64$ and $n_\mu = 32$ mesh points were chosen in the $v_\parallel$ and μ directions respectively.

The time-averaged amplitudes of k_y Fourier modes obtained from the electrostatic potential, $\langle|\hat{\phi}_{k_y}|\rangle_t(x,z)$, are shown for the non-linear ITG-2 case simulations in Fig. 4. Also shown are the corresponding z-averages $\langle|\hat{\phi}_{k_y}|\rangle_{z,t}(x)$. For each mode $k_y > 0$, these plots clearly point out the presence in the fully kinetic results of fine radial structures located at the associated MRS's, positioned at $x = \Delta m/(k_y\hat{s})$, Δm integer, and high-lighted with tick marks in the corresponding plots. Similar fine structures are absent in the hybrid simulations, confirming that the non-adiabatic passing electron structures identified in the linear simulations actually persist in the non-linear, turbulent regime. The clean procedure used in the linear study for measuring the radial width of non-adiabatic structures, based on subtraction of fully kinetic and hybrid results, is not applicable in any straightforward way for analyzing the non-linear data. One can nonetheless estimate the widths Δx^{num} by directly measuring on $\langle|\hat{\phi}^{\text{full.kin.}}_{k_y}|\rangle_{z,t}(x)$ the FWHM of fine structures appearing at each corresponding MRS's. Corresponding results show that the fine structures, although having survived in the non-linear regime, are significantly broadened. For example, at $k_y\rho_i = 0.35$, the linear results (for $\mu = 400$) present a width of order $\Delta x^{\text{num}}/\rho_i = 0.45$, while measuring structures on the corresponding mode in the non-linear simulations provides an average value $\Delta x^{\text{num}}/\rho_i \simeq 1.7$.

Considering in Fig. 4 the plot depicting $\langle|\hat{\phi}_{k_y}|\rangle_{z,t}(x)$ for $k_y\rho_i = 0$, there seems to be no apparent fine structure in this mode. Naturally, $k_y = 0$ itself has no associated MRS. However, when estimating the time-averaged shearing rate $\omega_{E\times B} = (d/dx)v_{E\times B} = -(1/B)(d^2/dx^2)\langle\phi\rangle_{\text{f.s}}$ of $E\times B$ flows associated to the zonal component $\langle\phi\rangle_{\text{f.s}} = \int \hat{\phi}_{k_y=0}(x,z)\,J\,dz/\int J\,dz$, clear fine structures with a scale length of a few ion Larmor radii ρ_i centered around the lowest order MRS's appear in the fully kinetic electron simulation, as clearly illustrated in Fig. 5. Such structures are again essentially absent in the results obtained with the hybrid model. The plot for $\langle\omega_{E\times B}\rangle_t$ in Fig. 5 thus is related to the second order derivative of the radial profile $\langle|\hat{\phi}_{k_y=0}|\rangle_{z,t}(x)$ in Fig. 4, the derivatives having accentuated the underlying fine structures in the kinetic runs. Remarkable in the corresponding radial profile of $\langle\omega_{E\times B}\rangle_t(x)$ is the regular pattern of fine structures that appear, periodically repeated over each lowest order MRS interval of length $1/(k_{y,\text{min}}\hat{s})$ (positions of lowest order MRS's are again high-lighted in the plot with tick marks). Structures in $\langle\omega_{E\times B}\rangle_t(x)$ are thus clearly aligned with the lowest order MRS's (related to $k_y = k_{y,\text{min}}$), presenting an *odd* parity with respect to these radial positions. Similar sub-structures can also be identified at the second order MRS's ($k_y = 2\,k_{y,\text{min}}$) as well as third order ones ($k_y = 3\,k_{y,\text{min}}$).

In a similar way as has been done for ϕ, one can investigate the flux-surface- and time-average of density and ion/electron temperature fluctuation fields. Considering for example the density fluctuation n_1 (same for ions and electrons as quasi-neutrality is verified), the radial derivative of the corresponding averaged profile, $(d/dx)\langle n_1\rangle_{\text{f.s.,t}}(x)$, is also shown in Fig. 5. As for the shearing rate profile $\langle\omega_{E\times B}\rangle_t(x)$, similar structures aligned with the MRS's are found for the fully kinetic model, again absent for the hybrid results. Note however that for $\langle n_1\rangle$, the structures present an *even* parity with respect to the positions of MRS's (to lighten notations one writes here $\langle n_1\rangle$ in place of $\langle n_1\rangle_{\text{f.s.,t}}$). In fact, the profile $(d/dx)\langle n_1\rangle(x)$ systematically presents *maxima* at the MRS's. This reflects the tendency towards stationary *flattenings* in the corresponding effective profiles. Indeed, the total stationary density profile reads $n = n_0 + \langle n_1\rangle$ and the condition $dn/dx = 0$ for a flattened region becomes $(L_n/n_0)(d/dx)\langle n_1\rangle = 1$. Values of $(L_n/n_0)(d/dx)\langle n_1\rangle \simeq 0.2$ appearing in Fig. 5 at lowest order MRS's thus correspond to partial flattenings of the density profile. Flattenings are observed as well at the next order MRS's, although weaker on average. Conversely, the profiles are *steepened* between MRS's, reflected by $(L_n/n_0)(d/dx)\langle n_1\rangle$ taking on negative values in the corresponding regions, thus resulting in an effectively step-like profile. The development of similar step-like profiles is observed for the ion/electron profiles as well (not shown here). These results thus confirm the corrugated profiles already observed in Ref. [3] using the GYRO code.

7. Conclusions

A simple theoretical estimate has been derived for the radial width Δx of non-adiabatic passing electron regions in which fine structures appear on the linear ITG/TEM eigenmodes with $k_y \neq 0$ at corresponding MRS's. This estimate has been successfully validated against numerical results obtained with linear, flux-tube simulations using the GENE code. The theoretical relation for Δx is in particular useful for determining the numerical resolution required for spatially resolving these non-adiabatic structures.

Preliminary results obtained from non-linear, flux-tube simulations in an ITG-dominated scenario have shown that the structures on the different $k_y \neq 0$ modes persist, albeit broadened, in the turbulent regime. Furthermore, through non-linear coupling to the $k_y = 0$ mode, fine structures located around lowest order MRS's also appear in the radial profiles of time- and flux-surface-averaged fluctuating fields, such as density n_1 and ion/electron temperature $T_{1,i/e}$. These structures are such that the associated total profiles show the tendency towards a flattening near low order MRS's and a steepening in between, leading to effective step-like density and temperature profiles, confirming results obtained in Ref. [3]. It should be emphasized that this phenomena of profile flattening near MRS is analyzed here for essentially electrostatic perturbations, and thus involves no electromagnetic fluctuations such as *e.g.* microtearing modes.

The modulation of the effective profile gradients can thus be interpreted as a clear sign that transport of particles and heat is, at least locally, influenced by the particular dynamics of non-adiabatic passing electrons near MRS's. In particular, stationary structures appearing as well on the radial shearing rate profile $\omega_{E\times B}(x)$, associated to the zonal flows, seem to play a key role in sustaining the step-like profiles. Indeed, regions of steepened gradients happen to coincide with regions where $|\omega_{E\times B}|$ is large. This is consistent with the fact that sheared flows effectively break up turbulent eddies, which results in reduced associated transport. Conversely, regions of flattened gradients coincide with low values of $|\omega_{E\times B}|$.

Further investigation of the role of non-adiabatic passing electrons in the vicinity of MRS's on profile evolutions and overall transport processes is clearly required and currently being carried out for both ITG and TEM dominated regimes, and shall be reported on in a future publication.

Acknowledgments

This work was partly supported by the Swiss National Science Foundation. GENE simulations have been carried out on the HPC-FF computer at the Jülich Forschungszentrum, Germany.

References

[1] Falchetto G L, Vaclavik J and Villard L 2003 *Physics of Plasmas* **10** 1424–1436
[2] Chowdhury J, Ganesh R, Angelino P, Vaclavik J, Villard L and Brunner S 2008 *Physics of Plasmas* **15** 072117
[3] Waltz R E, Austin M E, Burrell K H and Candy J 2006 *Physics of Plasmas* **13** 052301
[4] Jenko F, Dorland W, Kotschenreuther M and Rogers B N 2000 *Physics of Plasmas* **7** 1904–1910
[5] Görler T, Lapillonne X, Brunner S, Dannert T, Jenko F, Merz F and Told D 2011 *Journal of Computational Physics* **230** 7053 – 7071 ISSN 0021-9991
[6] Lapillonne X, Brunner S, Dannert T, Jolliet S, Marinoni A, Villard L, Gorler T, Jenko F and Merz F 2009 *Physics of Plasmas* **16** 032308
[7] Brizard A J and Hahm T S 2007 *Rev. Mod. Phys.* **79**(2) 421–468
[8] Görler T 2009 Ph.D. thesis University of Ulm, Germany
[9] Lapillonne X 2010 Ph.D. thesis Ecole Polytechnique Fédérale de Lausanne, Switzerland

Thermodynamical and microscopic properties of turbulent transport in the edge plasma

Ph Ghendrih[1], C Norscini[1], F Hasenbeck[1,2], G Dif-Pradalier[1], J Abiteboul[1], T Cartier-Michaud[1], X Garbet[1], V Grandgirard[1], Y Marandet[3], Y Sarazin[1], P Tamain[1], D Zarzoso[1]

[1]CEA-IRFM, F-13108 Saint-Paul-lez-Durance, France
[2]IEK-4 Plasma Physics, Forschungszentrum Jlich GmbH, Jlich, Germany
[3]PIIM, CNRS/Aix-Marseille Univ., Marseille, France

E-mail: philippe.ghendrih@cea.fr

Abstract. Edge plasma turbulence modelled with 2D interchange is shown to exhibit convective transport at the microscale level. This transport property is related to avalanche like transport in such a flux-driven system. Correlation functions and source modulation are used to analyse the transport properties but do not allow one to recover the Fick law that must characterise the system at large scales. Coarse graining is then introduced to average out the small scales in order to recover the Fick law. One finds that the required space averaging is comparable to the system size while the time averaging is comparable to the confinement time. The system is then reduced to a single reservoir such that transport is characterised by a single scalar, either the diffusion coefficient of the Fick law or a characteristic evolution time constant.

1. Introduction

The widely used diffusion approximation of transport modelling traces back to Einstein's interpretation of the Brownian motion later followed by the Langevin's toy model addressing the same problem [1]. These lead to the picture of a Markov chain of random steps associated to a random time that governs the transport at the microscopic level. An important consequence has been the interpretation it provides for the Fick law of proportionality between the transported flux and the particle gradients, the proportionality coefficient being the diffusion coefficient. However, it must be stressed that the Fick law must be satisfied at the macroscopic level even when the microscopic transport is not associated to diffusion. There appears therefore a scale separation between the microscopic physics that drive the transport and the macroscopic scale where thermodynamics require a Fick law to be satisfied. The key issue is therefore to determine those macroscopic scales such that the system, at these scale, is sufficiently close to thermodynamical equilibrium to recover the proportionality between the flux and the thermodynamical force. We address this problem here in the framework of turbulent plasma in a model where the microscopic transport is governed by turbulence, Section 2, and is found to be predominantly convective, Section 3. A coarse graining process is then introduced to recover the Fick law and to determine the macroscale of the problem, Section 4.

2. Steady state conditions with 2D interchange turbulence in the edge plasma

The present most advanced models of plasma turbulence are addressed in the gyrokinetic framework, hence determining the plasma distribution function f with global full-f and most importantly *flux-driven* codes [2, 3]. It has been reported that flux driven gyrokinetic simulations exhibit avalanche-like transport [4, 5], thus confirming the results obtained with fluid codes [6]. In this paper we use fluid simulations prior to analysing gyrokinetic simulations. The model is the same as in the seminal paper [6]. It is a simple 2-fields and 2D interchange model describing the pressure nT and vorticity W in the plane transverse to the magnetic field. Its structure is identical to that of the Rayleigh-Bénard system that describes convective instabilities. The buoyancy driven transport is the vorticity equation is governed by the magnetic curvature effect in the plasma model while the stream function is replaced by the electric potential. In the plasma model we consider for simplicity a constant temperature $T = T_e$, the electron temperature, so that one actually addresses particle transport in the cold ion limit and not thermal energy as in Rayleigh-Bénard. A key difference between the neutral fluid and the plasma model is the existence, in the latter, of a loss term in the vorticity equation that linearly stabilises the large structures. Consequently, the $E \times B$ convective cells do not reach the size of the system and boundary layer issues are less stringent than in Rayleigh-Bénard experiments and simulations. The a-dimensional equations take the following form:

$$\partial_t n \; + \; [\phi, n] \; - D\Delta n \qquad = -\sigma_n(n,\phi)\, n + S \tag{1}$$

$$\partial_t W \; + \; [\phi, W] - \nu \Delta W + g\partial_y n = \sigma_\phi(\phi) \tag{2}$$

The vorticity W is directly connected to the electric potential ϕ, normalised to T_e/e, by a Poisson equation $W = \nabla_\perp^2 \phi$. Six control parameters are obtained that govern the drive and loss terms. The linear instability is governed by the existence of the g-term of the order of ρ_s/R, where ρ_s is the hybrid Larmor radius that is introduce to normalise the transverse scales, and R the major radius of the device that drives the magnetic field curvature. The system is driven out of equilibrium by the particle source term S. In the present setting, the source is constant in the y-direction, akin to the poloidal direction, with Gaussian-peaking in the x-direction, akin to the radial direction. The loss term $\sigma_n(n,\phi)\, n$ is characteristic of the SOL region and corresponds to the parallel particle loss term to the wall. The structure of the control parameter $\sigma_n(n,\phi)$, and its possible dependence on ϕ to account for the sheath effects, is of little relevance in the present paper. A key aspect of this term is that this particle loss is distributed in the whole domain, it is therefore a volumetric sink. Indeed, each point in the 2D plane of interest is connected via the third dimension to a limiter surface. The finite confinement time along this direction, and plasma neutralisation at the wall thus governs the particle loss. When steady state conditions are achieved, this volumetric particle loss balances the particle source. Turbulent transport is thus organised from the source region, along the direction of symmetry of the source, the x-direction. The linear interchange instability sets in for negative density gradients, hence in the region localised on the right hand side of the source. Two diffusive transport coefficients are introduced, D and ν. These are normalised to the Bohm value $\rho_s c_s$ (where c_s is the sound velocity) and are small, equal to 10^{-2} in the present work. They are important in damping the small scales that build-up in the turbulent regime.

The last control parameter is the parallel current loss term $\sigma_\phi(\phi)$. Its exact structure, including or not sheath effects, is not crucial here. However, it must be stressed that it takes into account parallel current exchange that can be complex, and will depend on the electric potential, but that does depend explicitly on the vorticity $W = \nabla_\perp^2 \phi$. Consequently, as stated above, it will appear as $-\sigma_\phi/k_\perp^2$ in the calculation of the linear growth rates. Here $k_\perp$ is the wave vector in the plane orthogonal to the direction of the parallel current losses and stems from inverting the vorticity in Fourier space to recover the potential. Conversely, σ_ϕ can depend on the parallel wave vector $k_{\|}$. The term σ_ϕ thus drives a stabilisation the large mode structures $k_\perp \to 0$. In the present work, it is chosen independent on the density n so that this mechanism is homogeneous throughout the system. Consequences are two fold, first one can consider that it constrains the zonal flows, $k_y = 0$ where y is the poloidal direction. These must then have sufficient radial structure (x-direction) to yield finite values of $k_\perp$ and exhibit weak damping, σ_ϕ being a small parameter. This tends to preserve the shearing properties of the zonal flows but smears out global radial structures of the electric field. Another consequence is that the large scales of the system are stable with respect to turbulence. This favours the existence of macroscales where thermodynamics can apply.

The code solving equations (1, 2) is periodic in both x and y directions. Although the x-periodicity can be considered as unphysical, it allows one to investigate the plasma response in three regions, see figure 1. Due to the symmetry breaking set by the linear criterion for the turbulence onset, most of the simulation domain is governed by turbulent transport, while the region without turbulence, labeled region-III, in a narrow poloidal band that plays a crucial role numerically by preventing cross talk of turbulence in the radial direction, typically this region is characterised by $x \geq 439$ at the minimum of the density profile. Where turbulence governs transport, one labels region-I the region with nearly constant density gradient length $L_n \approx \lambda_n$, and region-II the region where L_n varies radially increasing from $L_n \approx \lambda_n$ to ∞ in the crossover region between the region with turbulent transport and region-III without. A fourth region should be mentioned, that of the particle source, located at $x = 1$ up to $x \approx 16$.

The curvature of the magnetic field and the geometry of the source break the symmetry between the x and y directions. Indeed, the source is constant in the y-direction, so that transport is organised along the x-direction. In this direction, translation invariance is thus broken by the source term. Conversely, the y-direction still exhibits translation invariance that allows one to average in the y-direction to gain in statistics (this is equivalent to a flux surface average). The curvature of the magnetic field drives the instability and leads to the g-term with an explicit dependence on y. In this model, it does not depend on x so that it does not generate extra thermodynamical forces. The average density and particle flux are plotted on figure 1. These are averaged on y and time for the whole simulation duration t_s. It is interesting to note that both density $\langle n\rangle_{y,t_s}$ and radial particle flux $\langle\Gamma_x\rangle_{y,t_s}$ exhibit an exponential fall-off as expected in the SOL plasma. However, it must be stressed that such an exponential fall-off is consistent with both purely diffusive and convective transport and thus cannot be identified as a signature of diffusive transport.

Statistics of these fields, at given x-value, lead to heavy tails Probability Density Functions as shown for Γ_x and $< \Gamma_x >_y$ on figure 2. Remarkable features are attached to these PDF. Let us first consider the PDF of Γ_x, black curve with open o-symbols, attached to the lower scale. First negative values of the particle flux account for a significant part of the data, hence driving particles from the sink towards the source. This result is in fact consistent with the properties of $E \times B$ eddy convection that does not discriminate directions. More unexpected is the fact that the most likely particle flux is negative, one the left hand side of the zero line (vertical dash-dot line). Second, one finds a heavy tail towards the large positive values of the particle flux. Since the small magnitude events are both positive and negative with high but nearly symmetric probability density, they tend to cancel out when addressing the mean value of the radial particle flux. From that point of view, one can state that the effective particle flux (the mean value) is essentially sustained by the heavy tail part of the PDF. This property is specific of flux driven system where the only constraint on the system is to ensure a balance, on average only, between the source and the sink. The heavy tails structures also appear to exhibit avalanche-like

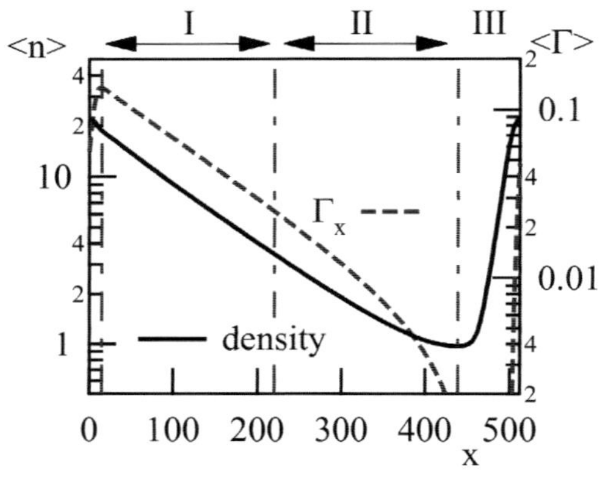

Figure 1. Averaged density and particle flux profiles, regions I, II and III are indicated.

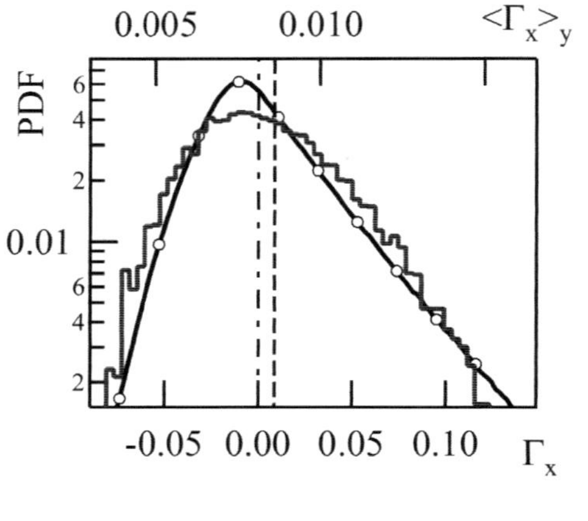

Figure 2. PDF of the particle flux, black curve open o symbols Γ_x lower scale, blue histogram $< \Gamma_x >$ upper scale.

behaviours with 1/f frequency spectra [6]. In the case of SOL plasmas, it can be connected to the so-called blob phenomenology later advocated in ref.([7]). Regarding the negative particle flux events, only an averaging procedure allows one to recover the thermodynamically expected sign of the particle flux. In fact the y-average is found to be sufficient for this process as exemplified by the PDF of $< \Gamma_x >_y$, also on figure 2, blue histogram upper scale. Only positive values are found for $< \Gamma_x >_y$ so that the PDF can be fitted with a lognormal PDF. More importantly is the strong change in the range of the fluctuations, typically a factor 10 between the lower and upper scale. For convenience the two PDF are centred at the same location of their mean value (which is of course identical) as indicated by the vertical dashed line. Despite the sharp drop in the fluctuation level, this PDF still exhibits a heavy tail. Consistently with the discussion of the PDF of Γ_x, one can consider that most of the data contributing to the PDF of $< \Gamma_x >_y$ must therefore stem from avalanche-like events.

The convergence towards a statistical steady state of this turbulence model is slow. To investigate this aspect we analyse the local convergence for each point in y, at the same x position of $\langle\Gamma_x\rangle_t$ when varying the duration t of the averaging process. We can then plot the distance $\left|\langle\Gamma_x\rangle_t - \langle\Gamma_x\rangle_{y,t_s}\right|$, see figure 3. One finds that the convergence takes place according to the scaling law $1/\sqrt{N}$ where N is the number of events. Fitting the data indicates that an event corresponds in fact to a finite duration τ_N, as indicated by the vertical dashed line on figure 3. This duration of a typical event must be compared to the confinement time of the system defined as $\tau_E = 1/\sigma_n$. Indeed, one can readily show that when switching off the source term, the total number of particles in the simulation domain will decay exponentially with e-folding time $\tau_E = 1/\sigma_n$. The simulation evidence thus leads to $\tau_N \approx \tau_E/6$. The events last a significant fraction of the confinement time. To complete the discussion on the characteristic times, another key time must be introduced, namely the turbulence correlation time τ_c. Data from the simulation yields $\tau_c \approx \tau_E/32$, figure 4. The typical event duration is thus in the range of 5 correlation times, a ratio also reported in gyrokinetic flux driven simulations [4]. When compared to the total simulation time $t_s \approx 40\tau_E$, one thus finds that typically 240 events take place. Together with the very irregular decay noticeable on figure 3, this large but not very large number of events underlines the difficulty in obtaining sufficient statistics to determine with precision the steady state regime.

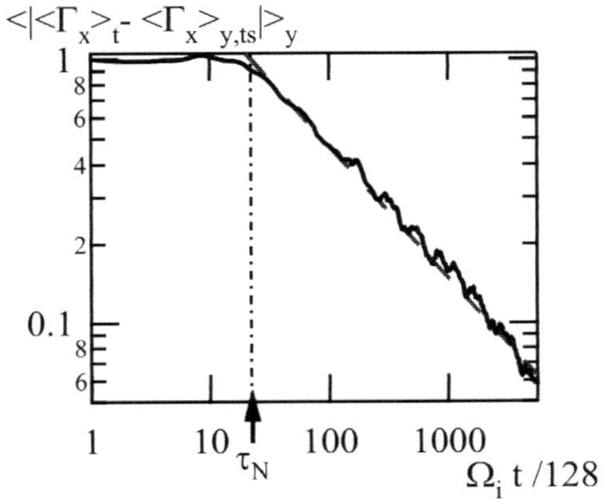

Figure 3. Convergence of the particle flux towards its mean value.

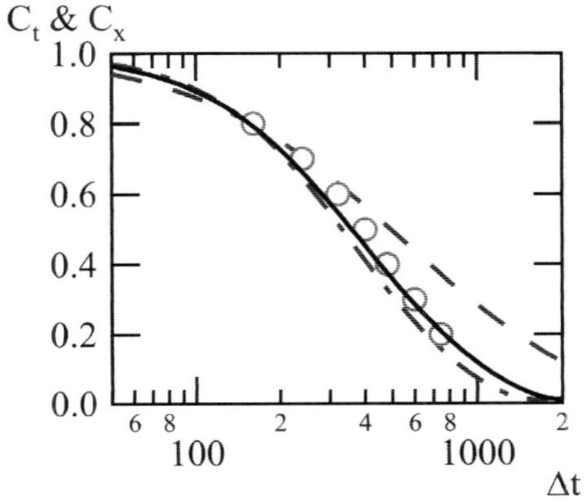

Figure 4. Time, $\mathcal{C}_t$ open circles, and space $\mathcal{C}_x$ correlation functions for $\alpha = 1.1$ plain, $\alpha = 1.5$ dashed and $\alpha = 1.0$ dash-dot lines.

3. The microscopic turbulent transport

The investigation of the microscopic turbulent transport has been undertaken with several approaches. Starting from the description of the transport in the set of equations (1, 2), one finds that in the turbulent regime the transport is governed by $E \times B$ convection. At the microscopic level, the issue is determining the sequence of events that bound this correlated convective motion, thus defining step sizes and step durations for the transport process. The PDF of these step characteristics will then govern the nature of the microscopic transport. In practice this approach is difficult to implement for turbulence since the decorrelation process that bounds the steps is not well defined. A standard alternative is to address the transport of test particles governed by the time dependent $E \times B$ motion. Investigating the time dependence of the mean distance between the test particles and their initial position with the present model in [6] indicates that the test particle transport is close to ballistic. A more demanding analysis is based on tracing the large flux events and determining their dynamics. One then finds that the events first experience an acceleration before propagating with a nearly constant velocity, hence behaving like fronts or avalanches. Since these large events dominate the average flux, it then appears that the microscopic transport is essentially ballistic [8]. A third approach proposed here is based on self similarity of transport such that the appropriate variable is x/t for ballistic transport and x^2/t for diffusive transport. Generalising this similarity constraint to turbulence transport, one expects transport to depend on x^α/t, the exponent α characterising the nature of transport. We then match the self-correlation function in time $\mathcal{C}_t(\Delta t)$ and self-correlation in space along the x-direction $\mathcal{C}_x(\Delta x)$ so that: $\mathcal{C}_t(\Delta t) \approx \mathcal{C}_x(c\ \Delta x^\alpha)$. In log scale for the abscissa, the constant c is determined as an offset while the parameter α governs the slope, figure 4. This procedure yields $\alpha \approx 1.1$ and $c \approx 15$, see figure 4, hence a value close to ballistic $\alpha = 1$ and clearly not diffusive $\alpha = 2$. This result is surprising since in the standard approach *à la mixing length* would stipulate the transport to be diffusive, the diffusion coefficient being $D_c = \lambda_c^2/\tau_c$ where $\lambda_c \approx 24$ is the correlation length. The correlation length and correlation time would then define the elementary event, and assuming a Markov process would govern the diffusive transport for scales larger than that of the elementary event. Given the values of λ_c and τ_c one then expects a diffusive coefficient $D_c \approx 1.11$ (normalised to the Bohm diffusion), typically 100 times larger than the small scale smoothing diffusion D. However, should one assume that the sign of the step is not random but always positive, the sequence of elementary events will lead to a ballistic transport of magnitude $V_c = \lambda_c\tau_c \approx 0.046$ (normalised to the sound velocity, and therefore effectively a Mach number) not too different from the upper bound value $1/c \approx 0.067$.

The different approaches used in analysing the microscopic transport properties indicate that the microscopic transport is close to ballistic in marked difference with that assumed when considering the Brownian motion.

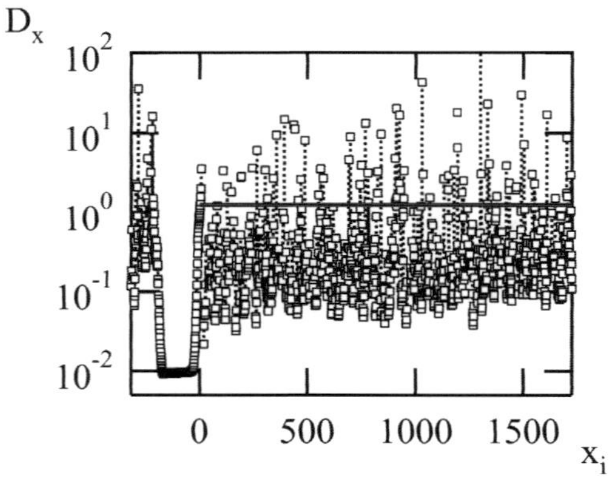

Figure 5. Diffusion coefficient D_x profile.

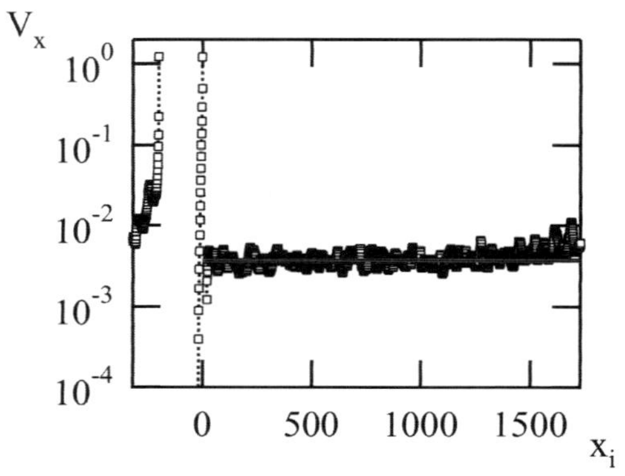

Figure 6. Velocity V_x profile.

4. Macroscopic transport analysis

Considering discretised transport operators, one can analyse the profile averaged in y and time either in terms of a profile of diffusive transport or in terms of a profile of convective transport, where the local diffusion D_x and convection V_x coefficients are defined according to the formulas:

$$D_x = \frac{\sigma_n n_i - S_i}{n_{i+1} + n_{i-1} - 2n_i}(\delta x)^2 \quad ; \quad V_x = \frac{2\left(\sigma_n n_i - S_i\right)}{n_{i+1} - n_{i-1}}\delta_x \tag{3}$$

Here the subscript i describes the discretised x-direction with a step δx. The result of the calculation is displayed on figure 5 and 6 clearly indicates that the values of the D_x exhibit a much larger fluctuation range than that of V_x. These fluctuations are clearly governed by the underlying turbulent transport since one properly recovers the value of $D = 10^{-2}$ in the stable region. Moreover, one finds that the values of D_x are reminiscent of a step like density profile with a sequence of small values followed by a large value. Based on these statistical properties, one would tend to consider that transport is more properly represented by a convection $< V_x >$ rather than the diffusion $< D_x >$ following therefore the microscopic properties. The macroscopic particle flux would then be of the form $< \Gamma_x > = < V_x > n$. However, from a thermodynamical point of view, the only force in the system is the density radial variation, hence the non-vanishing value of $-\nabla_x n$. Since the particle flux must be driven and opposite to the thermodynamical force, one expects recovering at lowest order a Fick law of the form $< \Gamma_x > = -\mathcal{D}\nabla_x n$. The mean density profile analysis does not appear to provide sufficient insight into the transport properties. One reason, already addressed in Section 2 could be the slow convergence of the statistics due to the occurrence of large and rare transport events that generate the departure from an exponential solution with a step like structure. In order to overcome this difficulty, and

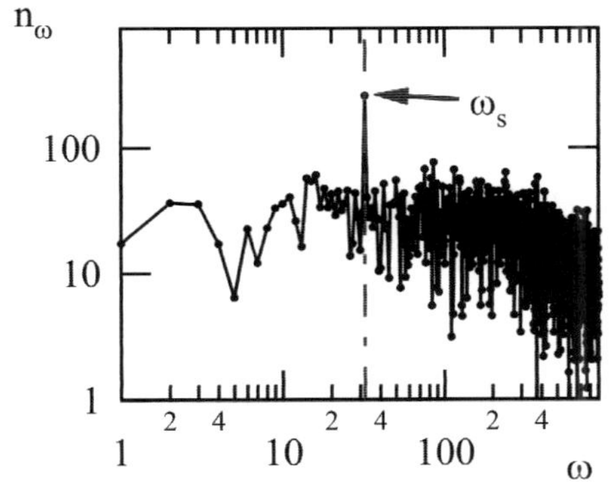

Figure 7. ω-spectrum of the density for a point in region I.

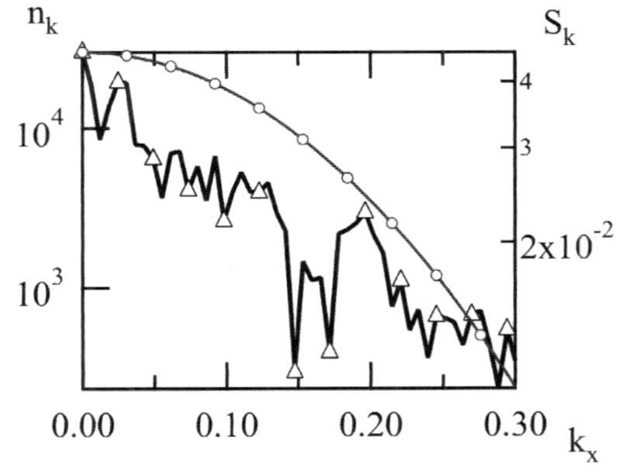

Figure 8. Source (open circles) and density (open triangles) k-spectra.

to investigate the possible coexistence of diffusion and convection, we have performed a series of simulations with a source modulation. The modulated source is localised in the same region as the main source, it is also constant along the y-direction and narrower in the x-direction, yielding a broad k_x spectrum. With the chosen value of the magnitude of the modulated source one identifies very well the density response at the source frequency ω_s, figure 7. One can then consider the transport equation in Fourier space:

$$\sigma_n - i\omega_s + \left(\mathrm{D}_{\mathrm{eff}} k_x^2 + i\mathrm{V}_{\mathrm{eff}} k_x\right) = \frac{\hat{S}}{\hat{n}} \tag{4}$$

where we have assumed the macroscopic, or effective, transport to be of the standard form $\Gamma = -\mathrm{D}_{\mathrm{eff}}\nabla_x n + \mathrm{V}_{\mathrm{eff}}\ n$. In such a framework the real part of the ratio $\hat{S}/\hat{n}$ should be a quadratic function of k_x while the imaginary part should be linear. The frequency ω_s must be small enough to allow the transport wave to propagate from the source to a sufficiently large distance, but not that large that it only drives an adiabatic response of the system at quasi-constant profile but slowly varying particle content. A value comparable to σ_n thus appears to be optimum. At any rate, as exemplified on figure 7, it is not possible to find a frequency outside the broad range of the turbulence spectrum. As expected for a system characterised by heavy tails of large transport events, there is no obvious time separation between turbulence and transport. No correlation between the source and density k_x-spectra has been observed, see figure 8. In particular, a signature of diffusive transport would govern a quadratic shape of the density spectrum at $k_x \to 0$ that must be more pronounced than that of the source, which is clearly not the case on figure 8 where the density spectrum is exponential like.

The two first transport analysis are based on a coarse graining process achieved by flux surface averaging, maximum coarse graining in y, and either averaging in time or selecting small frequencies both governing strong coarse graining in t.

We now analyse the impact of the coarse graining process by considering directly the particle flux $\langle\Gamma\rangle_{cg}$. The minimum coarse graining that is considered is a y-average to ensure that most of the data is consistent with the thermodynamical (macroscale) constraint of particle flux directed

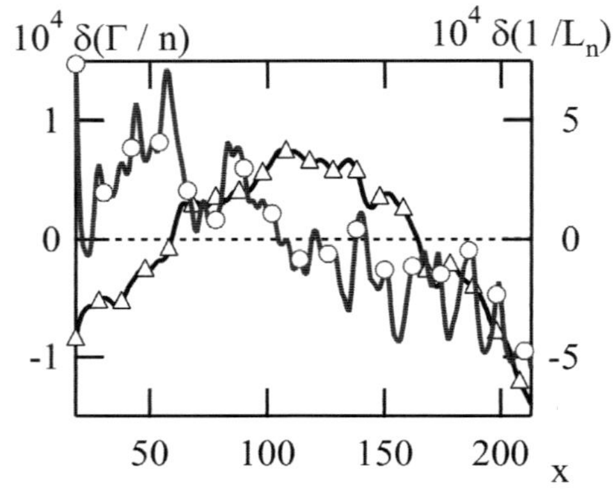

Figure 9. Blow up of the profiles of $\delta(\Gamma/n) = \Gamma/n - \langle\Gamma/n\rangle_x$, open triangles, and $\delta(1/L_n) = 1/L_n - \langle 1/L_n\rangle_x$, open circles.

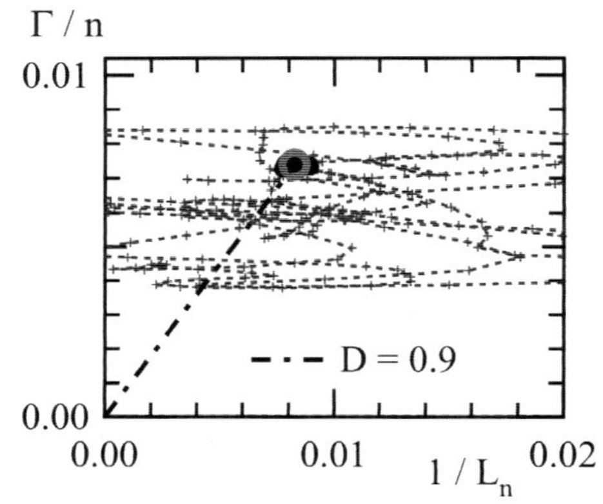

Figure 10. Γ/n as a function of $1/L_n$, y-average crosses and dashed lines, y and t average closed circles, x, y and t average open circle.

from the source towards the sink. One can expect $\langle\Gamma\rangle_{cg} / \langle n\rangle_{cg}$ to exhibit a polynomial expansion in $1/L_n$. In a linear expansion, consistent with weak departure from equilibrium the constant offset value would then be the convection velocity while the slope would be the diffusion coefficient. Data is selected from region I, figure 10, and does not appear to yield any trend, similarly to experimental evidence [9]. It is clear from this data that any relationship between $\langle\Gamma\rangle_{cg} / \langle n\rangle_{cg}$ and $1/L_n$ requires coarse graining in time. This is done by averaging over the duration of the simulation $t_s \approx 40\tau_E$. The data points exhibit more dispersion in $1/L_n$ than in $\langle\Gamma\rangle_{cg} / \langle n\rangle_{cg}$, figure 9. One then readily notices that the variation of $\langle\Gamma\rangle_{cg} / \langle n\rangle_{cg}$ is small and uncorrelated with the radial modulation of $1/L_n$. Again this data appears to be consistent with transport governed by convection. A response that is not possible since the system is governed by a single thermodynamical force measured by $1/L_n$. One is thus led to introduce coarse graining in x, here over region I to remove the profile effects, figure 9. The bottom line of this coarse graining process is that we have averaged over all possible direction to recover the Fick law, $\langle\Gamma_x\rangle_{cg} / \langle n\rangle_{cg} = -\mathcal{D}\,\langle 1/L_n\rangle_x$ represented by a single point on figure 10. In this process region I of the system is reduced to a single point, hence to a single reservoir interacting with other reservoirs. The single scalar characterising this reservoir, $\mathcal{D} \approx 0.9$, is equivalent to setting a characteristic evolution time $\tau = \mathcal{L}_x^2/\mathcal{D}$ where $\mathcal{L}_x \approx 190$ is the radial extent of region I, hence $\tau \approx 2.5\tau_E$. Some optimisation in the coarse graining process, different from the full time and radial coarse graining might be possible. Given the lack of correlation between the force and flux profiles in region I, figure 9, one can consider that the radial coarse graining cannot be modified while reducing the time coarse graining to τ might be possible.

5. Discussion and conclusion

We have shown in this paper that capturing the macroscale properties of a system driven by turbulence with self-organised transport events, avalanches or fronts, requires a strong coarse graining procedure. For the model we have implemented, this minimum coarse graining process includes a flux surface average (y-direction of the model), an average over the whole radial region and at least over times that correspond to a couple of confinement times. As a consequence, the total system appears to be reduced to only a few interacting reservoirs, the size of the latter being constrained by the requirement to have sufficient internal statistics to define an equilibrium thermodynamical state, these define therefore macroscale regions. With the available simulation run time, the evolution of such a reservoir is limited to a few steps. With such a level of coarse graining the Fick law is recovered, but the information regarding the reservoir is reduced to a single scalar, akin to a confinement time. The consequence would thus be that in the present system, the existence of the Fick law does not imply diffusive transport at the microscale. At the macroscale, where diffusion appears to be legitimate, the few interacting reservoirs makes a discussion of transport in terms of diffusion rather artificial since only a couple of steps separate the boundary conditions.

This result can be understood in the framework of avalanche like transport events. These events are localised in space but rather long lived in time, hence leading to a long range transport structure. They appear to propagate with a ballistic motion, with nearly constant velocity magnitude but varying direction in the plane orthogonal to the magnetic field. The order of magnitude of the poloidal component being comparable to that of the radial component. The heavy tail structure of transport events they encompass drives a very slow convergence of the statistics. The duration of the random events τ_N that governs this process appears to be large, so that only several such events occur during a confinement time. Present investigation of the time lag between the avalanches appears to be consistent with τ_N of the order of the shortest time lag of the events belonging to the heavy tail structure of the particle flux. These transport processes that are a key signature of global and flux driven systems cannot be captured by approaches that introduce scale separation, e.g. *à la quasilinear*. This is very clear in particular when considering a scale separation between a given gradient scale (assumed to be a macroscale quantity) and the underlying scale of the fluctuations. Indeed the avalanche events lead to fast transport over significant fractions of the radial extent of the domain, while remaining localised. This is shown to lead to broad spectra (partly as a consequence of the localisation) such that scale separation in time or in space cannot be justified.

Avalanche like events are also reported in global flux driven gyrokinetic simulations of the core plasma with ITG turbulence. In these systems the sink is not volumetric as in the present model but localised at the boundary layer. Although the latter point can modify some aspects of the present description, one can expect that the avalanche transport in gyrokinetics will also hinder scale separation and consequently require strong coarse graining to recover macroscale thermodynamical properties. This is the matter of our present effort in analysing gyrokinetic simulations performed with the GYSELA code.

Acknowledgments
This work was supported by ANR grants ESPOIR ANR-09-BLAN-0035-01 and GYPSY ANR-10-BLAN-941, as well as the European Community under the contract of Association between EURATOM, CEA, and the French Research Federation for fusion study. The views and opinions expressed herein do not necessarily reflect those of the European Commission.

References

[1] N G Van Kampen 1981 *Stochastic Processes in Physics and Chemistry* (Amsterdam: North Holland)
[2] Y Sarazin *et al.* 2010 *Nucl. Fusion* **50** 054004
[3] S Jolliet and Y Idomura 2012 *Nucl. Fusion* **52** 023026
[4] G Dif-Pradalier, P H Diamond, V Grandgirard, Y Sarazin, J Abiteboul *et al.* 2010 *Phys. Rev. E* **82** 025401(R)
[5] Y Sarazin, V Grandgirard, J Abiteboul, S Allfrey, X Garbet *et al.* 2011 *Nucl. Fusion* **51** 103023
[6] Y Sarazin and Ph Ghendrih 1998 *Physics of Plasmas* **5** 4214
[7] S I Krasheninnikov 2001 *Physics Letters* A **283** 368
[8] Ph Ghendrih , Y Sarazin, G Attuel *et al.* 2005 *J. Nucl. Mater.* **337-339** 347
[9] O E Garcia, R A Pitts, J Horacek, A H Nielsen, W. Fundamenski *et al.* 2007 *J. Nucl. Mater.* **363-365** 575

Journal of Physics: Conference Series **401** (2012) 012008 doi:10.1088/1742-6596/401/1/012008

Ion diamagnetic effects in gyrofluid collisionless magnetic reconnection

D Grasso[1], **F L Waelbroeck**[2], **E Tassi**[3],

[1]CNR-ISC, Dipartimento di Energia, Politecnico di Torino, Corso Duca degli Abruzzi, 24, 10129, Torino, Italy
[2]Institute for Fusion Studies, The University of Texas at Austin, Austin, TX 78712-1060, USA
[3]Centre de Physique Théorique, CNRS – Aix-Marseille Universités, Campus de Luminy, case 907, F-13288 Marseille cedex 09, France

E-mail: `daniela.grasso@infm.polito.it`

Abstract. Ion diamagnetic effects on collisionless magnetic reconnection are investigated by means of numerical simulations of a Hamiltonian gyrofluid model. The work is focused in particular on the effects of inhomogeneous density equilibria in the large Δ' regime. The linear growth rates predicted by asymptotic theory are recovered. Nonlinearly the island shape is strongly modified and the flow changes its typical four-cell structure into a simpler two cell one. Contrary to the resistive reconnection process, we find that in Hamiltonian (i.e. collisionless) reconnection the width of the final saturated island is independent from the diamagnetic effects and the magnetic island grows up to the equilibrium scale length.

1. Introduction

Due to its implication in many physical phenomena ranging from auroras in astrophysical plasmas to sawtooth crashes in Tokamak plasma [1, 2, 3], magnetic reconnection is widely studied in many different regimes and, consequently, with many different approaches. Here we are interested in high temperature and low β plasmas dominated by a strong guide field, characteristic of the Tokamak configuration. The high temperatures that characterize plasma close to ignition lead us to focus on the collisionless regime for the reconnection process, in which the electron inertia is the dominant mechanism for the decoupling between the plasma and the magnetic field motion. In the collisionless regime one of the main features of the reconnection process is the formation of narrow layers, whose width is well below the ion Larmor radius scale [4, 5]. This aspect raises also the question of which is the more appropriate physical description to be adopted. Here, as a compromise between the gyrokinetic description, accurate but computationally costly, and the fluid description that neglects all the sub-Larmor radius dynamics, we adopt the gyrofluid model described in [6]. In this framework we are interested in analyzing diamagnetic effects, generated by equilibrium ion pressure gradients, on the magnetic island evolution. In particular, among such diamagnetic effects, we focus on the island rotation and on the stabilization of the dominant mode when the diamagnetic frequency exceeds the linear growth rate [7, 8]. Moreover, we address the problem of nonlinear saturation, which results in an incomplete reconnection process and, hence, a saturated island. Indeed, one of the issues in sawtooth crashes is why the reconnection process ends before the sawtooth crash, as for instance shown in the ASDEX experiment [9]. A number of numerical simulations of the internal

Journal of Physics: Conference Series **401** (2012) 012008 doi:10.1088/1742-6596/401/1/012008

kink mode and sawtooth oscillations, with various degrees of sophistication, physical models and simplifying assumptions, have also been conducted in order to interpret the experimentally observed sawtooth dynamics. [10, 11, 12, 13]. While electron inertia is always invoked to account for the fast reconnection process at the start of the sawtooth crash, the possibility to reach a saturated island has been explored in connection with different effects.
Island saturation was observed in numerical simulations in the resistive regime in Ref. [11], where a relation between the island width and the diamagnetic frequency was found following a quasi-linear approach. While in the collisionless regime a rapid rotation of the plasma core itself was shown to be responsible for incomplete reconnection whenever the equilibrium profile of the plasma pressure is sufficiently steep Ref. [12]. Here we find that the final saturated island is independent from the diamagnetic effects and the magnetic island grows up to the equilibrium scale length. Unlike [8], in this paper we focus mainly on the ion dynamics and we neglect the electron temperature effects.
The paper is organized as follows. In Sec. 2 we review the model equations. Sec. 3 compares the ion density gradient and gyroradius effects on the stability and rotation frequency of the linear dominant mode with theoretical predictions. In Sec. 4 the nonlinear phase is analyzed. We conclude in Sec. 5.

2. The model

We consider the two-dimensional version of the Hamiltonian gyrofluid model described in Ref.[6] and adopted also in Ref.[8]. The model equations are

$$\frac{\partial n_i}{\partial t} + [\Phi, n_i] = 0, \tag{1}$$

$$\frac{\partial n_e}{\partial t} + [\phi, n_e] - [\psi, \nabla^2\psi] = 0, \tag{2}$$

$$\frac{\partial}{\partial t}(\psi - d_e^2\nabla^2\psi) + [\phi, \psi - d_e^2\nabla^2\psi] + \rho_s^2[\psi, n_e] = 0, \tag{3}$$

$$n_e = \Gamma_0^{1/2} n_i + (\Gamma_0 - 1)\phi/\rho_i^2. \tag{4}$$

The first two equations represent the continuity equations for the ion guiding centers and the electrons respectively. The third equation is the Ohm's law. The system is closed by the quasi-neutrality relation (4). In Eqs.(1)-(4) n_i is the ion guiding center density, n_e the electron density, ϕ the electrostatic potential, ψ the poloidal magnetic flux function, d_e the electron skin depth, ρ_s the sonic Larmor radius and $\Phi = \Gamma_0^{1/2}\phi$ is the gyro-averaged electrostatic potential. For our analysis the Padé approximant version $\Gamma_0^{1/2} = (1 - \rho_i^2\nabla^2/2)^{-1}$ of the gyro-average operator will be adopted, where ρ_i is the ion Larmor radius. Given a Cartesian coordinate system (x, y, z), we assume all the fields be translationally invariant along z and we define the canonical bracket between two generic fields f and g by $[f, g] = \hat{z}\cdot\nabla f \times \nabla g$. Eqs. (1-4)are normalized respect to the Alfvén time and to a magnetic equilibrium scale length L. Dependent variables are normalized in the following way:

$$n_i = \frac{L}{\hat{d}_i}\frac{\hat{n}_i}{n_0}, \qquad n_e = \frac{L}{\hat{d}_i}\frac{\hat{n}_e}{n_0}, \qquad \psi = \frac{\hat{A}_z}{BL}, \qquad \phi = \frac{\hat{\rho}_s^2}{L^2}\frac{L}{\hat{d}_i}\frac{e\hat{\phi}}{T_e},$$

where carets denote dimensional variables, d_i is the ion skin depth, n_0 a background density amplitude, A_z the magnetic potential, e the unit charge, B a characteristic toroidal magnetic field amplitude and T_e the electron temperature, which is assumed to be constant.

3. Growth rate and rotation frequency in the linear phase

The model equations are solved over the domain $\{(x, y) : -\pi \leq x < \pi, -a\pi \leq y < a\pi\}$, where we prescribe the value of the number a at each simulation. The grid is composed by 1024×128 points and double periodic boundary conditions are imposed on the field perturbations, which are the quantities that the code advances in time according to a third order Adams-Bashfort scheme. The equilibrium configuration is specified by the following expressions:

$$n_{i_{eq}}(x) = n_0' x, \qquad n_{e_{eq}}(x) = n_0' x, \qquad \psi_{eq}(x) = \sum_{n=-11}^{11} a_n \exp(inx), \tag{5}$$

where n_0' is a constant and the a_n are the Fourier coefficients of the function $f(x) = A/\cosh^2 x$, where A is a parameter determining how strong is the guide field respect to the poloidal magnetic field.

The equilibrium (5) is perturbed in n_i with a four-cell pattern disturbance of the form $\tilde{n}_i \propto \cos(x + y/a) - \cos(x - y/a)$. The field ϕ is also perturbed according to (4), in such a way that the initial perturbation on n_e is zero.

Linearly we assume the fields, ψ, ϕ, n_e and n_i to behave as $\mathrm{f}(x)\exp(iky + \lambda t)$, where $\mathrm{f}(x)$ is the amplitude of the perturbation, $k = 1/a$ the wave vector, and $\lambda = \gamma + i\omega$ (here we adopt a different sign convention respect to Ref. [8]). The growth rate, γ, and the rotation frequency, ω, then correspond to the real and imaginary part, respectively, of the complex frequency λ. The set of equations, once linearized, yield the definitions: $\omega_{*e} = kn_0'\rho_s^2$ and $\omega_{*i} = -\omega_{*e}\rho_i^2/\rho_s^2$.

In Ref. [8] we presented linear results including both ion and electron diamagnetic effects, while in this paper we focus only on the ion effects. Hence, we set ρ_s and ω_{*e} to 0 and only the ρ_i scale length will be retained in our set of equations. The linear theory in Ref. [7] gives a dispersion relation (see eq. (9) in Ref. [7]), valid in the asymptotic limit $\rho_i \gg d_e$ and $\Delta' d_e >> (d_e/\rho_i)^{1/3}$, where Δ' is the standard tearing stability parameter [14]. This relation, once split into its real and imaginary part, gives the rotation frequency and the growth rate for the dominant mode according to:

$$\gamma^2 \approx \gamma_0^2 - \left(\frac{\omega_{*i}}{2}\right)^2, \tag{6}$$

$$\omega \approx \frac{\omega_{*i}}{2}, \tag{7}$$

where $\gamma_0 = 2k_y(2d_e\rho_i^2/\pi)^{1/3}$. In Fig. 1 the numerical growth rates and the rotation frequencies are plotted versus their analytical values for a simulation campaign carried out adopting $A = 1$, $a = 2$, (yielding $\Delta' = 59.9$), $d_e = 0.1$, $\rho_i = 0.2$ and n_0 ranging in the interval [0,11], which corresponds to $-0.24 < \omega_{*i} < 0$. We can see that, while for moderate values of ω_{*i} the difference between analytical and numerical values remains constant, close to the region of marginal stability such difference increases. In particular numerical simulations indicate that the stabilization is reached with a value of ion diamagnetic frequency smaller than that predicted by the theory. We remark, however, that relations (6) and (7) are valid in an asymptotic regime, and for the choice of parameters corresponding to Fig. 1, we had ρ_i only slightly grater than d_e, and $\Delta' d_e$ exceeding only five times the value of $(d_e/\rho_i)^{1/3}$. Since diminishing d_e is computationally costly, we increased the box size to better satisfy the asymptotic conditions. In Fig. 2 we show the results reporting data from a simulation campaign with $A = 1$, $a = 4$, which corresponds to $\Delta' = 240.1$, $d_e = 0.15$, $\rho_i = 0.25$ and same values of n_0 corresponding now to values of ω_{*i} in the range [-0.172,0]. For these runs $\Delta' d_e$ is 40 times the value of $(d_e/\rho_i)^{1/3}$, and the agreement with the analytical theory is considerably improved also when approaching marginal stability.

Journal of Physics: Conference Series **401** (2012) 012008 doi:10.1088/1742-6596/401/1/012008

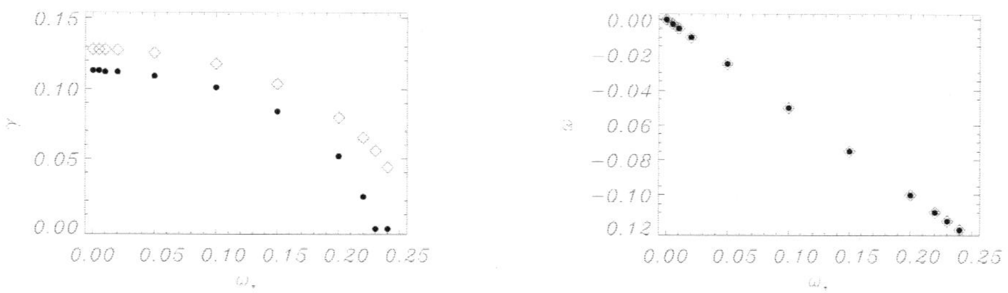

Figure 1. Comparison between the growth rates (right panel) and the rotation frequencies (left panel) obtained from the numerical simulations (circles) and the analytical values (diamonds) obtained from eqs. (6) and (7). The simulations have been carried out assuming $A = 1$, $a = 2$, $d_e = 0.1$, $\rho_i = 0.2$ and n_0' ranging from 0 to 11.

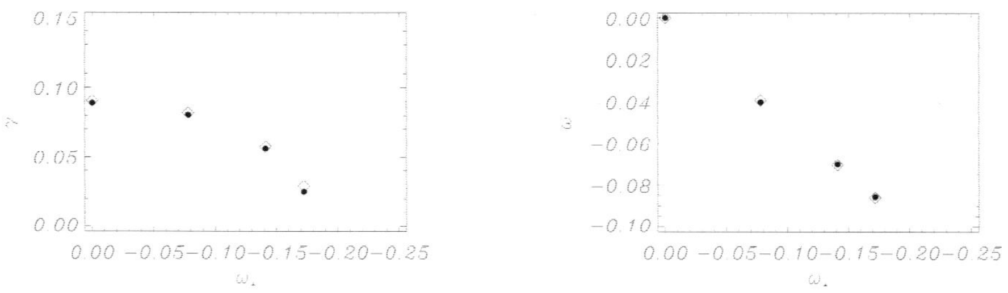

Figure 2. Comparison between the growth rates (right panel) and the rotation frequencies (left panel) obtained from the numerical simulations (circles) and the analytical values (diamonds) obtained from eqs. (6) and (7). The simulations have been carried out assuming $A = 1$, $a = 4$, $d_e = 0.15$, $\rho_i = 0.25$ and n_0' ranging from 0 to 11.

4. Nonlinear evolution

Saturation of the magnetic island induced by diamagnetic effects has been found in the resistive regime by Biskamp [11], where, following a quasi-linear approach, the island width is found to saturate at small amplitudes depending on the values of the diamagnetic frequency ω_{*i} and of the resistivity parameter η. In Ref. [12] the authors show that in the collisionless regime a partial reconnection process with island width larger with respect to the one determined in Ref. [11], is possible provided the value of the poloidal β parameter is large enough and the diamagnetic frequency is larger than the nonlinear growth rate. In this section we analyze the long term evolution of the reconnection process in order to investigate the possibility to reach a finite amplitude saturated island width resulting in partial reconnection of the plasma core. In our simulation we always find that the reconnection process evolves until the magnetic island is of the order of the equilibrium magnetic field scale length and proceeds up to the integration domain. So, no evidence of a partial reconnection process is found.

We believe that this difference is due to the intrinsic quasi-explosive behavior of magnetic reconnection in the collisionless regime, first identified in Ref. [4]. In this regime, a typical feature is that after the linear phase the growth rate accelerates in such a way that the magnetic island reaches a macroscopic amplitude on a time that is of the order of the inverse linear growth rate [15]. In Fig. 3 the growth rates for a collisionless and a resistive case are compared. The two runs have been performed in order to have similar linear growth rate, replacing the electron inertia into the Ohm's law with a value of resistivity such that $\eta \approx \gamma d_e^2$. From this figure it is clear that while the collisionless case has a growth rate which increases after the linear phase, the resistive case goes smoothly towards saturation. In this respect, we are not in contradiction with Ref. [12]. Rather, we observe that the condition that ω_{*i} is greater than the maximum growth rate attained in the nonlinear phase, prescribed in Ref. [12] is only feebly satisfied for our parameters.

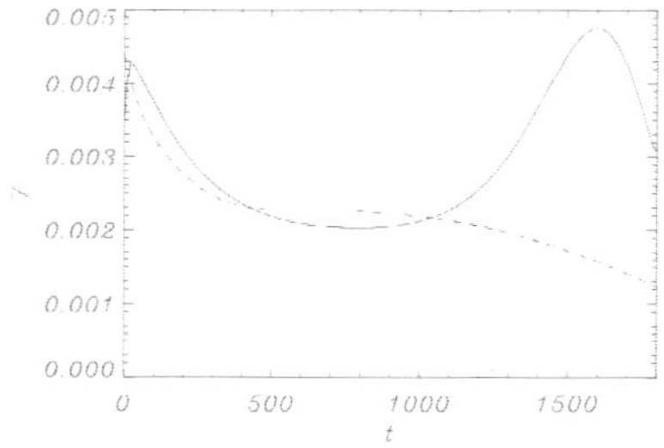

Figure 3. Comparison between the growth rate of a collisionless (solid line) and a resistive case (dashed line) having approximately the same linear growth rate. The collisionless case has a quasi-explosive behavior right after the linear phase. The simulations have been carried out setting $A = 0.03$, $a = 2$, $\rho_i = 0.2$, $n_0' = 0.4$, $d_e = 0.15$ for the collisionless case and $\eta = 0.0001$ for the resistive case.

The nonlinear evolution of the reconnection process is described in Fig. 4, where the stream function ϕ for three different values of ω_{*i} are shown. Superimposed to the plasma flow are the contour lines of the magnetic flux. For each run we have chosen to show a time step well into the nonlinear phase right after the maximum growth rate has been reached. From Fig. 3, besides

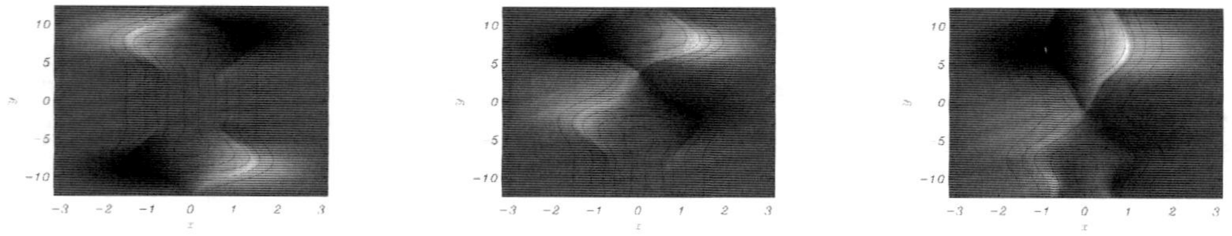

Figure 4. Comparison between the plasma flow at three different values of the diamagnetic frequency. From the left to the right $\omega_{*i} = 0, -0.078, -0.172$, corresponding to $n_0' = 0, 5, 11$ respectively. The simulations have been carried out assuming $a = 4$, $d_e = 0.15$, $\rho_i = 0.25$.

observing that the magnetic island has reached a macroscopic amplitude comparable with the scale length of the equilibrium magnetic field, we can also appreciate two different aspects of the reconnection process in this regime. First of all we observe that there is a strong nonlinear modification of the equilibrium magnetic field, due to the growth of the $k_y = 0$ wave number. Second, the plasma flow changes its structure from the typical four-cell one at $\omega_{*i} = 0$ into a two-cell one when increasing ω_{*i}. There is a progressive loss of symmetry in the y-direction related to the increasing value of the ion density equilibrium gradient.

5. Conclusions

We have investigated numerically a gyrofluid model for magnetic reconnection in collisionless regimes including ion diamagnetic effects. According to the linear theory of Ref. [7], we find stabilization for sufficiently high values of ω_{*i} and a propagation of the magnetic island in the ion direction drift. The agreement with the linear theory improves providing the asymptotic conditions for which the theory is valid are adequately satisfied. In contrast to the work presented in Ref. [8], here we focused on the difference between the saturation mechanism in the resistive regime considered in Ref. [11] and in the collisionless regime. We find that while in the resistive regime it is possible to achieve a saturated state, characterized by a finite island amplitude, corresponding to a partial reconnection process, in the collisionless regime, the magnetic island can grow up to the magnetic flux equilibrium scale length and to the integration domain, corresponding to a full reconnection process. Nonlinearly, the flow changes its topology from a symmetric four-cell structure into an asymmetric two-cell structure, when increasing the ion density equilibrium gradient.

Acknowledgments

This work was supported by the European Community under the contracts of Association between EURATOM and ENEA and between EURATOM, CEA, and the French Research Federation for fusion studies. The views and opinions expressed herein do not necessarily reflect those of the European Commission. ET acknowledges financial support from the Agence Nationale de la Recherche (ANR GYPSI)

References

[1] Priest E R and Forbes T G 2000 *Magnetic Reconnection* (Cambridge, Cambridge University Press)
[2] Biskamp D 2000 *Magnetic Reconnection in Plasmas* (Cambridge, Cambridge University Press)
[3] Wesson J A 1986 *Plasma Phys. Control. Fusion* **28** 243–248
[4] Ottaviani M and Porcelli F 1993 *Phys. Rev. Lett.* **71** 3802–3805
[5] Grasso D, Califano F, Pegoraro F and Porcelli F 2000 *Plasma Physics Reports* **26** 512–518
[6] Waelbroeck F L, Hazeltine R D and Morrison P J 2009 *Phys. Plasmas* **16** 032109
[7] Porcelli F 1991 *Phys. Rev. Lett.* **66** 425–428
[8] Tassi E, Waelbroeck F L and Grasso D 2010 *Journal of Phys. Conf. Series* **260** 012020
[9] Letsch A, Zohm H, Ryter F, Suttrop W, Gude A, Porcelli F, Angioni C and Furno I 2002 *Nuclear Fusion* **9** 1055–1059
[10] Aydemir A Y 1992 *Physics of Fluids B* **4** 3469
[11] Biskamp D 1981 *Phys. Rev. Lett.* **46** 1522–1525
[12] Biskamp D and Sato T 1997 *Phys. Plasmas* **4** 1326–1329
[13] Rogers B and Zakharov L 1996 *Phys. Plasmas* **3** 2411–2422
[14] Furth H P, Killeen L and Rosenbluth M N 1963 **6** *Physics of Fluids* 459
[15] Porcelli F, Borgogno D, Califano F, Grasso D, Ottaviani M and Pegoraro F 2002 *Plasma Physics and Controlled Fusion* **44** B389

Journal of Physics: Conference Series **401** (2012) 012009 doi:10.1088/1742-6596/401/1/012009

A dielectric response model for FEM solutions of ICRF wave fields

T Hellsten[a], A Hannan[a] , T Johnson[a], L J Höök[a] and L Villard[b]

[a]*School of Electric Engineering, KTH, Association VR-Euratom, Sweden*
[b]CRPP, EPFL, Association Euratom-Confédération Suisse, Lausanne, Switzerland

E-mail: thel@kth.se

Abstract. Modelling of fast wave ICRF heating in large machines with high density such as DEMO is challenging because of the short wave lengths. Therefore, fast, efficient global wave solvers are necessary. A major difficulty with calculating the wave field in a spatial dispersive medium is that the dielectric tensor becomes a function of the local wave vector, which in its turn depends on the solution. Furthermore, the solution may consist of several waves co-existing at the same location subjected to separate response functions. In order to model upshift of the parallel wave vector, higher order FLR-effects on the cyclotron absorption and TTMP damping for the electron absorption methods based on iteration, suitable for FEM codes, are proposed.

1. Introduction

Modelling fast wave ICRF (ion cyclotron range of frequencies) heating in large machines with high density such as DEMO is a challenging task because of the spatial dispersive effects in a hot plasma and the short wave lengths. A major difficulty when solving the wave equation for spatial dispersive media is that the dielectric tensor depends on the wave vector, which in its turn depends on the solution. The predominant effect of spatial dispersion of the fast magnetosonic wave comes from the parallel wave number. Other important spatial dispersive effects are higher order FLR (finite Larmor radius) terms, TTMP (transit time magnetic pumping) damping and linear mode conversion. Geometrical optics within the applicability of the theory describes how the wave vector changes in a transparent way, but misses out on interference. The interference can be taken into account by using global wave solvers based on local discretisation by finite differences and finite elements or spectral methods. However, the co-existence of several waves complicates the problem because the dielectric tensors for the different waves differ, which requires that the waves are separated into different components having their own response. This is important for waves being upshifted after a reflection so that they are not damped by upshift before the reflection has taken place. The straightforward way to tackle this problem is to solve an integro-differential wave equation [1]. For global wave solvers the problem is handled by Fourier decomposition the wave field into planar waves, which has the advantage that the wave numbers are known for each component. Such an approach allows the response to be constructed consistently with that for planar waves in quasi-homogenous plasma, but requires that the coupling between the modes is taken into account. For routine calculations fast efficient wave solvers being able to resolve the wave field are required. Thus, there is a need for approximating the response function allowing simpler methods to be used. When solving the wave field with a FEM code like LION [2, 3] the field is correctly calculated for an approximate response function. If the wave field consists of a single wave, the response can be corrected by iterations taking into account the wave field from the previous iteration when calculating the response. Here methods based on iteration are proposed, suitable for FEM codes, to model upshift of the parallel wave vector, higher order FLR-effects on the cyclotron absorption and TTMP damping for the electron absorption.

2. Upshift

The up or down shift of the parallel wave number, $k_{\|} = n_{\phi}B_{\phi}/BR + m_{\theta}B_{\theta}/Br$, appears because the poloidal mode number, m_{θ}, is not an invariant in a torus. The common approximation of the parallel wave number by $k_{\|} = n_{\phi}/R$ considerably simplifies the calculations, and is appropriate when the relative changes of $k_{\|}$ due to the poloidal mode number is small. The change in the poloidal mode number is taken into account in the geometrical optics in a transparent way. In global wave codes it can be taken into account by Fourier decomposing the wave field, which gives a spectrum of coupled modes. Note that the wave field corresponding to a narrow beam has a wide spectrum of poloidal Fourier components that is required to describe the cancellation of the wave field outside the beam. The wave field within the beam may well be described by a slowly varying single wave vector as in geometrical optics. Here we assume the damping to be strong and the wave field to consist of one wave with a local well defined wave vector, which is the typical case for most ICRH scenarios for DEMO.

Wave-particle interactions in inhomogeneous plasma, like a torus, differ fundamentally from that of a homogeneous plasma. In homogeneous plasma only a small fraction of the particles are resonant. In an inhomogeneous plasma the cyclotron frequency varies and, in general, all particles passing the cyclotron resonance are resonant, but at different locations, and the resonance occurs only at a short time, but still over many wave periods. Here we assume the wave-particle interactions to be fully decorrelated, and do not consider bounce resonances [4]. The fully decorrelated interactions are localised around the resonance, $\omega - \omega_c - k_{\|}v_{\|} = 0$, and only the change of the poloidal mode number around the resonance needs to be taken into account, where ω_c denotes the cyclotron resonance and $v_{\|}$ de parallel velocity of the ion. Fourier decomposition of the field in poloidal direction requires that all the poloidal modes are included with proper correlation between the interactions in order to cancel out the wave-particle interactions in regions where the wave field vanishes. Here we calculate the upshift consistent with the stationary phase method. The change in energy ΔW due to wave-particle interactions is obtained by integrating the local change in energy of the particle along the drift orbit

$$\Delta W = Ze\int_0^{\tau_b} \mathbf{v}\cdot\mathbf{E}dt\,. \tag{1a}$$

Expanding around the resonance, where $\omega_c = \omega_{c0}$, yields

$$\Delta W = Ze\int\left(v_{\perp}E_{+}\exp\left[i\left(\omega_{c0}-\omega\right)t + i\int_0^t dt' v_{\|}\left\{\frac{\partial\omega_{c0}}{\partial l}t' + \frac{\partial}{\partial l}\left[\mathrm{Im}\left(\ln E_{+}\right)\right]\right\}\right]\right)dt \tag{1b}$$

For simplicity we assume the waves interacting with passing particles or trapped far away from the turning point so that $v_{\|}$ and $k_{\|}$ can be regarded as constants. For a wave $\mathbf{E} = \mathbf{E}_0\ \exp i(k_{\|}l\text{-}\omega t)$ the integral in Eq. (1) oscillates rapidly as the particle moves away from the resonance, $\omega\text{-}\omega_c - k_{\|}v_{\|} = 0$, because of the variation of ω_c. To calculate the width of the resonance region we assume $E_+ = E_0\ \exp i(k_{\|}l\text{-}\omega t)$, which gives $k_{\|} = \mathrm{Im}(\ \partial(\ln E_+)/\partial l\ )$ and integrate the phase up top $\pi/2$ from the resonance, which defines one half of the resonance region

$$\frac{\pi}{2} = \int_0^{t_r}\left(\omega - \omega_{c0} - \int_0^{t''}\frac{\partial\omega_{c0}}{\partial l}v_{\|}dt' - k_{\|}v_{\|}\right)dt'' \tag{2}$$

The width of the resonance region along the magnetic field becomes $\Delta l_r = 2t_r\upsilon_\parallel$, or in the poloidal direction $r\Delta\theta_r = 2t_r\partial\theta/\partial l$, where $2t_r = \sqrt{4\pi qR^2/\upsilon_\parallel\omega_{c0} r\sin\theta}$. Here we have used $\int_0^t \frac{\partial\omega_{c0}}{\partial l}\upsilon_\parallel dt' = -\frac{\omega_{c0}}{2qR^2} r\sin\theta\upsilon_\parallel t^2$ and assumed $B(r,\theta) = B_0(1 + r\cos\theta/R_o)$. The width of the resonance depends on the parallel velocity of the particle. For trapped particles having their turning points near the resonance, for which case $\upsilon_\parallel \to 0$ and $t_r \to \infty$, the variation of $\upsilon_\parallel$ has to be taken into account [5]. For tangential resonance, for which $\partial\omega_{c0}/\partial l \to 0$, the change in ω_c has to be expanded up to second order. For a single wave the local wave vector can be obtained by taking the averaged value of the imaginary part of the logarithmic derivative of the electric field over the resonant region

$$k_l = \left\langle \mathrm{Im}\,\partial \ln E_+/\partial l \right\rangle_r \tag{3a}$$

or

$$k_\theta = \left\langle \mathrm{Im}\,\partial \ln E_+/\partial \theta \right\rangle_r, \tag{3b}$$

where $\left\langle \ldots \right\rangle_r = \frac{1}{2\Delta\theta_r}\int_{\theta_r-\Delta\theta_r}^{\theta_r+\Delta\theta_r} (\ldots)\,d\theta$. The region, over which one has to average $k_\parallel$, depends on the parallel velocity of the resonant ions. Further away from the resonance only particles with higher energy can satisfy the resonance condition and the region, over which one has to average increases.

For strong single pass damping when the wave field can be regarded as arising from one wave and the finite k_θ only makes a perturbation of $k_\parallel$, the wave field can be obtained by means of iterations replacing $k_\parallel = n_\phi/R$ in the susceptibility tensor with $k_\parallel = n_\phi B_\phi/BR + k_\theta B_\theta/B$ when k_θ is given by Eq. (3b). As a verification of that the solution consists of one propagating wave rather than a standing wave, one can compare k_θ calculated from $k_\theta = \left\langle \mathrm{Im}\,\partial \ln E_+/\partial\theta \right\rangle_r$ and from $k_\theta^2 = -\left\langle E_+^{-1}\,\partial^2 E_+/\partial\theta^2 \right\rangle_r$.

3. Higher order FLR terms

The dielectric response for hot homogeneous plasma can be found in e.g. Ref. [6]. The spatial dispersive effects caused by FLR depend in leading order only on $k_\perp$. The higher order terms depend also on the direction of $\mathbf{k}_\perp$, and are important for describing how the absorption by the E_+ and the E_- components add up. The perpendicular components of the susceptibility tensor in an arbitrary coordinate system can be written as

$$\chi_\perp = \begin{pmatrix} T + \sigma\cos 2\vartheta & -i\chi_{xy} - \sigma\sin 2\vartheta \\ i\chi_{xy} - \sigma\sin 2\vartheta & T - \sigma\cos 2\vartheta \end{pmatrix}, \tag{4}$$

where $T = (\chi_{xx} + \chi_{yy})/2$, $\sigma = (\chi_{xx} - \chi_{yy})/2$. The angle ϑ depends on the direction of $\mathbf{k}_\perp$. The susceptibility tensor components χ_{xx}, χ_{xy} and χ_{yy} are calculated in a coordinate system for which $k_y = 0$. In the limit of small Larmor radius $\sigma \to 0$ and $T \to \chi_{yx}$ the absorption depends only on $|E_+|$. For finite Larmor radius the E_- component also contributes to the absorption.

When defining $E_+ = 0.5\,(E_n + i\,E_b)$ and $E_- = 0.5\,(E_n - i\,E_b)$, where E_n and E_b are the normal and bi-normal components of the electric wave field, respectively, the absorbed power can be written as

$$\begin{aligned} p = &-2\left|E_+\right|^2 \mathrm{Im}\left(T + \chi_{xy}\right) - 2\left|E_-\right|^2 \mathrm{Im}\left(T - \chi_{xy}\right) - \\ &4\,\mathrm{Im}\,\sigma\left(\mathrm{Re}\left(E_+^* E_-\right)\cos 2\vartheta - \mathrm{Im}\left(E_+^* E_-\right)\sin 2\vartheta\right) \end{aligned} \tag{5}$$

The contributions to the absorption from the $|E_+|^2$ and $|E_-|^2$ terms do not depend on how the coordinate system is rotated, but the contributions proportional to $E_+^*E_-$ do. In a local coordinate system (x, y, z) with $k_y = 0$ in which we denote the wave field by $\tilde{\mathbf{E}}$ the absorbed power becomes

$$p = -2\left|\tilde{E}_+\right|^2 \operatorname{Im}\left(T + \chi_{xy}\right) - 2\left|\tilde{E}_-\right|^2 \operatorname{Im}\left(T - \chi_{xy}\right) - 4\operatorname{Re}\left(\tilde{E}_+^*\tilde{E}_-\right)\operatorname{Im}\sigma, \tag{6}$$

where $\tilde{E}_+ = 0.5\left(\tilde{E}_x + i\tilde{E}_y\right)$ and $\tilde{E}_- = 0.5\left(\tilde{E}_x - i\tilde{E}_y\right)$. Since the absorbed power is independent of the coordinate system it follows that

$$\operatorname{Re}\left(E_+^*E_-\right)\cos 2\vartheta + \operatorname{Im}\left(E_+^*E_-\right)\sin 2\vartheta = \operatorname{Re}\left(\tilde{E}_+^*\tilde{E}_-\right). \tag{7}$$

Let α_+ and α_- be defined by $e^{i\alpha+}|E_+| = 0.5\,(E_n + i\,E_b)$ and $e^{i\alpha-}\,|E_-| = 0.5\,(E_n - i\,E_b)$, the angle ϑ then becomes $\vartheta = (\alpha_- - \alpha_+)/2$. Self-consistent calculations of the wave field and the susceptibility tensor can be obtained by means of iteration using a local susceptibility tensor on the form of Eq. (4) calculated from the wave field components E_n and E_b obtained from the previous iteration, starting the initial calculation with $\vartheta = 0$.

4. Electron damping

The electron damping by the fast magnetosonic wave arises from TTMP and ELD (electron Landau damping). The TTMP damping arises from the acceleration of the electrons by the compression of the magnetic field by the wave due to the finite magnetic moment of the electrons. Here we denote the corresponding force as the mirror force. The force is proportional to the parallel gradient of the parallel component of the compressional magnetic wave field, $\tilde{B}_\parallel$. The power density caused by TTMP damping is given by

$$p_{TTMP} = 2\left(\frac{\omega}{k_\parallel w_\parallel}\right)^2 \frac{\sqrt{\pi}}{m_e\left|k_\parallel\right| w_\parallel} e^{-\left(\frac{\omega}{k_\parallel w_\parallel}\right)^2} n_e \left|\frac{m_e w_\perp^2}{2B} k_\parallel \tilde{B}_\parallel\right|^2, \tag{8}$$

by ELD

$$p_{ELD} = \left(\frac{\omega}{k_\parallel w_\parallel}\right)^2 \frac{\sqrt{\pi}}{m_e\left|k_\parallel\right| w_\parallel} e^{-\left(\frac{\omega}{k_\parallel w_\parallel}\right)^2} n_e \left|eE_\parallel\right|^2, \tag{9}$$

and the cross term by

$$p_m = \left(\frac{\omega}{k_\parallel w_\parallel}\right)^2 \frac{\sqrt{\pi}}{m_e\left|k_\parallel\right| w_\parallel} e^{-\left(\frac{\omega}{k_\parallel w_\parallel}\right)^2} n_e \frac{m_e w_\perp^2}{2B} k_\parallel \tilde{B}_\parallel^* eE_\parallel + cc, \tag{10}$$

where $w_\perp^2 = 2kT_\perp/m_e$ and $w_\parallel^2 = 2kT_\parallel/m_e$. The TTMP appears as a FLR effect because the damping by the mirror force depends on the magnetic moment. For planar propagating waves in a homogeneous plasma in a coordinate system with $k_y = 0$ and the magnetic field is in the z-direction, the TTMP damping appears in the electron susceptibility tensor in the $\chi_{yy}^{(e)}$ component acting on E_y. The electron damping from ELD appears in $\chi_{zz}^{(e)}$ component. The coupling between E_y and E_z is caused by the $\chi_{zy}^{(e)}$ and the $\chi_{yz}^{(e)}$ components. Coupling between the parallel electric wave field, E_z, and the E_x component occurs through the $k_x k_z$ term, when neglecting this coupling one obtains

$$E_\parallel \approx -\frac{m_e w_\perp^2}{2eB}\nabla_\parallel \tilde{B}_\parallel. \tag{11}$$

The parallel electric wave field component, caused by acceleration of the electrons along the magnetic field lines by the mirror force, counteracts the acceleration by the mirror force, and reduce the total damping to about half of that for TTMP alone [6]. To include the correction of the electron damping by the finite parallel electric field, when it is not included in the wave solution as in the LION code, only half of the electron susceptibility is then included.

$$\operatorname{Im}\chi_{yy}^{(eTTMP/ELD)} = \frac{k_\perp^2 w_\perp^2}{2\omega_{ce}^2}\frac{T_\perp}{T_\parallel}\frac{\omega_{pe}^2\sqrt{\pi}}{\omega k_\parallel w_\parallel} e^{-\left(\frac{\omega}{k_\parallel w_\parallel}\right)^2} \tag{12}$$

The above procedure is justified because the Hermitian part of the dielectric tensor component χ_{yy} is dominated by the ion contributions, and the parallel electric field has very little effect on the dispersion relation of the magnetosonic wave.

For incompressible waves, as the ideal Alfvén wave, the parallel electric field arises due to the coupling by $k_x k_z$-term, which we are not treated here.

Since the TTMP is a magnetic response and $\tilde{B}_\parallel$ is related to the electric wave field by $\partial\tilde{B}_\parallel/\partial t = -(\nabla\times\mathbf{E})_\parallel$, the response should be added to the derivatives of the electric wave field rather than the electric field itself. If one approximates $|\nabla\times\mathbf{E}|_\parallel^2 \approx k_\perp^2|E_b|^2$ the electron absorption can first be calculated by solving the wave equation with the anti-Hermitian part of the electron susceptibility tensor element $\chi_{bb}^{(e)}$, given by Eq. (12), where b denote the bi normal direction and then correct it by re-evaluating $\tilde{B}_\parallel$ after solving the wave equation by multiplying the electron absorption with $|\nabla\times\mathbf{E}|_\parallel^2/k_\perp^2|E_b|^2$ using $i\omega\tilde{B}_\parallel = (\nabla\times\mathbf{E})_\parallel$ with $\nabla\times\mathbf{E}$ given by the solution. This correction factor can be used in the tensor in the next iteration allowing the wave field and the dielectric tensor to be solved consistently.

The upshift for the TTMP damping should be calculated in the same way as in section 2. The parallel wave number, $k_\parallel$, for the transit time magnetic pumping should be evaluated from

$$\Delta W = \mu\int_0^{\tau_B} \mathbf{v}_{g\parallel}\cdot\nabla\tilde{B}_\parallel dt, \tag{13}$$

which gives

$$k_\parallel = \frac{B_\phi}{B}\frac{n_\phi}{R} + \frac{B_\theta}{rB}\operatorname{Im}\left[\left(\frac{\partial\tilde{B}_\parallel}{\partial l}\right)^{-1}\frac{\partial}{\partial\theta}\frac{\partial\tilde{B}_\parallel}{\partial l}\right]. \tag{14}$$

5. Conclusions and Discussions

The proposed methods of calculating effects of spatial dispersion for the fast magnetosonic wave are well suited for FEM codes, which only describe one single wave, and allow calculations of the upshift of the parallel wave number, the cyclotron absorption including higher order FLR terms, and TTMP damping. Higher order FLR terms are important for modelling harmonic damping. The width of the resonance region depends on the parallel wave number whereas the total harmonic damping depends only weakly on it. The FLR correction should therefore be a good approximation also for weakly damped modes as long as the magnitudes of their perpendicular wave vectors do not differ too much. The FLR model in section 3 gives a positive definite absorption for distribution function decreasing with energy, and differs from the frequently used so called "reduced FLR wave equation" [7, 8] by including the absorption due to the E_- component. It takes into account the decrease of the absorption at the fundamental cyclotron resonance due to the FLR terms and avoids the problem of creating negative absorption. The TTMP damping is important for fast wave current drive. The correction term is found to be most important near the magnetic axis reducing the

Journal of Physics: Conference Series **401** (2012) 012009 doi:10.1088/1742-6596/401/1/012009

electron damping. This correction is important for current drive, since in this region current drive has the highest efficiency because of the low fraction of trapped particles. The proposed method for calculating the upshift cannot handle weak single pass damping where the wave field consists of different waves (reflected wave). In this case one can only handle waves with large magnitudes of n_ϕ for which the upshift is negligible.

Acknowledgments
The work has been partly conducted within the framework of the EFDA Integrated Tokamak Modelling task force and the EFDA Goal Oriented Training Programme GOTiT.

References

[1] Jaeger E F *et al* 2001 *Physics of Plasma* **8** 1573
[2] Villard L *et al* 1986 *Computer Physics Reports* 4, 95
[3] Villard L *et al* 1995 *Nuclear Fusion* **35** 1173
[4] Kaufman A N 1972 Phys. Fluids **15** 1063
[5] Kerbel G D and McCoy M G 1985 Phys. Fluids **28** 3629
[6] Stix T H 1992 *Waves in Plasma Physics*, AIP, New York, USA
[7] Swanson D G 1981 *Physics of Fluids* **24** 2035
[8] Colestock P L and Kashuba R J 1983 *Nuclear Fusion* **23** 763

Joint Varenna-Lausanne International Workshop 2012 IOP Publishing
Journal of Physics: Conference Series **401** (2012) 012010 doi:10.1088/1742-6596/401/1/012010

Coupling JOREK and STARWALL Codes for Non-linear Resistive-wall Simulations

M Hölzl[1], P Merkel[1], G T A Huysmans[2], E Nardon[3], E Strumberger[1], R McAdams[4,5], I Chapman[5], S Günter[1], K Lackner[1]

[1]Max-Planck-Institute for Plasmaphysics, EURATOM Association, Boltzmannstr. 2, 85748 Garching, Germany
[2]ITER Organisation, Route de Vinon sur Verdon, St-Paul-lez-Durance, France
[3]CEA, IRFM, CEA Cadarache, F-13108 St Paul-lez-Durance, France
[4]York Plasma Institute, University of York, York, YO10 5DD, UK
[5]Euratom/CCFE Fusion Association, Culham Science Centre, Abingdon, OX14 3DB, UK

E-mail: `mhoelzl@ipp.mpg.de`

Abstract. The implementation of a resistive-wall extension to the non-linear MHD-code JOREK via a coupling to the vacuum-field code STARWALL is presented along with first applications and benchmark results. Also, non-linear saturation in the presence of a resistive wall is demonstrated. After completion of the ongoing verification process, this code extension will allow to perform non-linear simulations of MHD instabilities in the presence of three-dimensional resistive walls with holes for limited and X-point plasmas.

1. Introduction

Plasma instabilities growing on fast time-scales such as vertical displacement events, disruptions, external kink modes, or edge localized modes are connected with a time-dependent magnetic perturbation outside the plasma. The mirror currents induced in conducting structures around the plasma[1] by the varying field perturbations act back onto the instabilities. Thus, linear and non-linear dynamics of the plasma can be strongly influenced by the wall. An external kink mode driven by strong edge current densities, for instance, is converted into a resistive wall mode (RWM) in the presence of a conducting wall close to the plasma edge and grows on the timescale of resistive wall-current decay as described in great detail in Refs. [1, 2].

We describe the status of a resistive wall extension for the non-linear MHD code JOREK [3, 4] as well as first benchmark results. The implementation is done by coupling JOREK to a modified version of the STARWALL code [5] which determines the magnetic field in the vacuum region in the presence of three-dimensional walls with holes. The "response" of conducting structures to magnetic field perturbations is represented by response-matrices computed by STARWALL. These matrices are used in JOREK to keep track of wall-currents and calculate the tangential magnetic field required for the modified boundary conditions as discussed in Section 3.2.

The article provides some information about the STARWALL code in Section 2. The coupling equations are derived in Section 3. First benchmarks of the resistive wall extension are presented in Section 4. Conclusions and a brief outlook are given in Section 5.

[1] For simplicity, all conducting structures are denoted wall in the following.

Journal of Physics: Conference Series **401** (2012) 012010 doi:10.1088/1742-6596/401/1/012010

2. The STARWALL Code

STARWALL solves the vacuum magnetic field equation outside the JOREK computational domain in presence of a three-dimensional conducting wall with holes as a Neumann-like problem. The continuity of the magnetic field component normal to the boundary of the JOREK computational domain (called interface) is used as a boundary condition. The wall is represented by infinitely thin triangles which allows to approximate realistic wall structures very well using an effective resistance η_w/d_w where η_w and d_w denote the resistivity of the wall material[2] and its thickness, respectively. Wall currents are assumed constant within each triangle such that they can be described by current potentials Y_k at the triangle nodes [5][3].

In case of an ideally conducting wall, solving the vacuum field equation results in an algebraic expression for the magnetic field component tangential to the interface[4] in terms of the normal component. Wall currents don't need to be considered explicitly as they are instantaneously given by the normal field at the interface. Using the ideal wall response-matrix $\hat{M}^{\mathrm{id}}$ calculated by STARWALL, this can be written in the following way[5]:

$$B_{\mathtt{tan}} = \sum_i b_i \, B_{\mathtt{tan},i} = \sum_i b_i \sum_j \hat{M}^{\mathrm{id}}_{i,j} \, \Psi_j. \tag{1}$$

Here, b_i denotes a JOREK basis function consisting of a 1D Bezier basis function along the poloidal direction on the interface and Fourier basis function in toroidal direction. Ψ_j denotes poloidal flux coefficients. The poloidal flux at the interface is given by $\Psi = \sum_j b_j \, \Psi_j$.

When a resistive wall is considered, wall currents cannot be eliminated anymore. In this case, the tangential magnetic field is given by

$$B_{\mathtt{tan}} = \sum_i b_i \left(\sum_j \hat{M}^{\mathrm{ee}}_{i,j} \, \Psi_j + \sum_k \hat{M}^{\mathrm{ey}}_{i,k} \, Y_k \right), \tag{2}$$

where wall currents evolve in time according to

$$\dot{Y}_k = -\frac{\eta_w}{d_w} \, \hat{M}^{\mathrm{yy}}_{k,k} \, Y_k - \sum_j \hat{M}^{\mathrm{ye}}_{k,j} \, \dot{\Psi}_j. \tag{3}$$

Here, $\hat{M}^{\mathrm{ee}}$, $\hat{M}^{\mathrm{ey}}$, $\hat{M}^{\mathrm{ye}}$, and $\hat{M}^{\mathrm{yy}}$ denote resistive response matrices determined by STARWALL. Indices i and j run over all boundary degrees of freedom of the respective variable and k over all wall current potentials. For cross-checking, a relation between ideal and resistive response matrices can be derived by letting $\eta_w \to \infty$ (Appendix A).

Information about the STARWALL code coupled with CASTOR (sometimes called STARWALL_C) can be found in Refs. [5, 7]. It is similar to the STARWALL code coupled with JOREK (sometimes called STARWALL_J). An article describing more details is in preparation by the STARWALL author Peter Merkel.

3. Implementation in JOREK

For the coupling of JOREK and STARWALL, Eqs. (2) and (3) are discretized in time as described in Section 3.1. The vacuum response then enters into a natural boundary condition in JOREK as discussed in Section 3.2.

[2] Wall resistivity is normalized the same way as plasma resistivity (see Ref. [6] for JOREK normalizations).
[3] For simplicity, we will (inaccurately) speak about wall currents when wall current potentials are referred to.
[4] $B_{\mathtt{tan}}$ is the component of $(\mathbf{B} - \mathbf{B}_\phi)$ tangential to the interface, where $\mathbf{B}_\phi$ denotes the toroidal field component.
[5] For practical reasons, the algebraic expression is written in terms of the poloidal flux Ψ instead of the normal field component (which can be calculated from Ψ, of course).

3.1. Time-Discretization

Equation (2) is evaluated at the new timestep $n+1$ and discretizations $\Psi^{n+1} = \Psi^n + \delta\Psi^n$ and $Y^{n+1} = Y^n + \delta Y^n$ are used, where superscripts indicate evaluation at the given timestep. The tangential magnetic field is thus given by

$$B_{\mathbf{tan}}^{n+1} = \sum_i b_i \left[\sum_j \hat{M}_{i,j}^{\mathrm{ee}} \; (\Psi_j^n + \delta\Psi_j^n) + \sum_k \hat{M}_{i,k}^{\mathrm{ey}} \; (Y_k^n + \delta Y_k^n) \right]. \tag{4}$$

The general time-evolution scheme described in Ref. [8],

$$\left[(1+\xi) \left(\frac{\partial \mathbf{X}}{\partial \mathbf{u}} \right)^n - \Delta t \theta \left(\frac{\partial \mathbf{Z}}{\partial \mathbf{u}} \right)^n \right] \delta \mathbf{u}^n = \Delta t \, \mathbf{Z}^n + \xi \left(\frac{\partial \mathbf{X}}{\partial \mathbf{u}} \right)^{n-1} \delta \mathbf{u}^{n-1}, \tag{5}$$

is used for the JOREK equations written in the form $\partial \mathbf{X}(\mathbf{u})/\partial t = \mathbf{Z}(\mathbf{u})$, where $\mathbf{u}$ denotes the vector of JOREK physical variables like temperature or density, $\mathbf{X}$ and $\mathbf{Z}$ the left- respectively right-hand side terms of the time-evolution equations, and θ and ξ are numerical parameters[6]. The same general time-evolution method needs to be applied to the wall-current evolution (Eq. (3)) giving

$$\begin{aligned} &(1+\xi) \left[\delta Y_k^n + \sum_j \hat{M}_{k,j}^{\mathrm{ye}} \; \delta\Psi_j^n \right] + \Delta t \; \theta \; \frac{\eta_{\mathrm{w}}}{d_{\mathrm{w}}} \; \hat{M}_{k,k}^{\mathrm{yy}} \; \delta Y_k^n \\ &= - \; \Delta t \; \frac{\eta_{\mathrm{w}}}{d_{\mathrm{w}}} \; \hat{M}_{k,k}^{\mathrm{yy}} \; Y_k^n + \xi \left[\delta Y_k^{n-1} + \sum_j \hat{M}_{k,j}^{\mathrm{ye}} \; \delta\Psi_j^{n-1} \right]. \end{aligned} \tag{6}$$

Terms with δY_k^n are brought to the left hand side. The equation then reads

$$\begin{aligned} &\left(1 + \xi + \Delta t \theta \frac{\eta_{\mathrm{w}}}{d_{\mathrm{w}}} \hat{M}_{k,k}^{\mathrm{yy}} \right) \delta Y_k^n \\ &= - \; (1+\xi) \sum_j \hat{M}_{k,j}^{\mathrm{ye}} \; \delta\Psi_j^n - \Delta t \frac{\eta_{\mathrm{w}}}{d_{\mathrm{w}}} \hat{M}_{k,k}^{\mathrm{yy}} Y_k^n + \xi \delta Y_k^{n-1} + \xi \sum_j \hat{M}_{k,j}^{\mathrm{ye}} \; \delta\Psi_j^{n-1}. \end{aligned} \tag{7}$$

After solving for δY_k^n, one gets

$$\delta Y_k^n = \sum_j \hat{A}_{k,j} \; \delta\Psi_j^n + \hat{B}_{k,k} \; Y_k^n + \hat{C}_{k,k} \; \delta Y_k^{n-1} + \sum_j \hat{D}_{k,j} \; \delta\Psi_j^{n-1}, \tag{8}$$

where some of the "derived response matrices"

$$\begin{array}{lll} \hat{S}_{k,k} = 1 + \xi + \Delta t \theta \frac{\eta_{\mathrm{w}}}{d_{\mathrm{w}}} \hat{M}_{k,k}^{\mathrm{yy}} & \hat{D}_{k,j} = \xi \hat{M}_{k,j}^{\mathrm{ye}} / \hat{S}_{k,k} & \hat{H}_{i,j} = \hat{M}_{i,j}^{\mathrm{ee}} \\ \hat{A}_{k,j} = -(1+\xi) \; \hat{M}_{k,j}^{\mathrm{ye}} / \hat{S}_{k,k} & \hat{E}_{i,j} = \hat{M}_{i,j}^{\mathrm{ee}} + \sum_k \hat{M}_{i,k}^{\mathrm{ey}} \; \hat{A}_{k,j} & \hat{J}_{i,j} = \sum_k \hat{M}_{i,k}^{\mathrm{ey}} \; \hat{D}_{k,j} \\ \hat{B}_{k,k} = -\Delta t \frac{\eta_{\mathrm{w}}}{d_{\mathrm{w}}} \hat{M}_{k,k}^{\mathrm{yy}} / \hat{S}_{k,k} & \hat{F}_{i,k} = \hat{M}_{i,k}^{\mathrm{ey}} (1 + \hat{B}_{k,k}) & \hat{K}_{k,l} = -\Delta t \hat{M}_{k,k}^{\mathrm{yy}} \hat{S}_{l,k} \\ \hat{C}_{k,k} = \xi / \hat{S}_{k,k} & \hat{G}_{i,k} = \hat{M}_{i,k}^{\mathrm{ey}} \; \hat{C}_{k,k} & \hat{L}_{i,l} = \sum_k \hat{M}_{i,k}^{\mathrm{ey}} \; \hat{K}_{k,l} \end{array} \tag{9}$$

[6] Parameter values satisfying $\theta - \xi = 0.5$ are required to guarantee second-order accuracy of the time-evolution scheme. For instance, $(\theta = 0.5, \xi = 0)$ corresponds to a linearized Crank-Nicholson scheme and $(\theta = 1, \xi = 0.5)$ to the linearized Gears scheme.

Journal of Physics: Conference Series **401** (2012) 012010 doi:10.1088/1742-6596/401/1/012010

have been used. These matrices are computed at the beginning of a JOREK simulation from the STARWALL response matrices. They need to be updated if one or more parameters entering the definitions have changed, e.g., Δt, ξ, or $\eta_{\rm w}$. Plugging Eq. (8) into Eq. (4) gives

$$B_{\rm tan}^{n+1} = \sum_i b_i \Bigg[\sum_j \Big(\hat{M}^{\rm ee}_{i,j} + \sum_k \hat{M}^{\rm ey}_{i,k}\, \hat{A}_{k,j} \Big)\, \delta\Psi_j^n + \sum_k \hat{M}^{\rm ey}_{i,k} \left(1 + \hat{B}_{k,k}\right) Y_k^n + \sum_k \hat{M}^{\rm ey}_{i,k} \hat{C}_{k,k}\, \delta Y_k^{n-1} + \sum_j \hat{M}^{\rm ee}_{i,j} \Psi_j^n + \sum_j \sum_k \hat{M}^{\rm ey}_{i,k} \hat{D}_{k,j} \delta\Psi_j^{n-1} \Bigg] \tag{10}$$

and, making use of Eq. (9), we get

$$B_{\rm tan}^{n+1} = \sum_i b_i \left[\sum_j \hat{E}_{i,j}\, \delta\Psi_j^n + \sum_k \hat{F}_{i,k}\, Y_k^n + \sum_k \hat{G}_{i,k}\, \delta Y_k^{n-1} + \sum_j \hat{H}_{i,j}\, \Psi_j^n + \sum_j \hat{J}_{i,j}\, \delta\Psi_j^{n-1} \right]. \tag{11}$$

3.2. Boundary Integral

The fixed boundary conditions for poloidal flux and plasma current corresponding to an ideally conducting wall are removed from JOREK. As a consequence, a boundary integral in the current definition equation resulting from partial integration (which vanishes in fixed-boundary JOREK) needs to be considered now. It can be written in terms of the tangential magnetic field such that we can insert Eq. (11) as a natural boundary condition. Details are given in the following.

The current definition equation, $j = \Delta^*\Psi$, is written in weak form as

$$\int dV \frac{j_l^*}{R^2}\, (j - \Delta^*\Psi) = 0. \tag{12}$$

where j_l^* denotes the test-functions taken to be identical with the basis-functions. Using $\Delta^*\Psi \equiv R^2\, \nabla\cdot (R^{-2}\, \nabla\Psi)$, we get

$$\int dV\, \frac{j_l^*}{R^2}\, j - \int dV\, j_l^*\, \nabla\cdot \left(\frac{1}{R^2} \nabla\Psi \right) = 0, \tag{13}$$

where the second term of the resulting expression can be integrated by parts ($\int dV\ a\ \nabla\cdot\mathbf{b} = -\int dV\ \nabla a\cdot\mathbf{b} + \oint dA\ a\ \mathbf{b}\cdot\hat{\mathbf{n}}$) yielding

$$\int dV\, \frac{j_l^*}{R^2}\, j + \int dV\, \frac{1}{R^2}\, \nabla j_l^* \cdot \nabla\Psi - \oint dA\, \frac{j_l^*}{R}\, \underbrace{(\nabla\Psi\cdot\hat{\mathbf{n}}/R)}_{\equiv B_{\rm tan}} = 0. \tag{14}$$

Here, $\hat{n}$ denotes the unit vector normal to the interface and the tangential field is identified in the boundary integral (refer to Appendix B for details). Eq. (11) is inserted into Eq. (14) and after separating implicit and explicit terms, we have derived the form of the current equation implemented in JOREK,

$$\begin{aligned} &\sum_{i_{\rm elem}} \int \frac{dV}{R^2} \left(j_l^*\, \delta j^n + \nabla j_l^* \cdot \nabla\delta\Psi^n \right) - \sum_{i_{\rm bnd}} \oint dA \frac{j_l^*}{R} \sum_i b_i \sum_j \hat{E}_{i,j}\, \delta\Psi_j^n \\ &= - \sum_{i_{\rm elem}} \int \frac{dV}{R^2} \left(j_l^*\, j^n + \nabla j_l^* \cdot \nabla\Psi^n \right) \\ &+ \sum_{i_{\rm bnd}} \oint dA \frac{j_l^*}{R} \sum_i b_i \left[\sum_k \left(\hat{F}_{i,k}\, Y_k^n + \hat{G}_{i,k}\, \delta Y_k^{n-1} \right) + \sum_j \left(\hat{H}_{i,j}\, \Psi_j^n + \hat{J}_{i,j}\, \delta\Psi_j^{n-1} \right) \right], \end{aligned} \tag{15}$$

Journal of Physics: Conference Series **401** (2012) 012010 doi:10.1088/1742-6596/401/1/012010

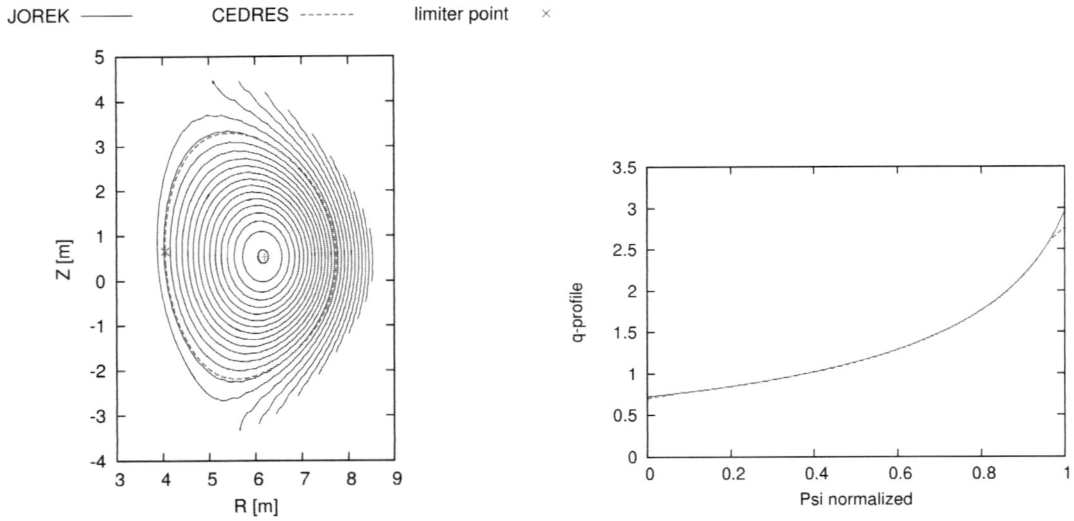

Figure 1
Left: *JOREK flux-surfaces of a limiter equilibrium (limiter point shown as blue cross) are compared to CEDRES++ last closed flux-surface and magnetic axis.*
Right: *Comparison of the respective q-profiles.*

with

$$\int dV = \sum_{i_{\text{elem}}} \int du\, dv\, d\phi\; J_2\, R, \quad \text{and} \quad \oint dA = \sum_{i_{\text{bnd}}} \int du\, d\phi\; R\sqrt{(R_{,u})^2 + (Z_{,u})^2}, \tag{16}$$

where the integral over u and v is carried out by Gauss-quadrature: $\int du \;\; X(u) = \sum_{m_G} w_{m_G}\; X(u_{m_G})$ with weights w_{m_G} and positions u_{m_G}. Coordinate u corresponds to the element-local coordinate of the 2D Bezier elements oriented poloidally along the interface. The other element-local coordinate is denoted v. J_2 denotes the 2D-Jacobian in the poloidal plane. Wall-currents are updated after each time-step according to Eq. (8) to guarantee consistency with the implicit time-stepping of JOREK.

For the $n = 0$ component, poloidal field coils need to be taken into account properly. This is done by replacing the expression for the tangential magnetic field, Eq. (2), by

$$B_{\mathtt{tan}} = \sum_i b_i \left(\sum_j \hat{M}^{\text{ee}}_{i,j}\, (\Psi_j - \Psi^{\text{coil}}_j) + \sum_k \hat{M}^{\text{ey}}_{i,k}\, Y_k + B^{\text{coil}}_{\mathtt{tan},i} \right). \tag{17}$$

Here, Ψ^{coil}_j and $B^{\text{coil}}_{\mathtt{tan},i}$ denote time-independent coil contributions to poloidal flux and tangential field at the interface, respectively. This leads to two additional terms in Eq. (15). A similar boundary integral also occurs in the Grad-Shafranov equation giving the plasma equilibrium. Here, only the $n = 0$ component needs to be considered (axisymmetric equilibria in JOREK) and no time-discretization is required.

4. First Benchmarks and Results

In this Section, we present first results of the ongoing benchmark effort for code validation.

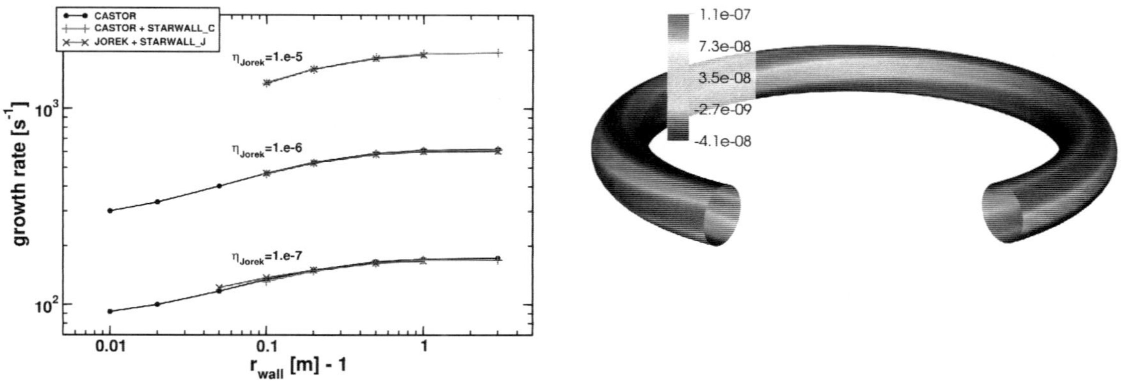

Figure 2
Left: *Linear growth-rates of a tearing-mode in a circular plasma with concentric ideal wall are compared between CASTOR using its own vacuum field module, CASTOR coupled with STARWALL, and JOREK coupled with STARWALL. Very good agreement is observed for a variety of different plasma resistivities and plasma-wall distances.*
Right: *The current potential distribution (arbitrary units) reflecting the 2/1 tearing mode structure is plotted during the linear phase ($\eta_{Jorek} = 1 \cdot 10^{-6}$, $r_w = 1.2$).*

4.1. Free-boundary equilibrium
Computing a free-boundary equilibrium with JOREK and comparing it to the results of another equilibrium solver allows to check some parts specific for $n = 0$ like poloidal field-coil contributions (see Sec. 3.2). We are considering an ITER-like limiter plasma. As seen from Figure 1, flux surfaces and q-profile of the JOREK equilibrium agree very well with the equilibrium computed by the CEDRES++ code [9]. Slightly different discretizations of the poloidal field coils might be responsible for small remaining differences.

4.2. Tearing Mode
A 2/1 tearing mode in a circular plasma with major radius $R = 10$, minor radius $a = 1$ and uniform plasma resistivity surrounded by an ideally conducting wall is considered. Figure 2 shows excellent agreement between JOREK and the linear CASTOR code [10]. Repeating the simulation for a resistive wall with zero resistivity is physically identical, of course, but numerically treated in a different way. The exact agreement we get here proves consistency.

4.3. Resistive Wall Mode
In this Section, results for 2/1 resistive wall modes (RWMs) in a circular plasma with minor radius $a = 1$ and major radius $R = 10$ surrounded by a resistive wall are presented. The left part of Figure 3 shows linear growth rates obtained for different wall radii r_w as a function of the wall resistivity. Clearly, the modes cannot be stabilized beyond a certain wall distance from the plasma even at very low wall resistivities. The right part of the Figure shows energy time-traces during the linear and saturation phase of a typical simulation.

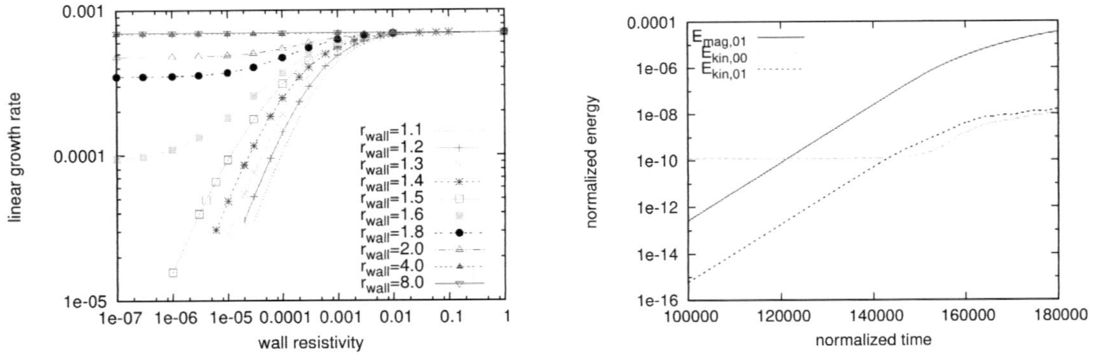

Figure 3
Left: *Linear RWM growth rates extracted from the simulations are shown for a variety of wall radii and wall resistivities.*
Right: *Energy time-traces are shown for linear and saturation phase of an RWM ($r_w = 1.2$, $\eta_{wall} = 1 \cdot 10^{-4}$). In saturation, wall currents on the high-field side have an amplitude comparable to the low-field side while they are smaller in the linear phase.*

5. Conclusions and Outlook

The ongoing implementation and verification of a resistive wall model in the non-linear MHD-code JOREK was summarized. Benchmarks for a freeboundary equilibrium and tearing mode cases show good agreement. Simulations of the linear and non-linear phase of RWMs were presented (comparison to analytical theory and linear codes ongoing).

Benchmarking will be continued and extended to non-linear comparisons. Realistic X-point geometries will be considered requiring a special treatment of grid corners. After completion, the code can be applied to a variety of MHD instabilities interacting with conducting structures like resistive wall modes, edge localized modes, vertical displacement events or disruptions.

Acknowledgements

Simulations were mostly carried out on the HPC-FF computing cluster in Jülich, Germany, and on the IFERC-CSC Helios computing cluster in Rokkasho-Mura, Japan.

References

[1] Chu M S and Okabayashi M 2010 *Plasma Physics and Controlled Fusion* **52** 123001
[2] Igochine V 2012 *Nuclear Fusion* **52** 074010
[3] Czarny O and Huysmans G 2008 *J. Comput. Phys.* **227** 7423 – 7445
[4] Huysmans G, Pamela S, Beurskens M, Becoulet M and van der Plas E 2010 *Proceedings of the 23rd IAEA Fusion Energy Conference* (Daejon, South Korea) THS/7-1
[5] Merkel P and Sempf M 2006 *Proceedings of the 21st IAEA Fusion Energy Conference* (Chengdu, China) TH/P3-8 URL `www-naweb.iaea.org/napc/physics/FEC/FEC2006/papers/th_p3-8.pdf`
[6] Hölzl M, Günter S, Wenninger R, Müller W C, Huysmans G, Lackner K, Krebs I and the ASDEX Upgrade Team 2012 *Physics of Plasmas (accepted)*
[7] Strumberger E, Merkel P, Tichmann C, and Günter S 2011 *Proceedings of the 38th EPS Conference on Plasma Physics* (Strasbourg, France) p5.082 URL `http://ocs.ciemat.es/EPS2011PAP/pdf/P5.082.pdf`
[8] Hirsch C 1989 *Numerical Computation of Internal and External Flows, Volume 1, Fundamentals of Numerical Discretization* (Wiley) ISBN 978-0-471-92385-5
[9] Hertout P, Boulbe C, Nardon E, Blum J, Brémond S, Bucalossi J, Faugeras B, Grandgirard V and Moreau P 2011" *Fusion Engineering and Design* **86** 1045–1048 Proceedings of the 26th Symposium of Fusion Technology (SOFT-26)
[10] Kerner W, Goedbloed J, Huysmans G, Poedts S and Schwarz E 1998 *J. Comput. Phys.* **142** 271 – 303

Journal of Physics: Conference Series **401** (2012) 012010 doi:10.1088/1742-6596/401/1/012010

Appendix A. Relation between ideal and resistive response matrices

Between the ideal and resistive response-matrices, the relation

$$\hat{M}^{\mathrm{id}}_{i,j} \equiv \hat{M}^{\mathrm{ee}}_{i,j} - \sum_k \hat{M}^{\mathrm{ey}}_{i,k}\, \hat{M}^{\mathrm{ye}}_{k,j} \tag{A.1}$$

must hold which becomes obvious when letting $\eta_{\mathrm{w}} \to 0$ and inserting Eq. (3) into Eq. (2). Also, the no-wall response can easily be identified as

$$\hat{M}^{\mathrm{nw}}_{i,j} \equiv \hat{M}^{\mathrm{ee}}_{i,j} \tag{A.2}$$

when letting $Y_k \to 0$. Eqs. (A.1) and (A.2) can be used to cross-check some parts of the implementation as both forms of the wall-treatment with eliminated wall-currents (Eq. (1)) and explicitly treated wall-currents (Eqs. (2) and (3)) are implemented in JOREK.

Appendix B. Magnetic field tangential to the interface

The magnetic field in the JOREK reduced-MHD model is given by

$$\mathbf{B} = \frac{F_0}{R}\hat{\mathbf{e}}_\phi + \frac{1}{R}\nabla\Psi \times \hat{\mathbf{e}}_\phi. \tag{B.1}$$

Its component tangential to the boundary of the JOREK computational domain (in the poloidal plane) can be determined from

$$B_{\tan} = (\mathbf{B} \times \hat{\mathbf{n}}) \cdot \hat{\mathbf{e}}_\phi, \tag{B.2}$$

where $\hat{\mathbf{n}}$ denotes the unit normal vector of the interface and $\hat{\mathbf{e}}_\phi$ the normalized toroidal basis vector. Inserting Eq. (B.1) and using vector identities, this can be written in the form

$$\begin{aligned} B_{\tan} &= -\frac{1}{R}\hat{\mathbf{e}}_\phi \cdot \left[\hat{\mathbf{n}} \times (\nabla\Psi \times \hat{\mathbf{e}}_\phi)\right] \\ &= -\frac{1}{R}\hat{\mathbf{e}}_\phi \cdot \left[\underbrace{(\hat{\mathbf{n}} \cdot \hat{\mathbf{e}}_\phi)}_{\equiv 0}\nabla\Psi - (\hat{\mathbf{n}} \cdot \nabla\Psi)\hat{\mathbf{e}}_\phi\right] \\ &= \frac{1}{R}\hat{\mathbf{n}} \cdot \nabla\Psi, \end{aligned} \tag{B.3}$$

which can be identified in the boundary integral of Eq. (14) and replaced by the STARWALL response.

Effect of plasma shaping and resonance location on minority ion temperature anisotropy in tokamak plasmas heated with ICRH

Ye O Kazakov[1], T Fülöp[1], I Pusztai[1] and T Johnson[2]

[1] Department of Applied Physics, Chalmers University of Technology, Euratom-VR Association, SE-41296 Göteborg, Sweden

[2] School of Electrical Engineering, Royal Institute of Technology, Euratom-VR Association, Stockholm, Sweden

E-mail: kazakov@chalmers.se

Abstract. Poloidal asymmetries of the impurity distribution, which are observed in tokamaks, may influence the impurity cross-field transport. Low field side ion cyclotron resonance heating (ICRH) often results in an inboard accumulation of impurities, which may in turn lead to an outward convective impurity flux. The temperature anisotropy of the ICRH-heated minority ions is identified to be one of the main parameters governing the impurity asymmetry strength. In the present work we analyze the effect of plasma shaping and the ICRH resonance location on the minority temperature anisotropy by means of the TORIC-SSFPQL modelling. We find that ellipticity reduces the anisotropy level due to the wave defocussing and broader absorption regions for the elongated plasmas. The temperature anisotropy decrease in case of the resonance layers located closer to the edge is caused by the significant reduction in heating power densities due to geometrical reasons.

1. Introduction

Ion cyclotron resonance heating (ICRH) is a widely used heating method in present-day fusion devices and is foreseen to be installed in ITER. In connection with ICRH the impurity density becomes often poloidally asymmetric [1–3]. The main reason for the asymmetry is an increase in the minority density at the outboard side of the magnetic surface due to ICRH. Preferentially perpendicular heating of the minority ions by ICRH traps excessive amount of positive charges at the smaller B locations and thus generates poloidal electrostatic electric field [4] that pushes impurities to the opposite (inboard) side. In recent experiments at the Alcator C-Mod tokamak it was conclusively demonstrated the link of the asymmetry level to ICRH [3]. In contrast to NBI-heated plasmas, where centrifugal effects usually lead to the outboard accumulation of impurities [3, 5], in ICRH-heated plasmas with the resonance layer located at the low magnetic field side (LFS) it is possible to reverse the side of the impurity accumulation from the outboard to the inboard. This effect has been recently recognized as a beneficial factor for controlling the impurity transport since the inboard accumulation of impurities can lead to the reduction of the impurity peaking factor in the core [6–8].

In a recent work [9], the poloidal asymmetry of the impurity ions in ICRH-heated tokamak plasmas has been analyzed. One of the most important parameters governing the impurity asymmetry strength has been identified to be the minority ion temperature anisotropy, $\alpha_T = T_\perp/T_\parallel$: the ratio of the perpendicular to the parallel temperature of the cyclotron heated minority population. The dependence of α_T on various plasma and ICRH parameters, such as ICRH power, minority and impurity concentrations, bulk plasma density and temperature, has been calculated. In the present paper we extend this work by studying the effects of the ICRH resonance location and plasma shaping on the minority temperature anisotropy.

2. Minority temperature anisotropy

In previous work [9], the poloidal impurity asymmetry strength has been shown to be proportional to the dimensionless parameter $k = \epsilon b_c(\alpha_T - 1)/[b_c + \alpha_T(1 - b_c)]$. Here, $\epsilon = r/R_0$ is the local inverse aspect ratio (r denotes the radial coordinate and R_0 is the plasma major radius), and $b_c = B_c/B_0$, where B_0 and B_c is the magnetic field at the plasma axis and at the minority cyclotron resonance layer, respectively. The parameter b_c, which depends on the location of the minority cyclotron resonance layer in the plasma, is controlled by the choice of B_0 and the radiofrequency (RF) generator frequency, f

$$b_c = \frac{f(\mathrm{MHz})}{15.25\, B_0(\mathrm{T})} \frac{A_\mathrm{m}}{Z_\mathrm{m}},$$

where A_m and Z_m are the atomic mass and the charge number of the resonant minority species. The parameter k also includes the minority temperature anisotropy α_T, which depends on many plasma and ICRH parameters in a complicated way, as was shown by numerical modelling in [9]. It has a maximum close to the resonance location and it increases approximately linearly with the ICRH power unless the heating power is too low. It is generally reduced with the minority concentration since the deposited RF energy per particle is lower. However, it has to be noted that α_T is not simply inversely proportional to the minority ion fraction, as the minority concentration affects the deposition profiles and ICRH power density reached. Minority temperature anisotropy as a function of plasma density scales roughly as $1/n_e^2$, but exhibits oscillations due to the sensitivity of the RF field pattern to multiple reflections and the interference undergone by the RF waves [10–15].

In this work we study the effect of the ICRH resonance location and plasma shaping (ellipticity κ and triangularity δ) on α_T. It is expected that the geometrical factors will affect the wave focussing and the ICRH power density, which determines the minority ion distribution function. The power density values are commonly smaller for the elliptical and D-shaped plasmas since the power absorption is spread over a larger region than in a circular case [16]. The ICRH resonance location is an additional parameter which is expected to affect significantly the impurity asymmetry strength. In Alcator C-Mod experiments, using B_0 scans to modify the location of the hydrogen minority resonance layer, clear 'in-out' asymmetry was observed only for the LFS heating. It is therefore instructive to determine the sensitivity of the reached temperature anisotropy to the ICRH resonance location.

In the general case, the RF field distribution and power deposition profiles in 2D tokamak geometry can only be determined numerically with a full-wave ICRH code. The analysis is further complicated by the fact that by modifying the resonance location one changes not only the minority absorption profile, but the local plasma parameters which determine the energetic minority distribution function (e.g. plasma density and temperature) are also different for different resonance locations. This motivates the use of numerical tools for the analysis of the effect of plasma shaping and cyclotron resonance location on the minority temperature anisotropy.

3. Numerical modelling and results

The temperature anisotropy is computed by the 2D full-wave TORIC code [17, 18] and the Fokker-Planck quasilinear solver SSFPQL [19] and the results are benchmarked to SELFO [20] simulations in the circular cross-section case. SELFO solves the wave field and the distribution function of the resonant ion species self-consistently, including finite-orbit width effects for the high-energy ions and RF-induced transport, but in its current status it cannot handle plasma shaping. TORIC solves Maxwell's equations in an axisymmetric but otherwise arbitrary toroidal magnetic configuration. Maxwellian distribution is assumed for all ion species, including minorities. The flux surface averaged power deposition profiles obtained by TORIC are exported to the quasilinear Fokker-Planck code SSFPQL [19], which evaluates the minority distribution function. SSFPQL provides information on the radial profiles of the high-energy tails generated by ICRH (distribution functions, perpendicular and parallel energies, number of fast ions), and on the collisional exchanges between the various species, but it does not account for toroidal effects (e.g. trapping or finite orbit width). SSFPQL calculates effective perpendicular and parallel temperatures for both the thermal and the tail parts of the distribution and an estimate for the fraction of fast minorities in the tail. Although the absorption and the local temperatures are different in TORIC-SSFPQL and SELFO, the average temperature is comparable, as it has been shown by comparison with self-consistent SELFO simulations presented in [9]. Therefore we expect that the averaged temperature anisotropy obtained by TORIC-SSFPQL gives a reasonable approximation in spite of the simplifying assumptions.

As a reference case we consider hydrogen minority heating in deuterium plasma with the minority concentration $X[\mathrm{H}] = 5\%$ for typical JET-like parameters: $R_0 = 3.0\,\mathrm{m}$, $a = 0.9\,\mathrm{m}$ (plasma minor radius), central electron density $n_{e0} = 3.6\times10^{19}\,\mathrm{m}^{-3}$, central plasma temperature $T_0 = 5.0\,\mathrm{keV}$ (background ions and electrons are assumed to have equal temperatures), the RF generator frequency $f = 41.2\,\mathrm{MHz}$, fast wave toroidal wavenumber $n_{\mathrm{tor}} = 27$, which is dominant for the dipole phasing of the JET A2 ICRH antenna [21,22], $B_0 = 3.0\,\mathrm{T}$, the safety factor at the center and the edge, $q_0 = 1.1$ and $q_1 = 4.5$, respectively. The model density and temperature profiles used in the simulations are $n_e(r) = n_{e0}[1 - 0.5(r/a)^2]$, $T(r) = T_0\,[1 - 0.5(r/a)^2]^{1.5}$. The ICRH power coupled to the plasma is assumed to be $P_{\mathrm{ICRH}} = 3.0\,\mathrm{MW}$. This choice of B_0 and f sets the fundamental hydrogen resonance layer to be located 33 cm to the LFS from the magnetic axis ($R_{\mathrm{H}} = 3.33\,\mathrm{m}$). To study the effect of plasma shaping on α_T for the reference heating scenario the ellipticity has been varied from $\kappa = 1$ to 1.6, and the triangularity from $\delta = 0$ to 0.4. The effect of the ICRH resonance location on α_T has been assessed by studying two additional cases $B_0 = 2.9$ T and $B_0 = 3.1$ T that correspond to $R_{\mathrm{H}} = 3.22\,\mathrm{m}$ and $R_{\mathrm{H}} = 3.44\,\mathrm{m}$, respectively.

For the reference case considered the flux surface averaged profile of the RF power absorbed by minority ions is presented in Fig. 1(a) with the red solid line. The heating maximum is reached at $r/a = 0.29$ and the radial width of the absorption profile (full width at half maximum, FWHM) is $\sim 11.5\,\mathrm{cm}$. Hydrogen minority ions absorb $p_{\mathrm{H}} = 86.6\%$ of the ICRH power, deuterium majority ions – $p_{\mathrm{D}} = 8.5\%$, and the rest of the incident ICRH power is deposited on electrons, $p_{\mathrm{e}} = 4.9\%$. Figure 1 shows the minority power deposition profiles and temperature anisotropy for different values of the ellipticity of the last closed magnetic surface (LCFS), κ_1. We assume the fixed ellipticity, $\kappa_0 = 1.0$ at the magnetic axis. The fraction of the ICRH power, which is absorbed by hydrogen minority ions, is practically unchanged when varying κ_1: $p_{\mathrm{H}} = 87.1\%$ for $\kappa_1 = 1.3$ and $p_{\mathrm{H}} = 87.6\%$ for $\kappa_1 = 1.6$. Figure 2 depicts the power deposition by minority ions in the poloidal cross-section. Clearly, the wave focussing is reduced for an elliptical plasma in comparison with the circular case. For elongated plasmas the power absorption is deposited in a larger volume that results in smaller power density values (Fig. 1(a)). Accordingly, the minority temperature anisotropy decreases for the elongated plasma as Fig. 1(b) illustrates. In the circular case the maximum temperature anisotropy is $\alpha_{\max} = 8.0$, whereas in the elliptic case $\alpha_{\max} = 6.8$ and

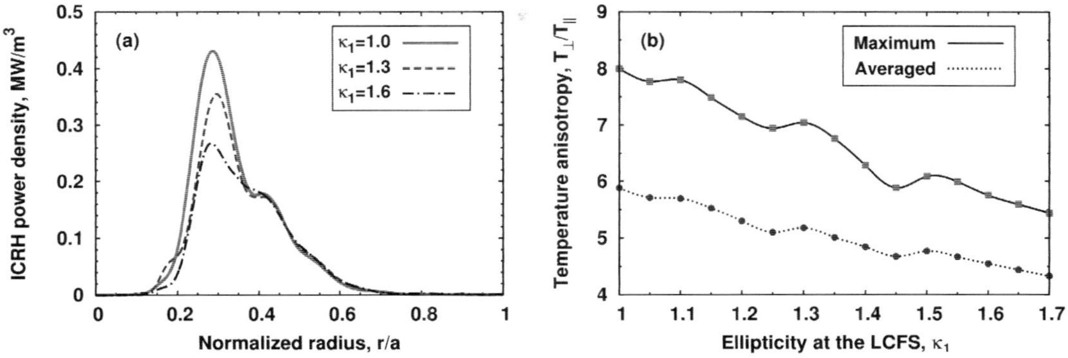

Figure 1. Flux surface averaged absorption profiles (a) and minority temperature anisotropy (b) evaluated by the TORIC-SSFPQL for various values of the ellipticity at the LCFS. The input ICRH power is $P_{\rm ICRH} = 3.0\,$MW.

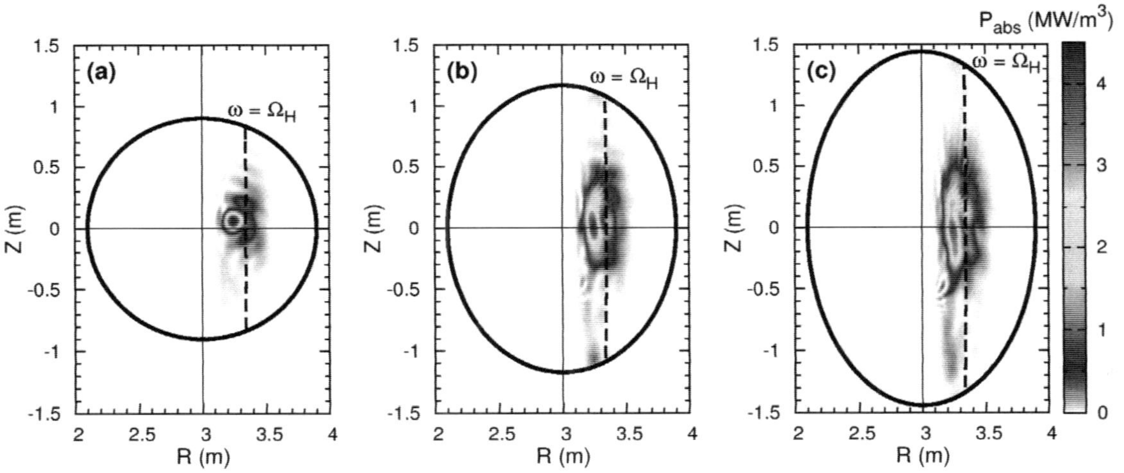

Figure 2. Contour plots of the RF power deposition to hydrogen minority ions in the poloidal cross-section from the TORIC simulations for different values of the edge ellipticity: (a) κ_1=1.0, (b) κ_1=1.3, (c) κ_1=1.6.

$\alpha_{\rm max} = 5.8$ for $\kappa_1 = 1.3$ and $\kappa_1 = 1.6$, respectively. The same conclusion applies to the averaged temperature anisotropy $\bar{\alpha}$, which is evaluated by averaging the temperature anisotropy over the radial range between FWHM points: $\bar{\alpha} = 5.9$, $\bar{\alpha} = 5.1$, $\bar{\alpha} = 4.6$ for $\kappa_1 = 1$, $\kappa_1 = 1.3$ and $\kappa_1 = 1.6$, respectively. As a result, more ICRH power needs to be coupled to an elongated plasma to achieve the same minority temperature anisotropy than that in a circular case.

For a fixed plasma elongation, the triangularity of the magnetic surfaces has much weaker effect on the power deposition profiles and, consequently, on the minority temperature anisotropy as shown in Fig. 3, which was calculated for $\kappa_1 = 1.6$. Figure 3(b) demonstrates that α_T slightly increases with the triangularity: for $\delta = 0.3$ $\alpha_{\rm max}$ and $\bar{\alpha}$ reach the values 6.9 and 5.1, respectively. For the heating scenario considered the Shafranov shift is also found to be of marginal importance on the values of the minority temperature anisotropy. Note that the

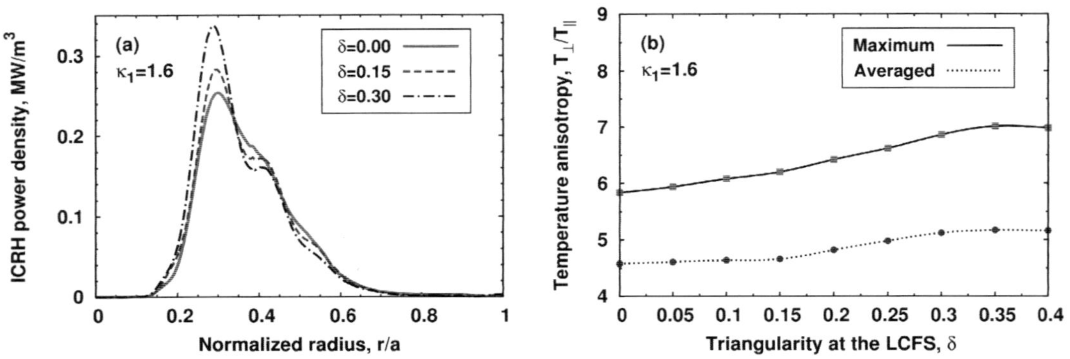

Figure 3. Flux surface averaged absorption profiles (a) and minority temperature anisotropy (b) evaluated by the TORIC-SSFPQL for various values of the plasma triangularity. The LCFS ellipticity is $\kappa_1 = 1.6$.

triangularity of the LCFS has much greater effect on the coupling efficiency [23, 24]. When the LCFS has the same curvature as the ICRH antenna facing the plasma, one commonly gets better coupling of the RF energy from the ICRH antenna to the plasma that results in the improved heating performance.

Figure 4 shows the profiles of the RF power absorbed in the poloidal cross-section (circular case) by hydrogen minority ions for $B_0 = 2.9\,\mathrm{T}$ and $B_0 = 3.1\,\mathrm{T}$. Figures 4(a,c) were calculated with TORIC assuming the Maxwellian background for all plasma species, while Figs. 4(b,d) were calculated with SELFO accounting self-consistently for the change in the minority distribution function. The corresponding radial absorption profiles are depicted in Fig. 5. One should note the difference in the absorption profiles predicted by TORIC and SELFO. While the TORIC profiles are typically peaked, SELFO predicts more radially smoothed absorption. This difference is caused by the simplifying assumption of the Maxwellian distribution for the heated hydrogen minorities underlying the TORIC modelling. Though it is possible to account for the change in the minority distribution function and its effect on the wave propagation and absorption in the TORIC-SSFPQL package [25, 26], the simulations made in the default TORIC mode allows one to get quick qualitative estimates for the temperature anisotropy and its parametric dependences since the averaged power densities and, accordingly, averaged perpendicular energies reached by the heated minorities are comparable for the two different approaches. The fraction of the RF power absorbed by minority ions is also in a fair agreement: for $B_0 = 3.1\,\mathrm{T}$ TORIC predicts $p_{\mathrm{H}} = 89.3\%$, while SELFO yields $p_{\mathrm{H}} = 91.6\%$ for the Maxwellian hydrogen distribution and $p_{\mathrm{H}} = 94.5\%$ in a steady-state.

For the fixed RF generator frequency, by increasing/decreasing the central magnetic field one causes the cyclotron resonance layer and, consequently, the power absorption region to be shifted towards the outer/inner side of the torus (see, Fig. 6(a)). According to TORIC results, for $B_0 = 3.1\,\mathrm{T}$ the heating maximum is observed closer to the edge ($r/a = 0.42$), while for $B_0 = 2.9\,\mathrm{T}$ it is located closer to the magnetic axis ($r/a = 0.15$). The flux surface averaged ICRH power density, which determines the perpendicular minority tail temperature, decreases with increasing B_0 due to the geometrical reasons. Since the RF power is deposited in a smaller volume as r decreases, the absorbed power density increases as the wave absorption takes place closer to the axis. For example, the maximum power density for $B_0 = 2.9\,\mathrm{T}$ is 1.6 times higher than that for $B_0 = 3.0\,\mathrm{T}$. Although the temperature anisotropy scales linearly with the ICRH power density, $\alpha_{\max}$ and $\bar{\alpha}$ are only 1.3 times larger for $B_0 = 2.9\,\mathrm{T}$ than that for the reference

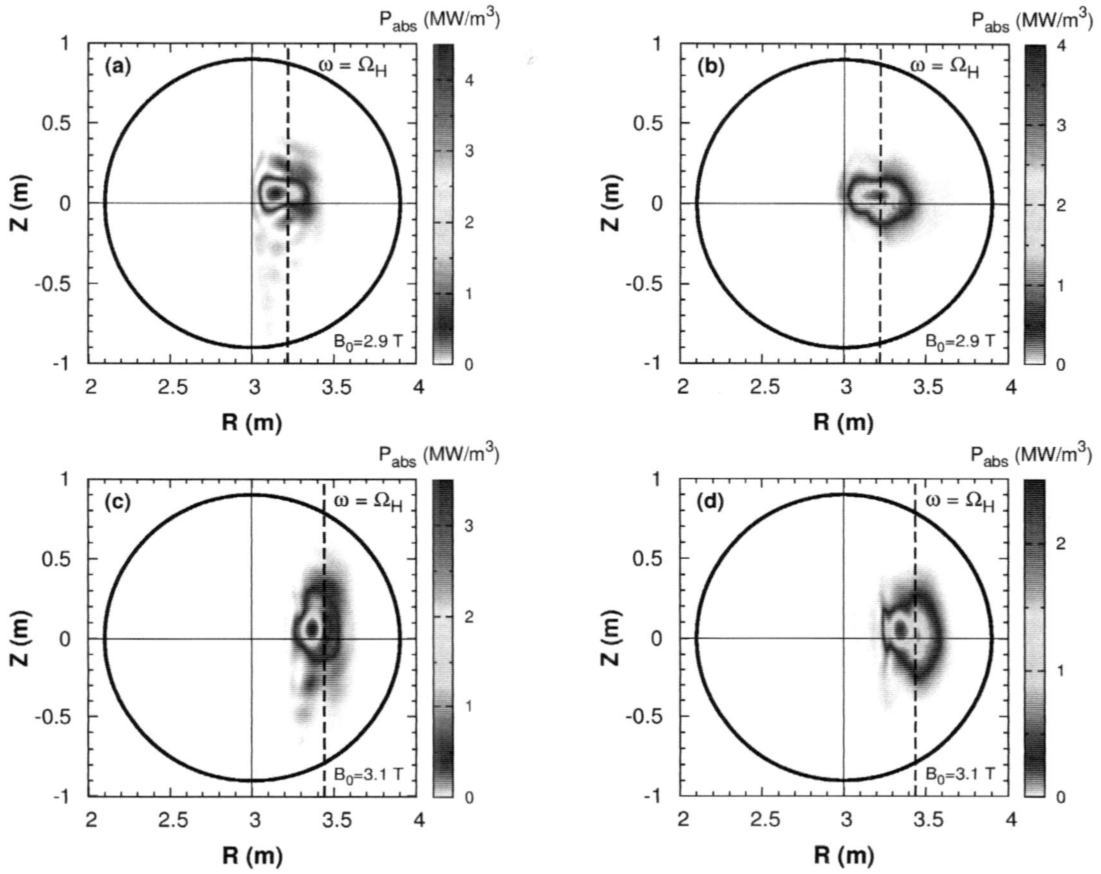

Figure 4. Contour plots of the RF power deposition to hydrogen minority ions in the poloidal cross-section calculated with the TORIC (a,c) and the SELFO (b,d) codes for $B_0 = 2.9\,\mathrm{T}$ and $B_0 = 3.1\,\mathrm{T}$.

case. This is due to the fact that the local plasma density at the absorption region changes when one shifts the ICRH resonance location. Since the temperature anisotropy scales roughly as $\alpha_T \propto 1/n_e^2$, it counteracts the anisotropy increase due to the increase of the local ICRH power density. The decrease of the minority absorption coefficient, p_{H} in case of the power deposition closer to the center (p_{H}=83.5% for $B_0 = 2.9\,\mathrm{T}$ and p_{H}=89.3% for $B_0 = 3.1\,\mathrm{T}$) is caused by the higher plasma temperature within the deposition region that favours the deuterium second harmonic and electron absorption mechanisms. Note, that the higher temperature anisotropy which is predicted for the case of the deposition closer to the axis does not necessarily lead to the higher impurity asymmetry strength since the latter parameter also includes dependence on the inverse aspect ratio and resonance magnetic field (see, Eq. (5) in Ref. [9]).

The non-monotonic dependence of the maximum temperature anisotropy seen in Fig. 1(b) and Fig. 6(b) is likely to be caused by the fact that the wave focussing and RF field pattern are intrinsically sensitive to the actual magnetic field geometry and multiple wave reflections in a plasma. By changing the central magnetic field, one changes the relative location of the absorption region and the high field side fast wave cutoff and thus affects the interference conditions that results in a modulation of the curves. However, it is clearly seen that for the averaged anisotropy such a modulation is substantially moderated. Thus, accounting for the different waves in the excited antenna spectrum, this effect is unlikely to be observed experimentally.

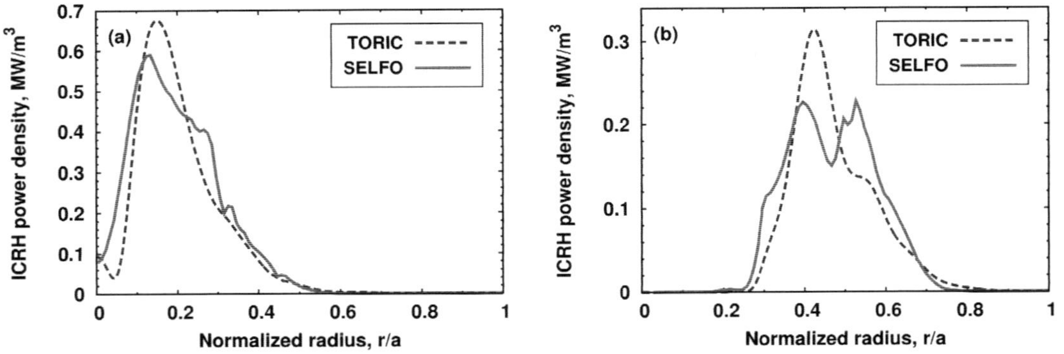

Figure 5. Flux surface averaged deposition profiles to hydrogen minority ions calculated with the TORIC (dashed) and the SELFO codes (solid) for $B_0 = 2.9\,\mathrm{T}$ (a) and $B_0 = 3.1\,\mathrm{T}$ (b).

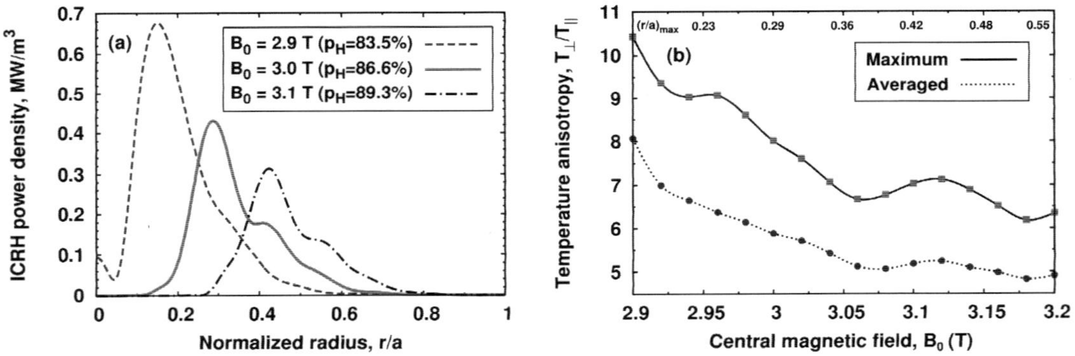

Figure 6. Flux surface averaged absorption profiles (a) and minority temperature anisotropy (b) evaluated by the TORIC-SSFPQL for different central magnetic field B_0.

4. Conclusions

In plasmas heated with the LFS ICRH, the inboard impurity accumulation is often observed. Such an asymmetry in impurity density has been recently considered as a possible mechanism for controlling the impurity transport. The asymmetry arises due to the accumulation of the cyclotron heated minority ions at the outboard side of the magnetic surface leading to the generation of the poloidal electrostatic electric field that pushes impurities to the opposite side. The minority temperature anisotropy has shown to be one of the main factors, which determines the impurity asymmetry strength. The temperature anisotropy depends on most of plasma and heating parameters, and in general case it should be evaluated numerically using the full-wave ICRH and Fokker-Planck codes. The difference in the power deposition profiles predicted by the TORIC and the SELFO codes is found to be appreciable, but it is not significant for the averaged anisotropy.

By means of the TORIC-SSFPQL modelling we analyzed the effect of plasma shaping and ICRH resonance location on the minority temperature anisotropy in tokamak plasmas. Accounting for the finite ellipticity of the flux surfaces a gradual decrease in the temperature anisotropy is observed which is caused by the wave defocussing and radially broader absorption regions for the elongated plasmas. The temperature anisotropy decrease in case of the resonance layers located closer to the LFS edge is shown. The lower values of the perpendicular minority temperatures are caused by the significant reduction in local surface averaged ICRH power densities due to geometrical reasons.

Acknowledgments
The authors are grateful to Dr. M. Brambilla and Dr. R. Bilato (IPP-Garching) for providing the TORIC-SSFPQL code. Computations for this article have been performed on the HPCFF cluster at the Jülich Supercomputing Center (JSC), Germany. This work was funded by the European Communities under Association Contract between EURATOM and *Vetenskapsrådet*. The views and opinions expressed herein do not necessarily reflect those of the European Commission.

References

[1] Ingesson L C, Chen H, Helander P and Mantsinen M J 2000 *Plasma Phys. Control. Fusion* **42** 161.
[2] Marr K D, Lipschultz B, Catto P J, McDermott R M, Reinke M L and Simakov A N 2010 *Plasma Phys. Control. Fusion* **52** 055010.
[3] Reinke M L *et al.* 2012 *Plasma Phys. Control. Fusion* **54** 045004.
[4] Choe W, Chang C S and Ono M 1995 *Phys. Plasmas* **2** 2044.
[5] Alper B, Edwards A W, Giannella R, Gill R D, Ingesson C, Romanelli M, Wesson J and Zastrow K-D 1996 *Proc. 23rd EPS Conf. on Plasma Phys. and Contr. Fusion* (Kiev, Ukraine) ECA **20C** p. 163–166.
[6] Fülöp T and Moradi S 2011 *Phys. Plasmas* **18** 030703.
[7] Moradi S, Fülöp T, Mollén A and Pusztai I 2011 *Plasma Phys. Control. Fusion* **53** 115008.
[8] Mollén A, Pusztai I, Fülöp T, Kazakov Ye O and Moradi S 2012 *Phys. Plasmas* **19** 052307.
[9] Kazakov Ye O, Pusztai I, Fülöp T and Johnson T 2012 *Plasma Phys. Control. Fusion* **54** 105010.
[10] Majeski R, Phillips C K and Wilson J R 1994 *Phys. Rev. Lett.* **73** 2204.
[11] Fuchs V, Ram A K, Schultz S D, Bers A and Lashmore-Davies C N 1995 *Phys. Plasmas* **2** 1637.
[12] Van Eester D *et al.* and JET EFDA contributors 2009 *Plasma Phys. Control. Fusion* **51** 044007.
[13] Kazakov Ye O, Pavlenko I V, Van Eester D, Weyssow B and Girka I O 2010 *Plasma Phys. Control. Fusion* **52** 115006.
[14] Van Eester D *et al.* and JET EFDA Contributors 2012 *Plasma Phys. Control. Fusion* **54** 074009.
[15] Kazakov Ye O, Kiptily V G, Sharapov S E, Van Eester D and JET EFDA Contributors 2012 *Nucl. Fusion* **52** 094012.
[16] Van Eester D, Bhatnagar V P and Koch R 1985 *Plasma Phys. Control. Fusion* **27** 1015.
[17] Brambilla M 1999 *Plasma Phys. Control. Fusion* **41** 1.
[18] Wright J C, Bonoli P T, Brambilla M, Meo F, D'Azevedo E, Batchelor D B, Jaeger E F, Berry L A, Phillips C K and Pletzer A 2004 *Phys. Plasmas* **11** 2473.
[19] Brambilla M 1994 *Nucl. Fusion* **34** 1121.
[20] Hedin J, Hellsten T, Eriksson L-G and Johnson T 2002 *Nucl. Fusion* **42** 527.
[21] Mayoral M-L *et al.* and JET EFDA contributors 2006 *Nucl. Fusion* **46** S550.
[22] Noterdaeme J-M, Eriksson L-G, Mantsinen M, Mayoral M-L, Van Eester D, Mailloux J, Gormezano C and Jones T T C 2008 *Fusion Sci. Technol.* **53** 1103.
[23] Bilato R, Mayoral M-L, Rimini F, Brambilla M, Hartmann D A, Korotkov A, Lamalle P U, Monakhov I, Noterdaeme J-M, Sartori amd JET EFDA contributors 2004 *Proc. 31th EPS Conf. on Plasma Phys.* (London, UK) ECA **28G** P-5.164.
[24] Bobkov Vl V, Bilato R, Braun F, Dux R, Noterdaeme J-M, the ICRF team and the ASDEX Upgrade team 2006 *Nucl. Fusion* **46** S469.
[25] Brambilla M and Bilato R 2009 *Nucl. Fusion* **49** 085004.
[26] Bilato R, Brambilla M, Maj O, Horton L D, Maggi C F and Stober J 2011 *Nucl. Fusion* **51** 103034.

Journal of Physics: Conference Series **401** (2012) 012012 doi:10.1088/1742-6596/401/1/012012

Interaction of electromagnetic waves and suprathermal electrons in the near-critical electric field limit

A Kómár[1], G I Pokol[1] and T Fülöp[2]

[1] Department of Nuclear Techniques, Budapest University of Technology and Economics, Association EURATOM, H-1111 Budapest, Hungary

[2] Department of Applied Physics, Nuclear Engineering, Chalmers University of Technology and Euratom-VR Association, Göteborg, Sweden

E-mail: `komar@reak.bme.hu`

Abstract. The velocity-space anisotropy of suprathermal electron distributions is a source of free energy that may destabilize plasma waves through a resonant interaction between the waves and the energetic electrons. In this work we use a suprathermal electron distribution appropriate for the case when the accelerating electric field is near-critical and we investigate the frequencies, wave numbers and propagation angles of the most unstable waves using a general dispersion relation. It is shown that if the electric field is sub-critical, the anisotropy is not enough to drive electromagnetic waves unstable, as the Landau damping of the waves overwhelms the drive through the anomalous Doppler resonance. In the case when the electric field is super-critical, two types of electromagnetic waves will be destabilized, the electron-whistler and the extraordinary electron wave. The number of electrons for destabilization of the latter is several orders of magnitude lower than for the electron-whistler wave. Consequently, the threshold for destabilization of the extraordinary electron wave is much lower.

1. Introduction

Magnetically confined plasmas often have suprathermal electron populations accelerated by electric fields induced in magnetic reconnection. If the electric field is larger than a certain critical field (E_c), the accelerating force overwhelms the friction for high energy electrons, and these electrons can run away. In tokamak disruptions, the electric field is often well above the critical field and these fields can accelerate electrons to relativistic energies and may generate a runaway electron current exceeding 1 MA. During normal operation the electric field is not as high as in a disruption, but it may nevertheless transiently exceed the critical field and accelerate electrons. There are several experimental observations that suprathermal electrons are produced during the sudden plasma cooling at magnetic reconnection events in sawtooth crashes [1, 2]. Estimates show that in these cases the electric field induced by the sawtooth crash is much larger than the critical one, and generates a suprathermal electron distribution. The steady-state on-axis electric field is close to the critical in these cases and this prevents a deceleration of the electrons between the sawtooth crashes.

The distribution of the suprathermal electrons has an anisotropy in velocity space and through this, they may drive electromagnetic waves unstable via a resonant interaction, not only when the electric field is far above the critical [3, 4, 5] but also in the near-critical case [6, 7]. In this work we generalize and extend the formalism describing the wave-particle interaction in the near-critical case by using a general dispersion relation and investigating also a case when the electric field is close to critical but is not quite as large as that (subcritical case).

2. Suprathermal distribution function in the subcritical case

In recent work [8], analytical expressions have been derived for the electron distribution function in plasmas where the electric field is close to critical, $\alpha \sim 1$, where $\alpha = E/E_c = 4\pi\epsilon_0^2 m_e c^2 E/n_e e^3 \ln\Lambda$. Here, n_e is the thermal electron density, m_e is the electron rest mass, e is the electron charge, $\ln\Lambda$ is the Coulomb logarithm, ϵ_0 is the dielectric constant and c is the speed of light. The distribution function of the suprathermal electrons in the slightly subcritical case ($\alpha \lesssim 1$) is obtained by matching solutions of the Fokker-Planck equation in different regions of momentum space and can be written as

$$f_{\rm st} = \exp F = \exp\left(\epsilon^{-1}F^{(0)} + \epsilon^{-1/2}F^{(1)} + F^{(2)} + ... + C_{\rm st}\right), \tag{1}$$

where $\epsilon = E_c/E_D \ll 1$, E_D is the Dreicer field, $F^{(0)}$, $F^{(1)}$, $F^{(2)}$ are given by Eqs. (23), (37)-(38) and the normalization constant $C_{\rm st}$ is defined by Eqs. (39) and (9) of Ref. [8]. Figure 1a shows a contour-plot of this distribution for $\alpha = 0.9$. The distribution is only slightly anisotropic, which is to be expected because of the subcritical value of the electric field.

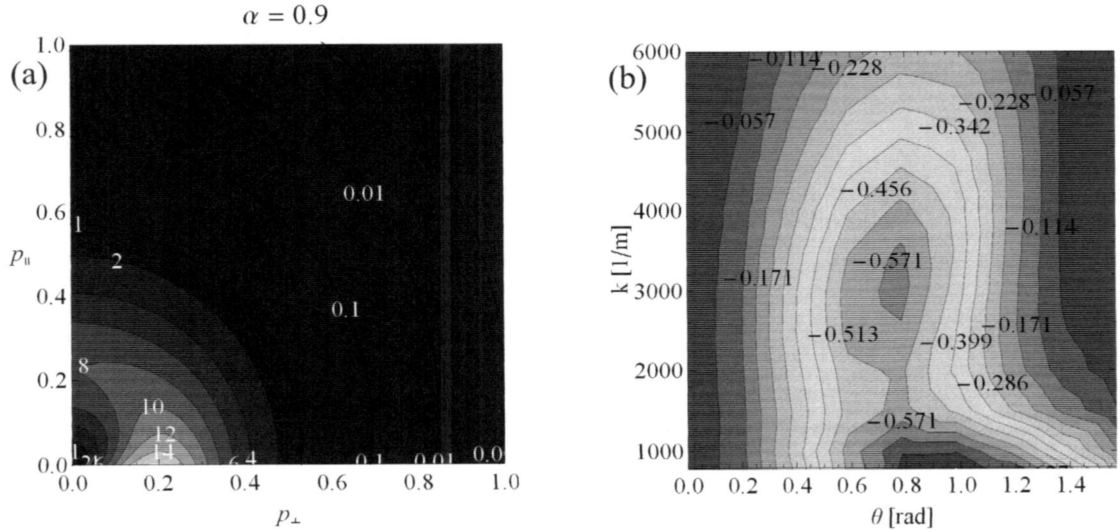

Figure 1. (a) Suprathermal electron distribution for $\alpha = 0.9$, plotted with respect to the parallel and perpendicular momentum normalized to $m_e c$. (b) Growth rate ($10^3\gamma_i/\omega_{ce}$) of the electron-whistler wave for suprathermal electrons, sum of the anomalous-Doppler and Cherenkov resonances. The parameters are $\alpha = 0.9$, $Z = 1$ effective ion charge, $n_e = 5 \cdot 10^{19}$ m^{-3} thermal electron density and $B = 2$ T magnetic field.

It should be noted that the $f_{\rm st}$ suprathermal distribution function is only valid for normalized relativistic momenta $p = \gamma v/c \sim \mathcal{O}(1)$.

Using this electron distribution we can calculate the linear growth rates for the high-frequency electromagnetic waves through a perturbative analysis similar to the one presented in [7]. Using the electromagnetic approximation, $\epsilon_{33}^{e+i} \gg k_\parallel k_\perp c^2/\omega^2$, leads to the dispersion relation $(\epsilon_{11} - k_\parallel^2 c^2/\omega^2)(\epsilon_{22} - k^2c^2/\omega^2) + \epsilon_{12}^2 = 0$, where ω is the wave frequency, k is the wave number, where the $\parallel$ index denotes its component parallel to the static magnetic field and $\boldsymbol{\epsilon}$ is the dielectric tensor of the plasma, consisting of the susceptibilities of the different plasma species: $\boldsymbol{\epsilon} = \mathbf{1} + \boldsymbol{\chi}^i + \boldsymbol{\chi}^e + \boldsymbol{\chi}^r$, the indices i, e and r denoting the ion, thermal electron and runaway population. Throughout this work we used the cold plasma approximation for the ion and thermal electron susceptibilities [9] which was seen to be accurate for temperatures up to 20 keV [7]. The dispersion relation can be further simplified using the approximation $\omega \gg \omega_{ce}\sqrt{m_e/m_i}$, for which the unperturbed dielectric tensor is $\epsilon_{11}^{e+i} = 1-\omega_{pe}^2/(\omega^2-\omega_{ce}^2)$, $\epsilon_{22}^{e+i} = 1-\omega_{pe}^2/(\omega^2-\omega_{ce}^2)$, $\epsilon_{12}^{e+i} = -i\omega_{pe}^2\omega_{ce}/[\omega(\omega^2-\omega_{ce}^2)]$, where ω_{pe} is the electron plasma- and ω_{ce} is the electron cyclotron frequency. This dispersion defines three different electromagnetic waves, of which it was shown that if $\alpha \gtrsim 1$ the lowest frequency branch, the so-called 'electron-whistler wave', was most unstable. The perturbative effect of the runaway electrons is taken into account by including the runaway susceptibility [7] in the dielectric tensor. This results in the changing of the wave frequency by a $\delta\omega$ term, the imaginary part of which is the linear growth rate of the wave.

Figure 1b shows that the linear growth rates excited by suprathermal electrons with the distribution (1) (valid for $\alpha \lesssim 1$) are negative on the whole wave number and propagation angle plane, where $\theta = k_\parallel/k$. This was somewhat unexpected, as it was shown in [6, 7] that for runaway electrons in a slightly higher electric field (near-critical case with $\alpha \gtrsim 1$) the corresponding growth rates are positive. The only part of the calculation different from the $\alpha \gtrsim 1$ case was the distribution function, which appears only in the anisotropy term of the runaway susceptibilities: $\mathcal{A} \equiv n\omega_{ce}\partial f/\partial p_\perp + kc\cos\theta\, p_\perp \partial f/\partial p_\parallel$. In the case of $\alpha \gtrsim 1$, this term was positive for both $n = 0$ (Cherenkov resonance) and $n = -1$ (anomalous-Doppler resonance), while in the $\alpha \lesssim 1$ case, the term is negative for $n = 0$ and positive for $n = -1$. The anomalous-Doppler term is negligible compared to the negative growth rate due to the Cherenkov resonance and therefore the waves in the subcritical case are stable.

3. Runaway distribution function

As we have established that the suprathermal populations that are formed in a sub-critical case are stable, we now turn our attention to the runaway distribution, the expression for which is also given in Ref. [8] as

$$f_r(p_\parallel, p_\perp) = \frac{A}{p_\parallel^{(C_s-2)/(\alpha-1)}} \exp\left(-\frac{(\alpha+1)p_\perp^2}{2(1+Z)p_\parallel}\right) {}_1F_1\left(1-\frac{C_s}{\alpha+1}, 1; \frac{(\alpha+1)p_\perp^2}{2(1+Z)p_\parallel}\right), \quad (2)$$

where $C_s = \alpha - \frac{(1+Z)}{4}(\alpha-2)\sqrt{\frac{\alpha}{\alpha-1}}$, Z is the effective ion charge, ${}_1F_1$ is the confluent hypergeometric function [10] and A is a normalization constant. Because of the continuous acceleration of electrons due to the constant electric field more and more electrons will run away, thus the integral of the distribution is divergent. However, in reality the electric field is constant only for a finite time, then drops. Therefore, the number of runaway electrons and their maximum energy will have an upper limit they can possibly reach. The value of this limit, $p_{\max}$ depends on the exact value and time evolution of the accelerating field. In this paper we will use $p_{\max} = 5$ corresponding to the maximum energy 2.6 MeV.

The distribution is positive on all of the momentum space only if the first argument of ${}_1F_1$ is positive: $C_s < \alpha + 1$. Furthermore, the condition $f_r \to 0$ for $p_\parallel \to \infty$ requires that $C_s > 2$. This distribution was used in [6, 7] to study the destabilization of the electron-whistler and magnetosonic-whistler waves. However, if the electromagnetic approximation breaks down, other waves may be driven unstable. In the following we will use the full dispersion relation to study the interaction of runaway electrons with the electromagnetic waves.

4. Wave dispersion

The full dispersion relation is the following:

$$\left[\left(\epsilon_{11} - \frac{k_\parallel^2 c^2}{\omega^2}\right)\left(\epsilon_{22} - \frac{k^2c^2}{\omega^2}\right) + \epsilon_{12}^2\right]\left(\epsilon_{33} - \frac{k_\perp^2 c^2}{\omega^2}\right) - \frac{k_\parallel^2 k_\perp^2 c^4}{\omega^4}\left(\epsilon_{22} - \frac{k^2c^2}{\omega^2}\right) = 0. \quad (3)$$

In order to proceed with the calculation of the growth rate we rewrite this dispersion relation, by substituting the formulas for the dielectric tensor elements, and arrive at: $\omega^8 - \omega^6\left(2k^2c^2 + \omega_{ce}^2 + 3\omega_{pe}^2\right) + \omega^4\left[k^4c^4 + 2k^2c^2(\omega_{ce}^2 + 2\omega_{pe}^2) + \omega_{pe}^2(\omega_{ce}^2 + 3\omega_{pe}^2)\right] - \omega^2\left[k^4c^4(\omega_{ce}^2 + \omega_{pe}^2) + k^2c^2\omega_{pe}^2(3/2\omega_{ce}^2 + 2\omega_{pe}^2 + 1/2\omega_{ce}^2\cos 2\theta) + \omega_{pe}^6\right] = -1/2k^4c^4\omega_{ce}^2\omega_{pe}^2(1 + \cos 2\theta)$. This dispersion gives four different branches of electromagnetic waves, three of which have already been studied in Refs. [6, 7] and one new branch. These are plotted on Fig. 2, where the lowest frequency branch is the previously named 'electron-whistler wave', and the second lowest is the new branch, which we will call 'extraordinary electron wave', as this is described in Ref. [9] as a high-frequency extraordinary wave.

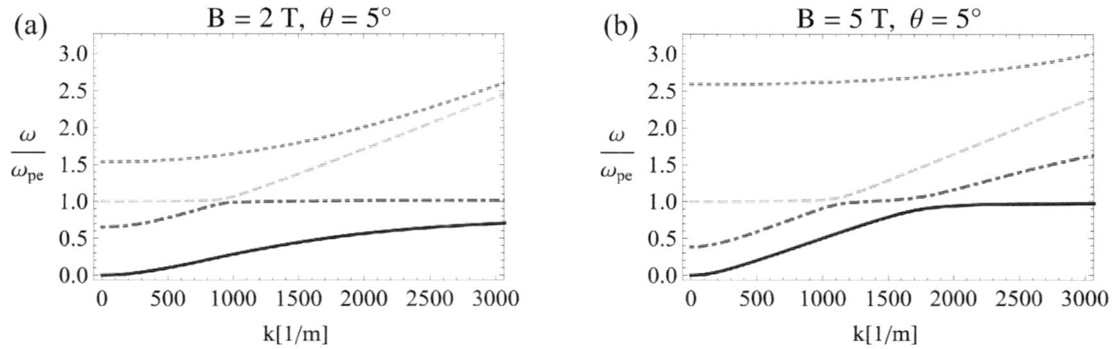

Figure 2. Wave dispersion of the electromagnetic waves at $\theta = 5$ degrees propagation angle for (a) $B = 2$ T and (b) $B = 5$ T. The thermal electron density is $n_e = 5 \cdot 10^{19}$ m^{-3}. The solid blue line corresponds to the electron-whistler wave and the red dashed line is the extraordinary electron wave. The branches with higher frequencies are not destabilized by the suprathermal electrons.

In the presence of runaway electrons, the linear growth rate of the waves can be calculated by perturbing the dispersion with the runaway susceptibility [7]. This results in an additional $\delta\omega$ term in the wave frequency, the imaginary part of which is the linear growth rate γ_i. The growth rate is then given by

$$\frac{\gamma_i}{\omega^6\left(\omega^2 - \omega_{ce}^2\right)} = \Im\left\{-\chi_{11}\left(\epsilon_{11} - k^2c^2/\omega^2\right)\left(\epsilon_{33} - k_\perp^2c^2/\omega^2\right) - 2\chi_{12}\epsilon_{12}\left(\epsilon_{33} - k_\perp^2c^2/\omega^2\right) - \chi_{22}\left[\left(\epsilon_{11} - k_\parallel^2c^2/\omega^2\right)\left(\epsilon_{33} - k_\perp^2c^2/\omega^2\right) - k_\parallel^2k_\perp^2c^4/\omega^4\right] - \chi_{33}\left[\epsilon_{11}^2 - \epsilon_{11}\left(k^2c^2/\omega^2 + k_\parallel^2c^2/\omega^2\right) + k^2k_\parallel^2c^4/\omega^4 + \epsilon_{12}^2\right]\right\}/P(\omega,k,\theta), \quad (4)$$

where $P(\omega,k,\theta) = 8\omega^7 - 6\omega^5\left(2k^2c^2 + \omega_{ce}^2 + 3\omega_{pe}^2\right) + 4\omega^3\left[k^4c^4 + 2k^2c^2(\omega_{ce}^2 + 2\omega_{pe}^2) + \omega_{pe}^2(\omega_{ce}^2 + 3\omega_{pe}^2)\right] - 2\omega\left[k^4c^4(\omega_{ce}^2 + \omega_{pe}^2) + k^2c^2\omega_{pe}^2(3/2\omega_{ce}^2 + 2\omega_{pe}^2 + 1/2\omega_{ce}^2\cos 2\theta) + \omega_{pe}^6\right]$.

The growth rates of the two highest frequency branches were already discussed in Ref. [7] and it has been concluded that these cannot be destabilized by the runaway population. Thus we turn our attention to the remaining two branches, the electron-whistler and the extraordinary electron wave.

The growth rate of the electron-whistler wave is plotted on Fig. 3 for two different magnetic fields. These growth rates do not differ much from the growth rate calculated with the simplified dispersion [6, 7], neither qualitatively nor quantitatively for $B = 2$ T, but as we will show later the parameters and growth rate of the most unstable wave differ for magnetic fields larger than

$B \gtrsim 2.5$ T. The parameters of the most unstable wave are determined by the maximum energy of the runaway electrons [7]. In Fig. 3 this quantity is chosen to be 2.6 MeV, corresponding to $p_{\max} = 5$, and the corresponding $k - \theta$ curve is plotted with a white dotted line on Fig. 3a, while it corresponds to such high wave numbers for $B = 5$ T that it cannot be seen on Fig. 3b. By decreasing the maximum energy, the wave number of the most unstable wave increases while its propagation angle decreases, resulting in the increasing frequency of the most unstable electron-whistler wave [7].

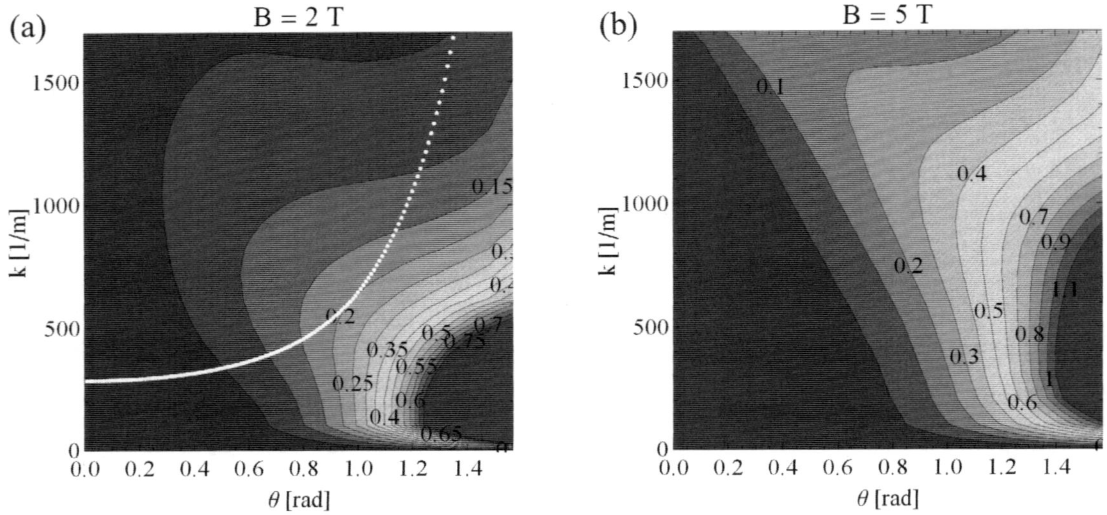

Figure 3. Growth rates ($10^2\gamma_i/\omega_{ce}$) for the electron-whistler wave with the full dispersion (contour lines) and the $k - \theta$ values corresponding to $p_{\max} = 5$ (white dotted line). Growth rate (a) for $B = 2$ T, (b) for $B = 5$ T magnetic field, sum of the anomalous-Doppler and Cherenkov resonances. The parameters are $\alpha = 1.3$, $Z = 1$, $n_e = 5 \cdot 10^{19}$ m^{-3}, $n_r = 3 \cdot 10^{17}$ m^{-3} runaway density and $p_{\max} = 5$.

By comparing the growth rate of the electron-whistler wave in the near-critical case to the growth rate of the magnetosonic-whistler wave in the high electric field case previously investigated [3, 4, 5] we can conclude that the growth rates are qualitatively similar in the two cases, having a maximum at low wave numbers and near-perpendicular propagation. For a high enough $p_{\max}$ this maximum of the growth rate would define the most unstable wave even in the near-critical case. Based on this and the qualitatively similar runaway distributions in the two cases, a smooth transition can be envisioned in between, for arbitrary electric fields. The quasilinear evolution of the whistler wave has already been investigated for high electric fields in Ref. [4] and it was found that the destabilization of the whistler wave led to the scattering of the runaway electrons in velocity space. Due to the similarities in the two cases, we expect that the effect of this interaction would be similar in the near-critical case, in the sense that a scattering of the runaway electrons would occur, while this effect would possibly be slower and less dominant.

The extraordinary electron wave can be destabilized only in a limited region of the $k - \theta$ parameter space and has a positive growth rate in this region, see Fig. 4. As in the case of the electron-whistler, the most unstable wave is determined by the maximum runaway energy, however, there is a difference in the two cases. While the electron-whistler wave can be destabilized on the whole $k - \theta$ parameter space, thus by runaways with arbitrary maximum energy, the extraordinary electron wave cannot be destabilized by runaways with energies higher

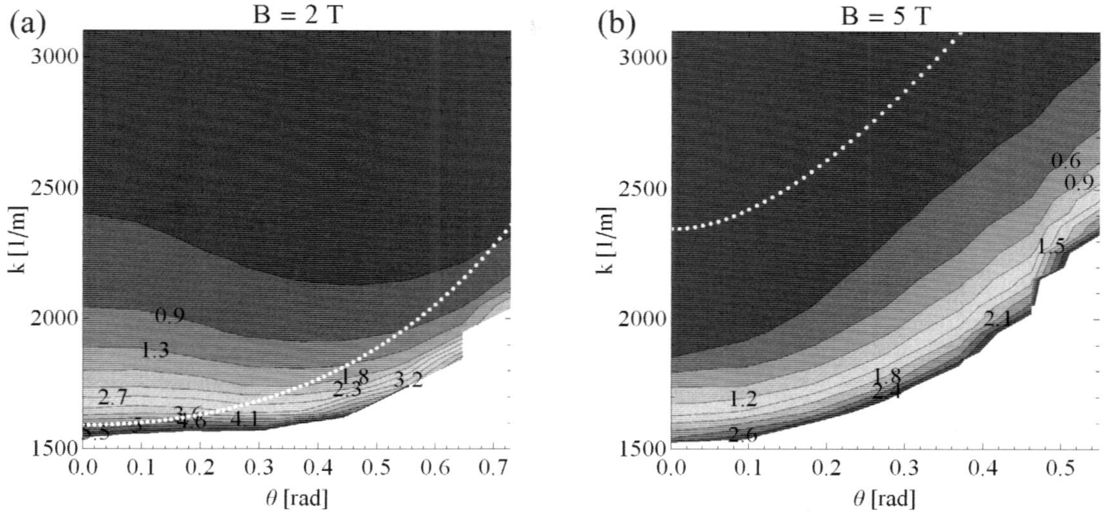

Figure 4. Growth rates ($10^2\gamma_i/\omega_{ce}$) for the extraordinary-electron wave with the full dispersion (contour lines) and the $k-\theta$ values corresponding to $p_{\max} = 5$ (white dotted lines). Growth rate (a) for $B = 2$ T, (b) for $B = 5$ T magnetic field, sum of the anomalous Doppler and Cherenkov resonances. The parameters are $\alpha = 1.3$, $Z = 1$, $n_e = 5 \cdot 10^{19}$ m^{-3}, $n_r = 3 \cdot 10^{17}$ m^{-3} and $p_{\max} = 5$.

than 2.6 MeV for the parameters in the figure caption (for $B = 2$ T). This is due to the restriction posed on the growth rate by the resonance condition: in order to have a real and positive resonant momentum certain limitations apply for the wave dispersion [7].

If we compare the growth rates of the electron-whistler and the extraordinary electron wave, we can conclude that for $p_{\max} = 5$ the most unstable extraordinary electron wave has a growth rate an order of magnitude higher than the electron-whistler wave.

5. Stability

By comparing the instability growth rate of the electron-whistler wave and the extraordinary electron wave to the various damping rates, we can determine a stability threshold for the destabilization of these waves. By taking into account collisional (γ_d) and convective damping (γ_v), a wave is unstable if $\gamma_l = \gamma_i - \gamma_d - \gamma_v > 0$. Collisional damping equals to $\gamma_d = 1.5\tau_{ei}^{-1}$ [11], where $\tau_{ei} = 3\pi^{3/2} m_{e0}^2 v_{Te}^3 \epsilon_0^2 / n_i Z^2 e^4 \ln\Lambda$ is the electron-ion collision time. This is the dominant damping mechanism in cold plasmas. The convective damping term takes into account the effect that wave energy is transported out of the runaway beam with the group velocity $\partial\omega/\partial k_\perp$, and equals to $\gamma_v \equiv (\partial\omega/\partial k_\perp)/(4L_r)$, where L_r is the radius of the runaway beam [5].

The growth rates of the most unstable waves driven by a near-critical electron distribution with $p_{\max} = 5$ were compared to the collisional and convective damping rates for $T_e = 20$ eV post-disruption electron temperature and for $T_e = 1$ keV temperature, and the obtained stability thresholds are shown in Fig. 5. If the runaway density is higher than the critical values plotted in Fig. 5, the corresponding wave is destabilized.

Fig. 5a shows the critical runaway densities for the destabilization of the electron-whistler wave. By comparing this to the stability thresholds calculated in Refs. [6, 7] with the simplified dispersion, it can be seen that the two cases differ both quantitatively and qualitatively for high magnetic fields. In Fig. 5a, the stability threshold decreases for increasing magnetic field for values higher than 3.5 T.

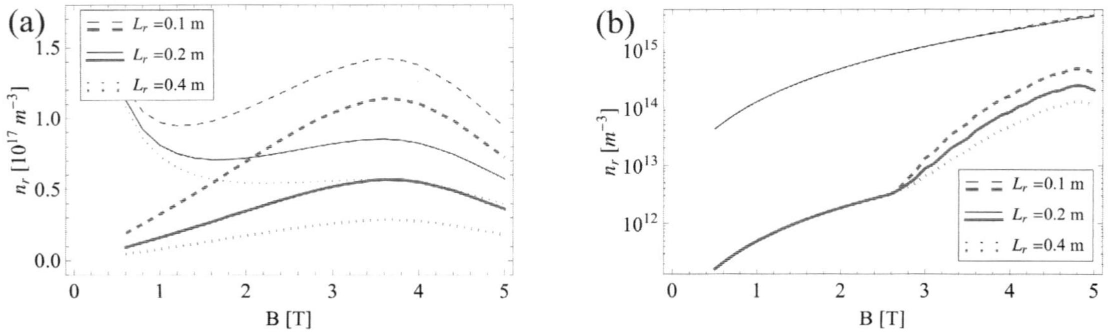

Figure 5. Stability thresholds in near-critical electric field. (a) Thresholds for the most unstable electron-whistler wave, (b) for the most unstable extraordinary electron wave as a function of the magnetic field. The electron temperatures are $T_e = 20$ eV (blue thin lines) and $T_e = 1$ keV (red thick lines). The runaway-beam radius is $L_r = 0.1$ m (dashed), $L_r = 0.2$ m (solid) and $L_r = 0.4$ m (dotted). The other parameters are $\alpha = 1.3$, $Z = 1$, $n_e = 10^{20}$ m^{-3} and maximum runaway energy of 2.6 MeV, corresponding to $p_{\max} = 5$.

On Fig. 6 the parameters and growth rate of the most unstable electron-whistler wave are shown for the (3) full dispersion and the simplified dispersion given in Section 2, as well as the convective damping of these waves. It can be seen that for higher magnetic fields these start to differ, and the most dominant effect behind the decrease in the stability threshold in Fig. 5a is that the growth rate of the wave (given by (3), the full dispersion relation) increases. This is due to two facts: (1) the parameters of the most unstable wave are different for high magnetic fields, while they are the same for low magnetic fields (see Fig. 6ab), and (2) the growth rate of the full dispersion increases faster with the magnetic field than the growth rate of the simplified dispersion.

On Fig. 5b stability thresholds are shown for the extraordinary electron wave. The critical density needed to destabilize this wave increases with the magnetic field. For 1 keV background temperature, the increase in the critical density suffers a break around 2.5 T, which is due to the corresponding break in the convective damping of the most unstable wave (see the dash-dotted line in Fig. 6d). This jump in the critical density becomes more dominant with increasing temperature, while it disappears at low temperatures, see the lines corresponding to 20 eV in Fig. 5b. By comparing the order of magnitude of the critical densities for the extraordinary electron wave to that of the electron-whistler, we can conclude that the extraordinary electron wave is more likely to be destabilized, as it has several order of magnitudes lower critical density.

6. Conclusions

Our results show that if the electric field is sub-critical, suprathermal electrons do not destabilize electromagnetic waves. In contrast, if the electric field is super-critical, both the electron-whistler wave and the extraordinary electron wave are destabilized. The extraordinary electron wave is a new branch, which is only present if the electromagnetic approximation is relaxed and the general wave dispersion is used. The wave number and propagation angle of the most unstable extraordinary electron wave is very different from its electron-whistler counterpart. It propagates almost parallel to the magnetic field if $B \lesssim 2.5T$. Above $B \simeq 2.5T$, the propagating angle becomes larger and this leads to a considerable increase in the convective damping. The extraordinary electron wave is much easier to destabilize than the electron-whistler branch. Interestingly, the stability threshold of the extraordinary electron wave exhibits a clear threshold at $B_{tr} \simeq 2.5$ T if the electron temperature is higher than 500 eV. Considerably fewer electrons

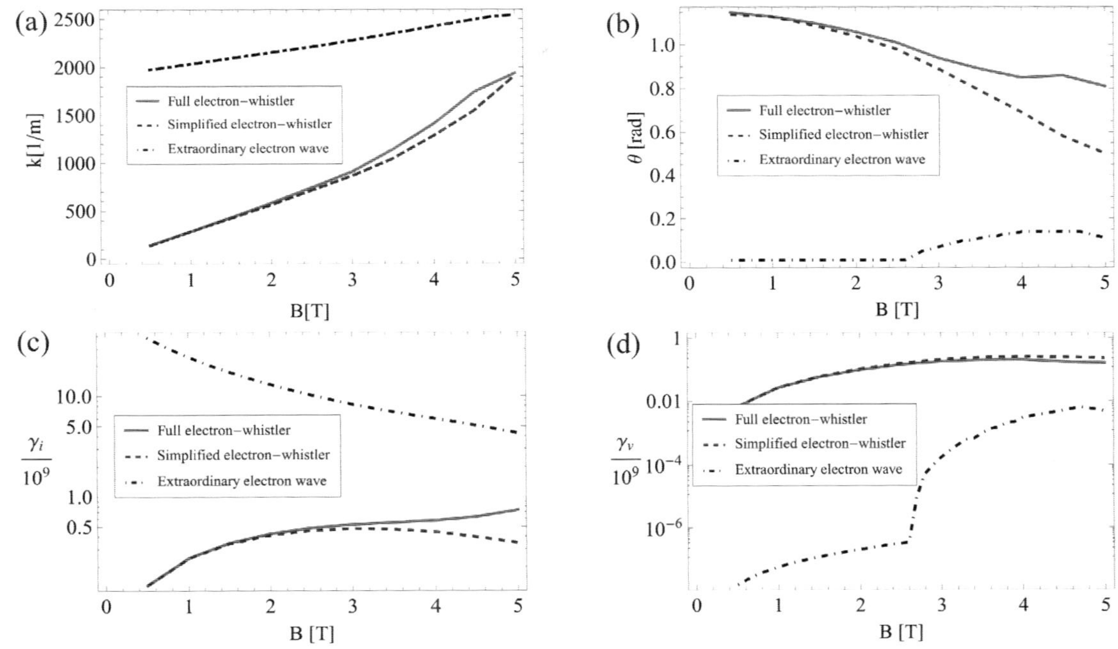

Figure 6. Parameters, growth rate and convective damping of the most unstable electron-whistler and extraordinary electron wave. (a,b) Parameters of the most unstable wave. (c,d) Growth rate and convective damping for the most unstable wave. The parameters are $\alpha = 1.3$, $Z = 1$, $n_e = 10^{20}$ m^{-3}, $n_r = 3 \cdot 10^{17}$ m^{-3}, $p_{\max} = 5$ and $L_r = 0.1$ m.

are needed for destabilization of this wave below B_{tr} than above. Destabilization of this wave is expected to lead to pitch-angle scattering of the runaway electron beam and therefore may be part of the reason for the observed magnetic field threshold for runaway beam formation in large tokamaks [12, 13].

Acknowledgments

This work was funded by the European Communities under Association Contract between EURATOM, HAS and *Vetenskapsrådet*. One of the authors acknowledges the financial support of the grant TÁMOP-4.2.2/B-10/1-2010-0009.

References

[1] Savrukhin P V, *Phys. Rev. Letters*, **86** (2001) 3036.
[2] Klimanov I, Fasoli A, Goodman T P, *Plasma Phys. Control. Fusion*, **49** (2007) L1.
[3] Fülöp T et al., *Phys. Plasmas* **13**, 062506 (2006).
[4] Pokol G et al., *Plasma Phys. Control. Fusion* **50**, 045003 (2008).
[5] Fülöp T et al., *Phys. Plasmas* **16**, 022502 (2009).
[6] Kómár A et al, in Proc. of the 39th European Physical Society meeting on Plasma Physics and 16th International Congress on Plasma Physics. P5.043 (2012);
[7] Kómár A et al, submitted to *Phys. Plasmas* (2012).
[8] Sandquist P et al., *Phys. Plasmas*, **13** 072108 (2006).
[9] Stix T H, Waves in plasmas, *American Institute of Physics*, New York (1992).
[10] Abramowitz M and Stegun I A, editors, *Handbook of Mathematical Functions*, chapter 13, page 503 (1972).
[11] Brambilla M, *Phys. Plasmas* **2**, 1094 (1995).
[12] Gill R D et al., *Nucl. Fusion* **42**, 1039 (2002).
[13] Yoshino R et al., *Nucl. Fusion* **39**, 151 (1999).

Coupling the beam tracing code TORBEAM and the Fokker-Planck solver RELAX for fast electrons

O Maj[1], E Poli[1] and E Westerhof[2]

[1] Max Planck Institute for Plasma Physics, EURATOM Association, 85748 Garching bei München, Germany.

[2] FOM Institute DIFFER - Dutch Institute for Fundamental Energy Research, Association EURATOM-FOM, Trilateral Euregio Cluster, Nieuwegein, The Netherlands.

E-mail: omaj@ipp.mpg.de

Abstract. In this paper the interface between the beam tracing code **TORBEAM** [Poli, Peeters and Pereverzev, Comp. Phys. Comm. 136, 90 (2001)] and the quasi-linear Fokker-Planck solver **RELAX** [Westerhof, Peeters and Schippers, Rijnhuizen Report No. RR 92-211 CA, 1992] is presented together with preliminary testing results for electron cyclotron waves in ITER plasmas and their effects on the electron distribution function. The resulting numerical package allows us to account for diffraction effects in the construction of the quasi-linear wave-particle diffusion operator. The coupling of the paraxial-WKB code **TORBEAM** to the ray-based code **RELAX** requires a re-interpretation of the paraxial wave field in terms of extended rays, which are addressed in details.

1. Introduction

Electron cyclotron resonance heating, which exploits the resonant interaction of electrons with a wave beam tuned in the electron cyclotron frequency range, is a major source of fast (suprathermal) electrons in tokamak plasmas. Such suprathermal electrons are essentially generated on the high-energy tail of the distribution function, as quasi-linear effects on bulk electrons are typically very weak [1]. The detailed knowledge of such high-energy tails is essential for the understanding of soft and hard X-ray diagnostics.

A precise study of the electron distribution function requires a description of both the power carried by the wave beam and the velocity-space diffusion of the distribution function due to collisions and wave scattering.

Typically, electron cyclotron wave beams are highly collimated/focused, in order to deposit the injected power "surgically" in the plasma, and, in such regimes, diffraction effects should be properly accounted for in the description of the wave propagation. This motivates the use of the beam tracing code TORBEAM [2], which, being based on the paraxial WKB approximation developed by Pereverzev [3, 4], can describe diffracting beams in a natural way.

The electron distribution function is computed by the relativistic bounce-averaged Fokker-Planck solver RELAX [5], which, however, cannot process the TORBEAM paraxial wave field directly, as it has been optimized for a ray-based description of the wave beam.

Our strategy has been obtaining a ray-based description of the paraxial WKB solution, which fully retains the effects of diffraction; this is made possible by making use of *extended rays*. A novel module has been implemented in TORBEAM for the calculation of extended rays

Journal of Physics: Conference Series **401** (2012) 012013 doi:10.1088/1742-6596/401/1/012013

defined as the field lines of the wave energy flux vector. This requires the integration of an additional system of ordinary differential equations with coefficients determined in terms of the main `TORBEAM` output. An appropriate scheme is applied in order to avoid interpolations of `TORBEAM` data. Extended rays are then passed to the interface with the Fokker-Planck solver `RELAX` for the calculation of the electron distribution function.

The interface has been written from scratch on the model of the existing interface of `RELAX` with the ray tracing code `TORAY-FOM` [6, 7]. For this part of the package, the language of choice is `Python`[1], which allows us to take advantage of object-oriented programming as well as of the scientific libraries `NumPy` and `SciPy`[2]. Time consuming procedures have been written in `Cython`[3], a `C` extension for `Python`, in order to avoid the performance shortcomings of interpreted programming languages such as `Python`.

The physics ideas at the basis of `TORBEAM` extended rays are discussed in section 2, while preliminary numerical results are presented in section 3 for the specific case of power deposition on the flux surface $q = 3/2$ in ITER scenario 2, which is a very well documented test bed [7, 8].

2. Extended ray tracing and diffraction effects.

The physics basis of coupling `TORBEAM` to `RELAX` without modifications of the latter relies on the representation of diffracting paraxial wave fields in terms of a bundle of trajectories that (i) reduce to geometrical optics rays when the beam is non diffracting and (ii) represent the wave energy flow to leading order in the high frequency limit. The second condition is particularly important for building the input data to the `RELAX` code, in which it is assumed that rays describe the energy flow. Trajectories satisfying requirement (i) are called extended rays and have been first obtained in the framework of the complex eikonal method [9-13].

2.1. Extended rays

In order to fix the notation, let us begin from the standard definition of geometrical optics rays [14] for the construction of asymptotic solutions of the Maxwell's wave equation, which, in normalized form, reads [4],

$$\nabla \times (\nabla \times E(\kappa, x)) - \kappa^2 \varepsilon(\kappa, x) E(\kappa, x) = 0, \quad \kappa = \omega L/c \to +\infty, \tag{1}$$

where L is the typical scale length of the medium inhomogeneity, $x = (x^1, x^2, x^3)$ is the spatial position normalized to L, $E(\kappa, x)$ is the wave electric field of a beam of frequency ω in a stationary spatially non-dispersive medium with dielectric tensor $\varepsilon(\kappa, x) = \varepsilon_0(x) + \kappa^{-1}\varepsilon_1(x)$ and with $\varepsilon_0(x)$ being Hermitian (weakly dissipative medium).

Geometrical optics rays are the spatial projection $x = x(t)$ of orbits $(x(t), N(t))$ of the Hamiltonian system

$$dx/dt = \nabla_N H(x, N), \quad dN/dt = -\nabla_x H(x, N), \tag{2}$$

with the constraint $H(x, N) = 0$, and $N = (N_1, N_2, N_3)$ has the physical meaning of the refractive index vector carried by the ray. The Hamiltonian function $H(x, N)$ is in the wave x-N phase-space is obtained as follows. First, one considers the dispersion tensor $D_0(x, N)$, given in components by

$$D_{0,ij}(x, N) = N^2 \delta_{ij} - N_i N_j - \varepsilon_{0,ij}(x), \tag{3}$$

which is the leading order term in the plane-wave response, namely,

$$e^{-i\kappa N \cdot x} \mathcal{L}(\kappa, x, \nabla) e^{i\kappa N \cdot x} = \kappa^2 D_0(x, N) + O(\kappa),$$

[1] http://www.python.org/
[2] http://numpy.scipy.org/
[3] http://www.cython.org/

$\mathcal{L}$ being the operator on the left-hand side of (1). With our assumptions, D_0 is Hermitian, and its eigenvectors determine the polarization of wave modes supported by the medium. Let $e(x, N)$ be the specific eigenvector corresponding to the considered wave mode (mode conversion is not discussed here), then the eigenvalue $H(x, N) = e^*(x, N) \cdot D_0(x, N)e(x, N)$ is the Hamiltonian in equation (2).

Upon computing (when possible) the integral of $ds = N \cdot dx$ along solutions of (2) one obtains the action function $s(x)$ which solves the Hamilton-Jacobi equation

$$H(x, \nabla s) = 0. \tag{4}$$

We can now define the normalized group velocity field

$$V(x) = \nabla_N H(x, \nabla s(x)), \tag{5}$$

which has the property of being tangent to geometrical optics rays, i.e., $dx(t)/dt = V(x(t))$. In addition, the vector field V determines the wave energy transport, which is governed by the amplitude transport equation

$$\nabla \cdot [V(x)|A(x)|^2] = -2\gamma(x)|A(x)|^2, \tag{6}$$

where $A(x)$ is the complex scalar amplitude of the wave field, and $\gamma(x) = e^*(x, \nabla s) \cdot \varepsilon_i^a(x)e(x, \nabla s)$, with ε_1^a being the anti-Hermitian part of ε_1. The wave electric field is then obtained in the form

$$E(\kappa, x) = e(x, \nabla s)A(x)e^{i\kappa s(x)}, \tag{7}$$

and one can check that this is an asymptotic solution of (1) in the limit $\kappa \to +\infty$.

Summarizing, in the standard geometrical optics theory, the wave energy density flux vector is proportional to the vector $V(x)$ defined in equation (5), and thus to geometrical optics rays.

The standard geometrical optics theory, however, does not allow us to describe diffractive beams, which is a severe limitation when it comes to the electron cyclotron beams currently employed. The complex eikonal theory [9-13] is an extension of geometrical optics which includes diffractive beams. The solution (7) is replaced by

$$E(\kappa, x) = e(x, \nabla s)A(x)e^{i\kappa s(x) - \kappa\phi(x)}, \tag{8}$$

where $e(x, N)$ is the same as in (7), but the amplitude $A(x)$ and the phase $s(x)$ are modified by the introduction of a new non-negative function $\phi \geq 0$, which is regarded as the imaginary part of the complex phase (eikonal) $\psi = s + i\phi$. As a consequence the Hamilton-Jacobi equation (4) is replaced by

$$H(x, \nabla s) - \frac{1}{2}\frac{\partial \phi}{\partial x^i}\frac{\partial \phi}{\partial x^j}\frac{\partial^2 H}{\partial N_i \partial N_j}(x, \nabla s) = 0, \tag{9}$$

together with the constraint $\nabla\phi \cdot \nabla_N H(x, \nabla s) = 0$. We note from (8) that the envelope of the wave field $e^{-\kappa\phi}$ has a maximum where $|\nabla\phi| = 0$, which is where the wave field is localized. We are therefore interested in a region of space where $|\nabla\phi|$ is actually small so that the second term in (9) acts as a perturbation of the standard Hamilton-Jacobi equation. The corresponding Hamiltonian orbits are the extended rays of complex geometrical optics. It is worth noting that Hamilton's equations for extended rays depend on $\nabla\phi$, which, in turn, satisfies a partial differential constraint.

We can still define the vector field $V(x)$ by equation (5), s being now a solution of (9), but, differently from geometrical optics rays, for extended rays we have,

$$\frac{dx}{dt} = V(x) - \frac{1}{2}\frac{\partial \phi}{\partial x^i}\frac{\partial \phi}{\partial x^j}\frac{\partial^3 H}{\partial N_i \partial N_j \partial N}, \tag{10}$$

i.e., $V(x)$ is no longer exactly tangent to ray trajectories. It is however approximately tangent to ray trajectories in the region of interest, where $|\nabla\phi|$ is small enough.

As for the wave energy flux it can be shown [15] that equation (6) holds in complex geometrical optics as well, with the only difference that the function s is now a solution of the modified Hamilton-Jacobi equation (9).

It follows that, according to the standard definition, extended rays satisfy the first requirement stated at the beginning of this section, i.e., (i) they reduce to geometrical optics rays when diffraction is negligible. The second requirement on the energy flux, however, is not exactly satisfied, cf. equation (10), though it is approximately satisfied.

One could define a second set of extended rays as the field lines of the vector field $V(x)$, which is proportional to the wave energy flux. Such rays satisfy condition (ii) on the energy flow by construction and, in addition, satisfy condition (i) as well, since the function s reduces to the corresponding geometrical optics solution for non diffracting beams.

Quantitatively, the differences between extended rays computed according to the foregoing two definitions is found to be small in the neighborhood of the beam trajectory [15], so that the distinction of the two different bundle of extended rays has no practical consequence.

In addition, one should note that the calculation of the field lines of the vector $V(x)$ requires the knowledge of the phase function $s(x)$, which, in turn, requires the calculation of standard extended rays.

2.2. The case of paraxial WKB wave fields and extended rays from `TORBEAM`

The paraxial WKB solution for the lowest order Gauss-Hermite mode [3, 4] is formally the same as equation (8), but with a different representation of the amplitude A, the phase s, and the field envelope ϕ, which we briefly recall. The trajectory of the beam is identified with the locus of points $\phi(x) = 0$ and it is found to be a particular geometrical optics ray $x = \bar{x}(t)$, called reference ray and carrying the refractive index $N = \bar{N}(t)$. In a neighborhood of the reference ray, one extends the parameter t to a curvilinear coordinate $t(x)$, and consider the Taylor expansion

$$A(x) \approx A(t(x)), \tag{11}$$

$$s(x) \approx s_0(t(x)) + \bar{N}(t(x)) \cdot \tilde{x} + \tfrac{1}{2}\tilde{x} \cdot \mathrm{S}(t(x))\tilde{x}, \tag{12}$$

$$\phi(x) = \tfrac{1}{2}\tilde{x} \cdot \Phi(t(x))\tilde{x}, \tag{13}$$

where $\tilde{x} = x - \bar{x}(t(x))$, while the amplitude $A(t)$ and the two beam tracing matrices $\mathrm{S}(t)$ and $\Phi(t)$ satisfy ordinary differential equations describing their evolution along the reference ray. (The first two terms in the expansion of ϕ vanish because of the definition of reference rays and the condition $\phi \geq 0$.) The physical meaning of the two symmetric matrices $\mathrm{S}(t)$ and $\Phi(t)$ has been extensively discussed [2-4].

Also in the paraxial WKB method [3, 4], the wave energy flux is proportional to the wave field $V(x)$ defined in equation (5) with $s(x)$ given in the form (12). In order to see this, let us note that equation (1) implies the continuity equation $\nabla \cdot F = -2E^* \cdot \varepsilon_1^a E$, where ε_1^a is defined after equation (6) and

$$F(\kappa, x) = \frac{2}{\kappa}\mathrm{Im}[E^*(\kappa, x) \times (\nabla \times E(\kappa, x))], \tag{14}$$

is the Poynting flux properly normalized. We substitute the wave object (8) for $E(\kappa, x)$ and, retaining the lowest order term only in the limit $\kappa \to +\infty$, we obtain

$$F(\kappa, x) = [2\nabla s - (e^* \cdot \nabla s)e - (e \cdot \nabla s)e^*]e^{-2\kappa\phi}|A|^2 + \mathcal{R}_1(\nabla\phi) + O(1/\kappa),$$

the remainder $\mathcal{R}_1(\nabla\phi)$ being proportional to $\nabla\phi$. The term in square brackets on the right-hand side reads

$$[2\nabla s - (e^* \cdot \nabla s)e - (e \cdot \nabla s)e^*] = e^*(x, \nabla s) \cdot \frac{\partial D_0(x, \nabla s)}{\partial N} e(x, \nabla s),$$

and, again, upon differentiating the eigenvalue equation $D_0 e = He$ with respect to the refractive index and making use of (9), one has

$$e^*(x, \nabla s) \cdot \frac{\partial D_0(x, \nabla s)}{\partial N} e(x, \nabla s) = \nabla_N H(x, \nabla s) + \mathcal{R}_2(\nabla \phi),$$

where the remainder $\mathcal{R}_2(\nabla\phi)$ is quadratic in $\nabla\phi$. In both the complex eikonal and the paraxial WKB theories, the estimate $|\nabla\phi|^n e^{-\kappa\phi} = O(\kappa^{-n/2})$ is proven for every non-negative integer $n \geq 0$, [4, 15], and that implies

$$F(\kappa, x) = V(x)e^{-2\kappa\phi(x)}|A(x)|^2 + O(1/\sqrt{\kappa}), \tag{15}$$

showing that the leading order term in the (normalized) Poynting flux is proportional to the vector field $V(x) = \nabla_N H(x, \nabla s)$. This asymptotic form of the Poynting flux vector is obtained regardless for the specific representation of the functions A, s, and ϕ, and, thus, it holds for both the complex eikonal method and the lowest-order Hermite-Gaussian mode of the paraxial WKB solution.

In the latter case, one can make use of equation (12) in order to compute $\nabla s(x)$ up to an $O(\tilde{x}^2)$ remainder, namely, $\nabla s(x) = \bar{N}(t(x)) + \mathrm{S}(t(x))\tilde{x} + O(\tilde{x}^2)$ and this approximation can be used in the calculation of the vector field $V(x) = \nabla_N H(x, \nabla s(x))$ in a neighborhood of the reference ray. This leads to the following system of ordinary differential equations for extended rays,

$$\frac{dx}{dt} = \nabla_N H(x, \nabla s(x)), \quad \nabla s(x) = \bar{N}(t(x)) + \mathrm{S}(t(x))[x - \bar{x}(t(x))], \tag{16}$$

where the vector $\bar{N}(t)$ and the symmetric matrix $\mathrm{S}(t)$ are obtained from the solution of the standard equations of paraxial WKB [2-4].

Equations (16) are used to compute the bundle of extended rays in `TORBEAM`. With this aim an appropriate numerical scheme has been developed on the basis of the second-order Heun's method for the solution of ordinary differential equation, which belongs to the family of Runge-Kutta methods. This is, indeed, a rather unconventional choice of Runge-Kutta method, but it allows us to combine nicely the calculation of extended rays with the main `TORBEAM` calculation, without making use of interpolations for the refractive index vector $\bar{N}(t)$ and the matrix $\mathrm{S}(t)$.

Let $x^n \approx x(t_n)$ be an approximation of the position of a ray at the discrete time-like parameter t_n, and, analogously, $(\nabla s)^n \approx \nabla s(x(t_n))$; then, the second-order Heun's integration step for the position of the rays at the next time-like point $t_{n+1} = t_n + \tau_n$, τ_n being the local step size, can be written as a predictor-corrector scheme,

$$x^* = x^n + \nabla_N H(x^n, (\nabla s)^n)\tau_n, \quad x^{n+1} = x^n + \frac{1}{2}[\nabla_N H(x^n, (\nabla s)^n) + \nabla_N H(x^*, (\nabla s)^*)]\tau_n, \tag{17}$$

where x^* is the (explicit Euler) predictor for the advanced position of the ray, $(\nabla s)^* \approx \nabla s(x^*)$, and x^{n+1} is the corrected advanced position of the ray. Under the assumption that the curvilinear coordinate $t(x)$ along the propagation of the beam satisfies the condition $t(x^n) = t_n$, at least approximately, the gradients of the phase are computed by

$$(\nabla s)^n = \bar{N}(t_n) + \mathrm{S}(t_n)[x^n - \bar{x}(t_n)], \quad (\nabla s)^* = \bar{N}(t_{n+1}) + \mathrm{S}(t_{n+1})[x^* - \bar{x}(t_{n+1})]. \tag{18}$$

All coefficients in the integration step (17) are then known from the standard `TORBEAM` output at the time-like points t_n and t_{n+1}, and the bundle of extended rays can be advanced in parallel to and without interfering with the main `TORBEAM` calculation. It is worth noting that, in contrast to the complex eikonal method, rays here are not used to compute the phase s and the beam

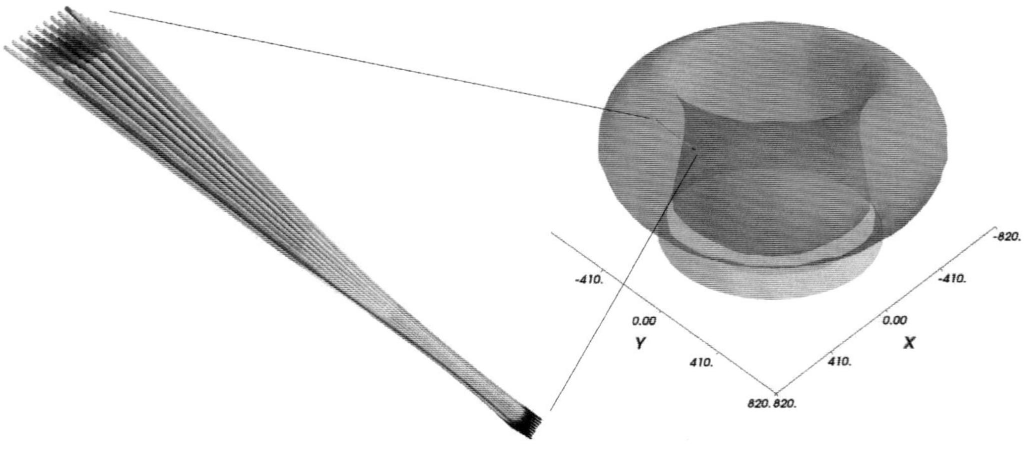

Figure 1. Three-dimensional representation of the extended rays obtained from the beam tracing code TORBEAM for the specific case of the ITER upper launcher, for the standard scenario 2 with power deposition on the flux surface $q = 3/2$. Rays are color-coded with the fraction of carried power (from red for high power to blue for low power) and the position of the beam with respect to the ITER separatrix is shown.

envelope ϕ, as those are already provided by the TORBEAM code; thus, reasonable results can be obtained with a limited number of extended rays.

Figure 1 shows an array of 9×9 extended rays computed by a new module implemented in the beam tracing code TORBEAM on the basis of the forgoing numerical scheme. Launching conditions are those relevant to the ITER upper launcher for power deposition on the flux surface $q = 3/2$ in the standard scenario 2; specifically the beam is assumed to be stigmatic and to have an initially circular cross section of width $w = 4.599$cm and phase-front radius of curvature $R = 193.4$cm; poloidal and toroidal injection angles are $\alpha = 53^\circ$ and $\beta = 20^\circ$, respectively; the coordinates of the launching position are $(688., 0., 418.)$ in centimeters. One can note that extended rays do not cross each other: diffraction effects appear to be properly accounted for.

3. Numerical results

In this section, preliminary numerical results of the Fokker-Planck calculation are presented, with the aim of testing the TORBEAM-RELAX package.

For ITER parameters, quasi-linear effects are completely negligible [1, 7] and thus linear power-deposition and current drive profiles, obtained from TORBEAM, should agree with the corresponding quasi-linear profiles obtained from RELAX.

In figure 2 the results for the ITER upper launcher scenario 2, with power deposition on the $q = 3/2$ flux surface, are reported; beam parameters are the same as in figure 1, the launched power is 1MW, and the array of 9×9 extended rays of figure 1 is used. (Qualitatively similar results can also be obtained with a 3×3 array.) The non-linearity parameter of Harvey *et al.* is $(dP/dV[\mathrm{W/cm^3}])/N[10^{13}\mathrm{cm^{-3}}]^2 = 0.0004$ [1], whereas quasi-linear effects start to be relevant for $(dP/dV[\mathrm{W/cm^3}])/N[10^{13}\mathrm{cm^{-3}}]^2 \approx 0.5$, which, for the considered case, is achieved for ≈ 1.1GW of injected power. One can note that the temperature profile is practically unchanged by the Fokker-Planck calculation and the anisotropy induced by the wave-particle interaction, quantified by the ratio of perpendicular and parallel energies, is extremely low. Both these results confirm that quasi-linear effects are totally negligible and, therefore, the current drive

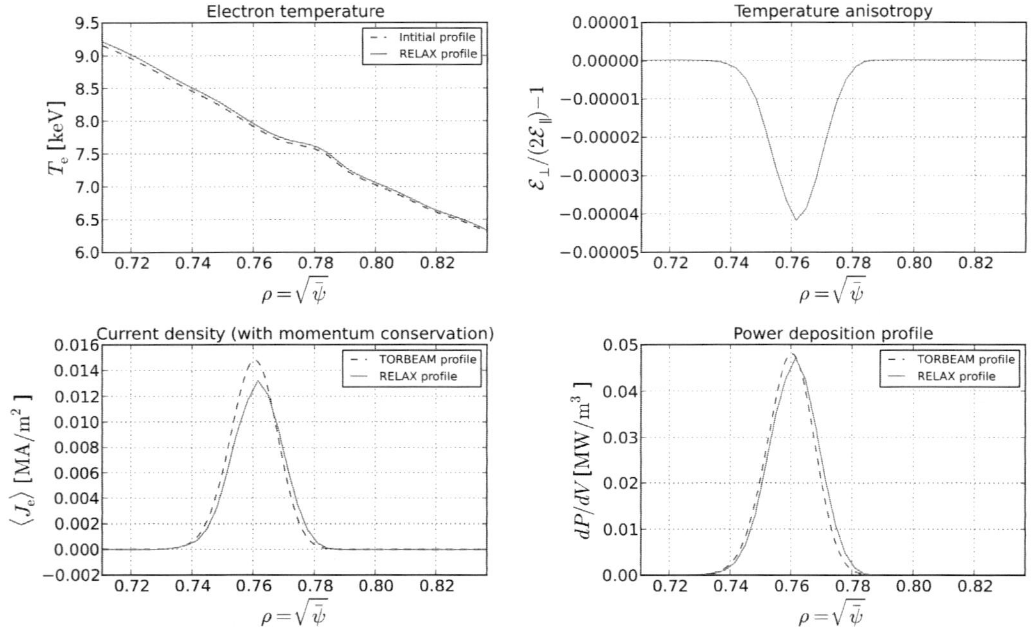

Figure 2. Temperature (upper-left panel), anisotropy (upper-right panel), driven current density (lower-left panel), and power deposition (lower-right panel) profiles obtained from the TORBEAM-RELAX package for the case of ITER scenario 2, power deposition on the $q = 3/2$ flux surface. The parameters of the beam are the same as in figure 1.

and the power deposition profile computed by TORBEAM and RELAX should be the same. The agreement shown in figure 2, lower panels, is qualitatively the same as that observed with the TORAY-RELAX package [7]; the total absorbed power in the two codes agrees within an error of $\approx 1.1\%$. Let us remark that both the linear code TORBEAM and the quasi-linear code RELAX compute the current drive and power deposition profiles independently and without any *ad hoc* renormalization of the absorbed power. Furthermore, a momentum conserving electron-electron collision operator is used in both TORBEAM and RELAX current drive calculations.

On going back to the anisotropy index, figure 2 upper right panel, we note that the perpendicular energy is actually reduced with respect to the parallel energy; this result might appear counter-intuitive as one would expect cyclotron wave to increase the perpendicular energy. In order to understand this point, we can increase the injected power holding constant the other parameters up to the level where quasi-linear effects are strong enough to be appreciated in the electron distribution function. For the case of the ITER tokamak, this takes an unrealistic value of the injected power (8GW, corresponding to the non-linearity parameter of Harvey *et al.* $(dP/dV[\mathrm{W/cm^3}])/N[10^{13}\mathrm{cm^{-3}}]^2 = 3.76$).

Figure 3 shows the contour plot of the electron distribution function in the $(p_\parallel, p_\perp)$-plane in logarithmic scale, at the crossing point of the beam on the surface where the maximum absorbed power density is achieved, i.e., $\rho = \sqrt{\psi} = 0.745$ (this value is significantly shifted with respect to the maximum of the deposition profile for the 1MW case). The black circle denotes the locus of points where the first-harmonic wave-particle resonance condition

$$\gamma - \frac{\omega_c}{\omega} - N_\parallel \frac{p_\parallel}{m_e c} = 0, \tag{19}$$

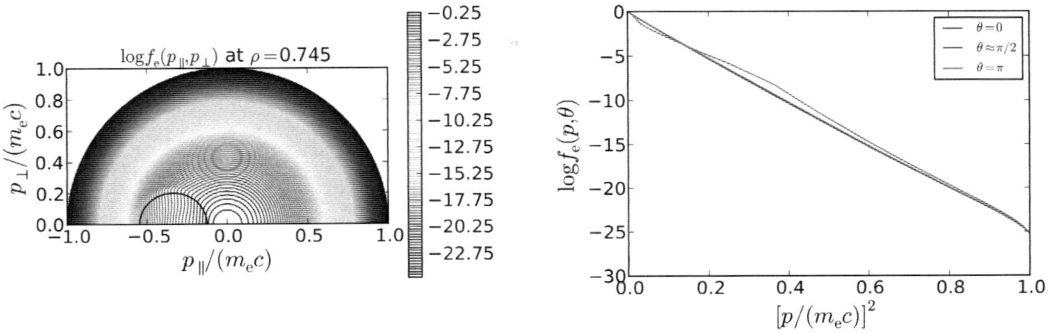

Figure 3. Electron distribution function in the quasi-linear regime obtained by pushing the injected power far beyond ITER operational parameters for testing purpose only.

is satisfied, with the electron cyclotron frequency ω_c and the parallel refractive index $N_{\|}$ being estimated by their average values over crossing points of rays through the considered magnetic surface. We can note that the deformation of the electron distribution function due to the interaction with the electron cyclotron beam occurs in the region of the velocity space delimited by the resonance condition as expected. For the current drive scenario under consideration such region is localized in the part of the velocity space for which $p_{\|} < 0$ and $p_\perp$ is small. The section of the distribution function at fixed velocity pitch-angle ϑ also shows a significant deformation for $\vartheta = \pi$ only, i.e., in the direction anti-parallel to the equilibrium magnetic field. This explains the anisotropy profile of figure 2.

Acknowledgments
We wish to thank Michael Kraus for his assistance with the `Cython` modules, and Nicola Bertelli for sharing with us his previous work on `TORBEAM` and `RELAX`. The output routines of `RELAX` have greatly benefited by the formatted output library `fmtout` by Roberto Bilato.

References
[1] R. W. Harvey, M. G. McCoy and G . D. Kerbel, Phys. Rev. Lett. **62**, 426 (1989).
[2] E. Poli, A. G. Peeters and G. V. Pereverzev, Comp. Phys. Comm. **136**, 90 (2001).
[3] G. V. Pereverzev, in *Reviews of Plasma Physics*, Vol. 19, edited by B. B. Kadomtsev, Consultants Bureau (New York, 1996) .
[4] G. V. Pereverzev, Phys. Plasmas **5**, 3529 (1998).
[5] E. Westerhof, A. G. Peeters and W. L. Schippers, *RELAX, a computer code for the study of collisional and wave driven relaxation of the electron distribution function in toroidal geometry*, Rijnhuizen Report No. RR 92-211 CA, 1992.
[6] E. Westerhof, *Implementation of TORAY at JET* National Technical Information Service Document no. PB91-114819INZ Rijnhuizen Report 89-183, 1989.
[7] N. Bertelli and E. Westerhof, Nucl. Fusion **49**, 095018 (2009).
[8] R. Prater, Nucl. Fusion **48**, 035006 (2008).
[9] S. Choudhary and L. B. Felsen, IEEE Trans. Antennas Propag. **AP-21**, 827 (1973).
[10] E. Mazzucato, Phys. Fluids B **1**, 1855 (1989).
[11] S. Nowak and A. Orefice, Phys. Fluids B **5**, 1945 (1993).
[12] A. G. Peeters, Phys. Plasmas **3**, 4386 (1996).
[13] D. Farina, Fusion Sci. Technol. **52**, 154 (2007).
[14] Yu. A. Kravtsov and Yu. I. Orlov, *Geometrical Optics of Inhomogeneous Media*, Springer (Berlin 1990).
[15] O. Maj, A. Mariani, E. Poli and D. Farina, *Extended rays and wave energy flux in complex geometrical optics*, to be submitted 2012.

Journal of Physics: Conference Series **401** (2012) 012014 doi:10.1088/1742-6596/401/1/012014

Gyrokinetic transport relations for gyroscale turbulence

BF McMillan, P Hill

Department of Physics, Warwick University, Coventry CV47AL, UK

E-mail: B.F.McMillan@warwick.ac.uk

S Jolliet, T Vernay, L Villard

Centre de Recherches en Physique des Plasmas, Association Euratom-Confédération Suisse, Ecole Polytechnique Fédérale de Lausanne, PPB, 1015 Lausanne, Switzerland.

A Bottino

Max Planck Institut für Plasmaphysik, Boltzmannstr. 2, D-85748 Garching, Germany

Abstract.
We derive explicit local transport relations for the global gyrokinetic formalism at arbitrary wavelength. This is an extension of the analysis in Scott et. al. 2010 where this was examined in the long-wavelength limit. Deriving a local expression for the fluxes requires that the gyroaveraging operator is symmetric, so that if point B is on the gyroring around A, point A is on the gyroring around B, for the same value of the magnetic moment. An algorithm for constructing a symmetric gyroring in a global code is described. Finally, using a simple 2D gyrokinetic code, we demonstrate the application of the momentum transport relation in a model problem with the full gyroaveraging operator without any long-wavelength approximation.

Symmetries and the related conservation properties underpin almost every domain of basic physics. Approximate models of a physical system which recapitulate the complete system's conservation properties are forced to embody the qualitative features of Hamiltonian dynamics. In turbulence physics, we are directly interested in the dynamics seen through the frame of these conserved quantities; we are primarily concerned with particle, heat and momentum transport through a system.

Global conservation of several important physical quantities, like particle number, momentum, and energy, is gauranteed in gyrokinetic formalisms based on a Lagrangian form[1]. In the full underlying Vlasov-Maxwell system, we are also able to write down transport equations which determine local time variation of these quantities in terms of fluxes evaluated using local values and gradients of the distribution function and electromagnetic fields; this is generally the case for systems with bounded speed for information propagation. It is *a priori* unclear whether we can do the same for gyrokinetic systems which involve nonlocal interactions across the gyroradius, due to rapid response on timescales which play an implicit role in the model.

Scott and Smirnov[2] give a general formalism for toroidal momentum transport, but only derive specific local transport relations for certain field dependences of the Lagrangian, with bilinear dependence on field gradients, appropriate for long-wavelength expansions of the gyrokinetic Poisson equation. Abiteboul et. al. provide an explicit, local, form of this toroidal momentum relation for the formalism used in Gysela, based on a Pade expansion of the gyroaveraging operator. Many codes (like GYRO, GENE and NEMORB) use the full gyroaveraging operator at some point in their formalism. We demonstrate here how an explicit polarisation term and gyro-momentum can be constructed when the exact gyroaveraging operator is used.

As might be expected, the finite-range non-locality due to gyroaveraging introduces a nonlocality into the flux-surface averaged transport relations, so that fluxes and toroidal momentum density need to be defined in terms of gyroaveraged quantities.

We perform calculations to evaluate local conservation laws for the exact abritrary wavelength gyroaveraging terms in the standard gyrokinetic equation, which were not evaluated explicitly in terms of local quantities in refs [2; 3]. Although the existence of a globally conserved toroidal momentum immediately allows the local momentum evolution to be written as the divergence of a global integral, this is not particularly useful for analysis. By deriving explicit, local, expressions for the fluxes, we show that the nonlocality in the fluxes is over the gyroradius length scale.

The existence of a local 1-D momentum transport equation is particularly vital for global gyrokinetic codes because they gaurantee that the slow evolution of flows and temperature profiles at long length scales is consistent with the divergences of fluxes in the systems. We might otherwise be concerned that non-conservation would result in small non-local transfers of energy, which might overwhelm these divergences, which are also small due to the long length and timescales involved.

1. Polarisation vector

The total toroidal momentum depends both on the toroidal component of the flow along the field line and the bulk $E \times B$ velocity. The dependence on the potential ϕ, is related via the gyrokinetic Poisson equation to the charge density n_i. We can use the conservation of charge relation to determine the dynamics of the $E \times B$ component of the toroidal momentum in terms of the time derivative of the polarisation vector.

We choose an electrostatic Lagrangian typically used for gyrokinetic simulations, which is close to that of ref. [4], but using a time-constant background density f_0 in the coupling to the squared potential (this leads to a linearised polarisation equation). We have

$$L = \int d^6 Z L_{pl} f + L_{pn} f_0 \tag{1}$$

with

$$L_{pl} = \frac{e}{c}(\mathbf{A} + p_z \mathbf{b}) \cdot \dot{\mathbf{R}} + \frac{mc}{e}\mu\dot{\theta} - \frac{p_z^2}{2m} - \mu B - e < \phi > \tag{2}$$

and

$$L_{pn} = -\frac{em}{2qB}\frac{\partial}{\partial \mu} < \tilde{\phi}^2 > . \tag{3}$$

Here, L is the Lagrangian density, which is integrated over the Lie-transformed phase space to form the Lagrangian functional. The particle coordinates, $\mathbf{Z}_p = R, p_z, \mu$, are the gyrocentre position, parallel canonical momentum and magnetic moment. α is the ignorable sixth coordinate representing the gyroangle, e and m are the species charge and mass, $\mathbf{A}$ and $\mathbf{b}$ the magnetic vector potential and unit vector, and $\mathbf{B} = \nabla \times \mathbf{A}$ the magnetic field. The angle brackets around the electrostatic potential ϕ denote a gyroaveraging about a ring $\mathbf{R} + \boldsymbol{\rho}(\alpha, \mu)$. The standard gyrocentre transform uses $\rho = (2\mu m/B)^{1/2}/q$, and $\boldsymbol{\rho}/\rho = cos(\alpha)\mathbf{e_1} + sin(\alpha)\mathbf{e_2}$, with $\mathbf{e_1}$ and $\mathbf{e_2}$ unit vectors defined perpendicular to the magnetic field. We also have:

$$\tilde{\phi} = \phi - < \phi > \tag{4}$$

Note that there is usually a sum over multiple species involved in the Lagrangian, which we supress here to simplify notation.

We then follow Scott et. el., defining H_p via

$$H(f) = (H_0 f + q\phi f) + H_p(f) \tag{5}$$

with H_0 collecting all the particle terms not dependent on ϕ. The definition of H_p implies

$$H_p(f) = \left[q < \phi > f - q\phi f + \frac{e}{2\Omega_{cs}} \frac{\partial}{\partial \mu} < \tilde{\phi}^2 > f_0 \right] \tag{6}$$

We define a polarisation,

$$\nabla \cdot \mathbf{P} = \frac{\partial}{\partial \phi} H_p(f), \tag{7}$$

which will be used to define charge fluxes through the system.

Let us separately consider $\mathbf{P_0}$ associated with the first two terms of the eq. 6, and $\mathbf{P_1}$ associated with the coupling to the background. We have

$$\nabla \cdot \mathbf{P_0} = e \int d^6 Z \frac{\partial}{\partial \phi} (< \phi > - \phi) H_P(f) = e \int d^6 Z \left[\delta \left(\mathbf{R} - \mathbf{R}' + \boldsymbol{\rho}(\mu, R, \theta) \right) - \delta(\mathbf{R} - \mathbf{R}') \right] H_P(f, \phi[\mathbf{R}']), \tag{8}$$

where the last expression exhibits the implicit dependence of the Hamiltonian on position.

We can insert a Greens' function-type solution to find an integral form for P which has the divergence specified above. Because we wish for P to be near-local, we choose a function which is non-zero only on the gyrodisc. The polarisation term can be constructed via 'lines of field' connecting the charge at the gyrocentre, with a ring of charge on the gyrocircle: this accounts for the gryoaveraging which was not included in the definition of H_0. We can define a $\mathbf{P_0}$ which satisfies this equation by introducing an integration from one end of this 'line of charge' to the other, for parameter $g \in [0, 1]$, via

$$\mathbf{P_0} = e \int d^3 R d\mu dv_{\|} d\alpha dg B *_{\|} (\mathbf{R} + \boldsymbol{\rho}) \boldsymbol{\rho} \delta(\mathbf{R} + g\boldsymbol{\rho} - \mathbf{R}') f(\mathbf{R} + \boldsymbol{\rho}). \tag{9}$$

To perform this calculation for $\mathbf{P_1}$ we need to define f_0, which will be chosen as $f_0 = n(x) \exp(-2\mu B/T + m v_{\|}^2 / 2T)$. To explicitly calculate $\mathbf{P_1}$ as an integro-differential function of ϕ, we use the weak form of the definition of the polarisation $\mathbf{P_1}$,

$$\int d^6 Z \nabla \cdot \mathbf{P_1} \delta\phi = \frac{\partial}{\partial \phi} \frac{e}{2\Omega_{ci}} \int d^6 Z f_0 \frac{\partial}{\partial \mu} < (\phi - < \phi >)^2 > \delta\phi \tag{10}$$

and integrate by parts by parts, using the μ-independence of the phase space volume element, to find

$$\int d^6 Z \nabla \cdot \mathbf{P_1} \delta\phi = \frac{\partial}{\partial \phi} \frac{eB}{2T\Omega_{ci}} \int d^6 Z f_0 < (\phi - < \phi >)^2 > \delta\phi \tag{11}$$

$$= \frac{\partial}{\partial \phi} \frac{eB}{2T\Omega_{ci}} \int d^6 Z f_0 (< \phi^2 > - < \phi >^2) \delta\phi. \tag{12}$$

Inserting explicit gyroaverages results in

$$\int d^6 Z \nabla \cdot \mathbf{P_1} \delta\phi(\mathbf{X}) \tag{13}$$

$$= \frac{\partial}{\partial \phi} \frac{eB}{2T\Omega_{ci}} \int d^6 Z f_0(\mathbf{X}) [d\alpha_1 d^3 X' \delta(\mathbf{X} - \mathbf{X}' - \boldsymbol{\rho}[\alpha_1, \mathbf{Z}]) \phi(\mathbf{X}')^2 \tag{14}$$

$$- d^3 X' d^3 X'' d\alpha_1 d\alpha_2 \delta(\mathbf{X} - \mathbf{X}' - \boldsymbol{\rho}[\alpha_1, \mathbf{Z}]) \delta(\mathbf{X} - \mathbf{X}'' - \boldsymbol{\rho}[\alpha_2, \mathbf{Z}]) \phi(\mathbf{X}') \phi(\mathbf{X}'')] \delta\phi(\mathbf{X}). \tag{15}$$

The right hand side is the variation of ϕ by an infinitestimal $\delta\phi$, which allows us to evaluate this equation as

$$\int d^6Z'\nabla\cdot\mathbf{P_1}\delta\phi(\mathbf{X}') = \frac{eB}{T\Omega_{ci}}\int d^6Zd^3X'd^3X''[d\alpha_1\delta(\mathbf{X}-\mathbf{X}'-\boldsymbol{\rho}[\alpha_1,\mathbf{Z}]))\phi(\mathbf{X}') \quad (16)$$
$$-d\alpha_1d\alpha_2\delta(\mathbf{X}-\mathbf{X}'-\boldsymbol{\rho}[\alpha_1,\mathbf{Z}])\delta(\mathbf{X}-\mathbf{X}''-\boldsymbol{\rho}[\alpha_2,\mathbf{Z}])\phi(\mathbf{X}'')]\delta\phi(\mathbf{X}')f_0(\mathbf{X}), \quad (17)$$

where we have changed the dummy variable on the LHS to allow us to identify the divergence of the polarisation

$$\nabla\cdot\mathbf{P_1}(\mathbf{X}') = \frac{eB}{T\Omega_{ci}}\int d^6Zd^3X''[d\alpha_1\delta(\mathbf{X}-\mathbf{X}'-\boldsymbol{\rho}[\alpha_1,\mathbf{Z}]))\phi(\mathbf{X}') \quad (18)$$
$$-d\alpha_1d\alpha_2\delta(\mathbf{X}-\mathbf{X}'-\boldsymbol{\rho}[\alpha_1,\mathbf{Z}])\delta(\mathbf{X}-\mathbf{X}''-\boldsymbol{\rho}[\alpha_2,\mathbf{Z}])\phi(\mathbf{X}'')]f_0(\mathbf{X}). \quad (19)$$

A polarisation vector satisfying this equation which is a local function of the field can be written using the 'lines of charge' described above approach,

$$\mathbf{P_1}(\mathbf{X}) = -\frac{eB}{T\Omega_{ci}}\int d^6Z''d\alpha_1d\alpha_2dg f_0(\mathbf{X}-\boldsymbol{\rho}[\alpha_1,\mathbf{Z}''])\hat{\boldsymbol{\rho}}_b\delta(\mathbf{X}-g\boldsymbol{\rho}_b-\mathbf{X}'')\phi(\mathbf{X}'') \quad (20)$$

where $\boldsymbol{\rho}_b = \boldsymbol{\rho}[\alpha_1,\mathbf{Z}''] + \boldsymbol{\rho}[\alpha_2,\mathbf{Z}'']$ and the integration domain for g is $[0,1]$.

2. Momentum equation

Following ref. [2], with the polarisation defined, the toroidal momentum equation can be written in the form

$$-\frac{\partial}{\partial t}\left\langle\frac{1}{c}P\cdot\nabla A_\psi\right\rangle + ... + \left\langle f\frac{\partial H}{\partial\psi}\right\rangle = 0 \quad (21)$$

We evaluate the last term which contains the derivative of the Hamiltonian with respect to the toroidal component.

$$f\frac{\partial H}{\partial\psi} = f\frac{\partial H}{\partial <\phi>}\frac{\partial <\phi>}{\partial\psi} + f\frac{\partial H}{\partial <\phi^2>}\frac{\partial <\phi^2>}{\partial\psi} \quad (22)$$
$$= -f\frac{\partial <\phi>}{\partial\psi} - \int\frac{eBf_0}{T\Omega_{ci}} <\phi> \frac{\partial <\phi>}{\partial\psi} + \int\frac{eBf_0}{T\Omega_{ci}} <\phi\frac{\partial\phi}{\partial\psi}>. \quad (23)$$

This needs to be written as the divergence of a flux to yield a local transport relation. The terms can be refactored as

$$f\frac{\partial H}{\partial\psi} = <f>\frac{\partial\phi}{\partial\psi} - \frac{eB}{T\Omega_{ci}}(<f_0<\phi>> - <f_0>\phi)\frac{\partial\phi}{\partial\psi} \quad (24)$$
$$-\left[<f>\frac{\partial\phi}{\partial\psi} - f\frac{\partial <\phi>}{\partial\psi}\right] \quad (25)$$
$$+\left[\frac{eB}{T\Omega_{ci}} <f_0<\phi>> \frac{\partial\phi}{\partial\psi} - \frac{eB}{T\Omega_{ci}}f_0 <\phi> \frac{\partial <\phi>}{\partial\psi}\right] \quad (26)$$
$$+\left[\frac{eB}{T\Omega_{ci}} <f_0>\phi\frac{\partial\phi}{\partial\psi} - \frac{eB}{T\Omega_{ci}}f_0 <\phi\frac{\partial\phi}{\partial\psi}>\right] \quad (27)$$

As in Scott and Smirnov, the first two terms are the functional derivative of Hf with respect to the potential, which is the Poisson equation, and therefore vanishes. The last two pairs of terms are of the form $f<g> - g<f>$:

$$\int d^6Z'f(\mathbf{Z})g(\mathbf{Z}') - f(\mathbf{Z}')G(\mathbf{Z}))\delta(\mathbf{R}-\mathbf{R}'-\boldsymbol{\rho}(\mathbf{R}')). \quad (28)$$

In order to produce a near-local convservation relation, this need to be written in the form of a divergence. As before, we will relate this to a line integral

$$\nabla \cdot \int dg d^6 Z' f(\mathbf{Z}') g(\mathbf{Z}' + \boldsymbol{\rho}) \delta(g(\mathbf{R}' + \boldsymbol{\rho}) + (1-g)\mathbf{R}' - \mathbf{R}) \hat{\boldsymbol{\rho}} \tag{29}$$

where $\boldsymbol{\rho} = \boldsymbol{\rho}(\mathbf{R}, \alpha)$, as long as the gyroaveraging procedure has 'pairing symmetry', so that

$$\mathbf{R}' = \mathbf{R} + \boldsymbol{\rho}[\mathbf{R}, \alpha] \implies \mathbf{R} = \mathbf{R}' + \boldsymbol{\rho}[\mathbf{R}', -\alpha]. \tag{30}$$

This relation is trivially true for standard flux tube codes, where $\boldsymbol{\rho}$ is defined to be on the radial and binormal plane ($\boldsymbol{\rho} \cdot z \equiv 0$), and is only a function of the direction along the field line, z ($\boldsymbol{\rho} = \boldsymbol{\rho}(z)$).

For global codes the gyroaveraging procedure is typically asymmetric, because ρ is a function of the magnetic field strength and direction, which varies across the simulation domain. Although the gyroaveraging will still be symmetric to first order in $\rho*$, there is a small component which is non-local, which will be associated with non-local energy and momentum transport in the model. Abiteboul (ref [3]) notes that this non-local component will be zero as long as the gyroaveraging operator is Hermitian: the pairing symmetry requirement is sufficient to ensure that the gyroaveraging is Hermitian.

Before we explain how to construct pair-symmetric gyroaverages in a global model, we will evaluate these terms in the momentum balance equation for a simple physical problem.

3. A demonstration of FLR polarisation terms in a simple gyrokinetic model.

To demonstrate the practical calculation of the FLR terms in the momentum relation, we set up a simple gyrokinetic model and solve it numerically.

We consider one of the simplest possible gyrokinetics models: the solution of the 2D GK-Vlasov Poisson equation in the plane perpendicular to the magnetic field. To simplify further, no gradients are assumed to exist in the system, and the perturbed distribution function is taken to be $\delta f = \delta(\mu - \mu_0)\delta(v_{\|})$. The background distribution f_0 is assumed Maxwellian. The magnetic field is taken to be of the form $\vec{B} = B(x)\hat{\vec{z}}$. Parallel dynamics are absent in the model, and only the spatial advection is non-zero, so the equations of motion are simply

$$\frac{d^3 R}{dt} = (mv_\perp^2/qB) B \times \nabla B/B^2 - B \times \nabla\phi/B^2. \tag{31}$$

To emphasise the FLR effects, electrons are assumed to act as a neutralising background, and do not respond adiabatically to the field. The Poisson equation, which describes the balance between the ion polarisation charge, and the gyrodensity is of the form

$$\int < f > d^6 Z = \int d^6 Z (< \phi > - \phi) \delta(\mathbf{R} - \mathbf{R}' + \boldsymbol{\rho}) f_0 d\mathbf{Z}. \tag{32}$$

At large scales, this model is equivalent to the short-wavelength limit of the Hasegawa-Mima equation.

Periodic boundary conditions are chosen in x and y, with $x \in [0, L_x]$, $y \in [0, L_y]$. The distribution function is described using the PIC method, and the Poisson equation is discretised via a finite element method using quadratic splines. Fourth order Runge-Kutta timestepping is used to solve the resulting coupled Vlasov-Poisson system. For the example problem treated here, 256 grid points here chosen in each direction, and 2^{20} markers were loaded into the simulation domain. Because the problem is symmetric in the y-direction, the Poisson solution is performed in Fourier space, as in [5], allowing variations of field strength in the x direction which will be considered in further work.

For an initial demonstration, a scenario with a homogeneous magnetic field was chosen. As the system does not contain spatial gradients in f_0, there are no linear instabilities. To produce interesting dynamics, initial states with substantial flows must be specified. We choose

$$\delta f = A \cdot \text{sign}[\sin k_y y] \tag{33}$$

to produce an initial state with banded flow zones. We took $k_y = 4\pi/L_y$.

This situation is unstable to a Kelvin-Helmholtz instablity, which develops nonlinearly to produce filamentation at short scales and the transfer of the initial energy to long scales. The predominant mode in the end state of the simulation is the largest which fits in the box, with $(N_x, N_y) = (1, 0)$.

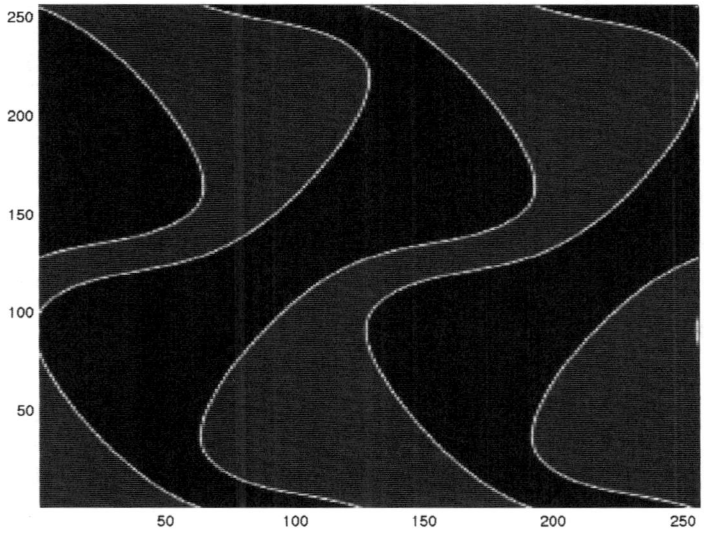

Figure 1. Color plot of gyrocentre density on the xy spatial domain early in the nonlinear period of the 2D gyrokinetic simulation.

For this situation, the momentum transport and vorticity transport equations are equivalent. We instrumented the code with the appropriate terms in the momentum balance equation. For ease of implementation, instead of directly evaluating the polarisation via the Green's function approach described above, the divergence was calculated instead, and the inverse of the divergence operator was found via a Fourier solver.

The terms in the momentum transport equation, and the overall numerical imbalance, are shown in fig. 3. During the development of the instability, the conventional Reynolds stress and the additional non-local FLR correction to the Reynolds stress are of similar magnitude, so that this test problem is a strong test of accurate conservation: the RMS error is around 0.5% at this late stage of the simulation, where the absolute error is largest.

This provides a clear demonstration that vorticity transport remains a practical approach to understanding turbulent dynamics, even when the transport is determined by somewhat nonlocal operators on the gyroscale. The same principles should be applicable to typical 3D codes with inhomogeneous magnetic fields, which are required for global tokamak modelling.

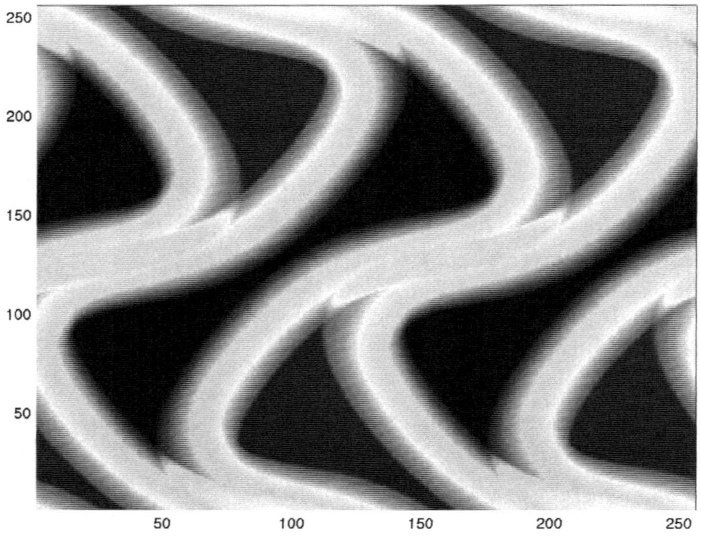

Figure 2. Color plot of gyroaveraged density on the xy spatial domain early in the nonlinear period of the 2D gyrokinetic simulation.

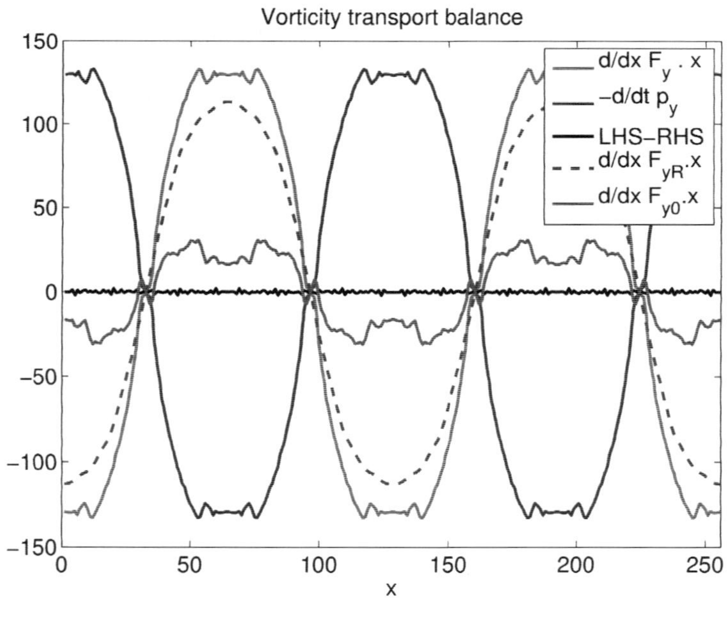

Figure 3. Terms in the 1D momentum balance equation well into the nonlinear regime in the simple 2D simulation.

4. Pair-symmetric gyroaveraging for global gyrokinetics

In global gyrokinetic codes, gyroaveraging is typically defined on planes not exactly perpendicular to the field line: usually these are surfaces of constant toroidal or poloidal angle.

For example, NEMORB defines gyroaveraging on poloidal planes, whereas the global GENE code uses surfaces of constant poloidal angle. Numerical gyroaveraging is an integral around an ellipse on these planes, which is the intersection of the plane with a cylinder around a line tangent to the magnetic field at the gyrocentre (for the case of poloidal planes, these are approximately circles). This is, to leading order in $\rho*$, equivalent to using gyroaveraging rings perpendicular to the field lines, because the perturbations are flute-like. The size of the ellipses is dependent on the magnetic moment μ and the field strength B at the gyrocentre $\mathbf{R}$.

In general, this leads to gyroaveraging points which are not conjugate. However, it is possible to redefine the gyroradius $\boldsymbol{\rho}(\theta, \mu, \mathbf{R})$ so that it has 'pairing symmetry', but is equal to the usual gyroradius at first order.

We define a pair $\mathbf{R}, \mathbf{R}'$ to be gyroconjugate if they lie at generalised distance ρ on straight lines on a plane K, with the distance defined by a metric $d\mathbf{R}'^2 = Bd\mathbf{R}_\perp^2 = Bd\mathbf{R}^2 - B(d\mathbf{R} \cdot \mathbf{b})^2$ in terms of the usual Euclidean distance $\mathbf{R}$. In algorithmic terms, this means we integrate from the gyrocentre along a straight line until $\int d\mathbf{R}' = (2\mu m)^{1/2}/q$.

Because the magnetic field scale length is long compared to the gyroradius, approximate expressions can be easily generated for numerical evaluation. For example, by iteratively evaluating the metric one-half of a gyroradius from the initial gyrocentre, gyro-conjugate points can be generated without extensive coding.

As an example, we construct a symmetrised gyroring around a single point in a strongly spatially varying magnetic field, and show that the gyrorings of points on this curve are tangent to the initial point. Figure 4 shows these symmetrised gyrorings, which are somewhat distorted circles, for a case where the ratio of the gyroradius to the magnetic field scale length is 0.7. A large ratio is chosen for the purposes of visually demonstrating the distortion, and the conjugacy of the gyropoints (although clearly the gyrokinetic ordering is only valid when this ratio is small).

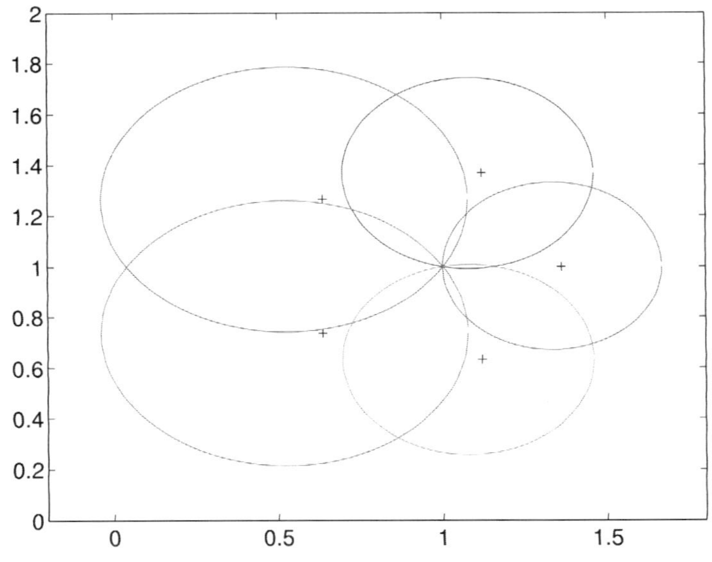

Figure 4. Five points on a symmetrised gyroring around the point $(1, 1)$ (addition symbols), and the five conjugate gyrorings around these points.

5. Discussion

We have demonstrated that local conservation principles can be applied in practice to gyrokinetic simulations capturing physics all the way from the gyroscale to the system scale. This serves to reinforce in a more explicit form the statements of recent papers[2; 3; 6] on momentum conservation in gyrokinetic systems.

It is as yet unclear what the implications of the slight non-Hermicity of gyroaveraging in some gyrokinetic simulations is. Although distortions in the gyroring to allow the gyroaveraging to remain Hermitian in finite-size tokamaks would be small, small modifications in polarisation due to system size effects are known to be significant in certain situations. For example, a detailed code comparison between the codes ORB5 and GENE showed that the zonal response of the Poisson operator was subtly different for system-scale perturbations because GENE treated some small $\rho *$ effects differently to ORB5[7].

References

[1] Brizard A and Hahm T S 2007 *Reviews of Modern Physics* **79** 421–468

[2] Scott B and Smirnov J 2010 *Physics of Plasmas* **17** 112302

[3] Abiteboul J, Garbet X, Grandgirard V, Allfrey S J, Ghendrih P, Latu G, Sarazin Y and Strugarek A 2011 *Physics of Plasmas* **18** 082503 (pages 11)

[4] Hahm T 1988 *Physics of Fluids* **31** 2670–2673

[5] McMillan B, Jolliet S, Bottino A, Angelino P, Tran T and Villard L 2010 *Computer Physics Communications* **181** 715 – 719 ISSN 0010-4655

[6] Brizard A J and Tronko N 2011 *Physics of Plasmas* **18** 082307 (pages 14)

[7] Lapillonne X, McMillan B F, Gorler T, Brunner S, Dannert T, Jenko F, Merz F and Villard L 2010 *Physics of Plasmas* **17** 112321 (pages 10)

Journal of Physics: Conference Series **401** (2012) 012015 doi:10.1088/1742-6596/401/1/012015

Fusion modeling approach for novel plasma sources

D Melazzi[1], A Cardinali[2], M Manente[1], D Pavarin[1]
[1]CISAS "G. Colombo" Centro Interdipartimentale Studi Attività Spaziali, University of Padova, Padova, Italy
[2]Associazione Euratom-ENEA sulla Fusione, C.P. 65 - I-00044 - Frascati, Rome, Italy
E-mail: davide.melazzi@gmail.com

Abstract. The physics involved in the coupling, propagation and absorption of RF helicon waves (electronic whistler) in low temperature Helicon plasma sources is investigated by solving the 3D Maxwell-Vlasov model equations using a WKB asymptotic expansion. The reduced set of equations is formally Hamiltonian and allows for the reconstruction of the wave front of the propagating wave, monitoring along the calculation that the WKB expansion remains satisfied. This method can be fruitfully employed in a new investigation of the power deposition mechanisms involved in common Helicon low temperature plasma sources when a general confinement magnetic field configuration is allowed, unveiling new physical insight in the wave propagation and absorption phenomena and stimulating further research for the design of innovative and more efficient low temperature plasma sources. A brief overview of this methodology and its capabilities has been presented in this paper.

1. Introduction

Analytical and numerical Ray-Tracing techniques, which are based upon the Wentzel-Kramers-Brillouin-Jeffreys (WKBJ) approximation and assume the wave pattern as a superposition of noninteracting modes, have been extensively employed to provide deep insight on wave propagation and absorption and to carry out heating modeling studies in reactor relevant Tokamak devices. These techniques have been proven essential for simulation of Lower Hybrid (LH)[1],[2] and Electron Cyclotron Heating (ECH)[3] and Current Drive (CD), for which the wavelengths are much smaller than the machine dimensions, and they have been also used for the simulation of Ion Cyclotron Resonance Heating (ICRH)[1],[4].

They are still of high interest in heating scenarios where mode conversion is important. Moreover, they play a key role in investigation of basic wave-plasma interaction physics like electrons heating, local transport coefficients behaviour, current drive and current profile tailoring, Internal Transport Barrier (ITB) creation, and Neoclassical Tearing Modes (NTM) stabilization. The application of these methods has been considered for the study of whistler mode propagation in linearly stratified plasma in magnetosheath and plasma sheet regions of the Earth's magnetosphere[5], the analysis of whistlers reflected by the magnetosphere as a source of plasmaspheric hiss[6] and the conversion of whistler waves to lower hybrid waves in the presence of short-scale field-aligned density irregularities (striations) in Earth's low-altitude ionosphere[7].

In Ray-Tracing techniques the wavevector satisfies the local dispersion relation, and the wave amplitude must not change much over a wavelength. This can fail in a number of situations, such as in the presence of strong wave focussing or reflection, in mode conversion regions, and inside

evanescent regions. However, Ray-Tracing has been proven a powerful modeling tool, providing much insight on wave propagation, especially for the simulation of wave related phenomena where associated wavelengths are much smaller than the machine dimensions and finite difference or finite element discretisations would require huge meshes, forbidding the treatment of 2D and 3D problems. Dependences of the plasma density and temperature, the frequency, the parallel wave number spectrum, and the antenna location can be taken into account as well.
Maxwell's equations are usually treated as Boundary Value Problem (BVP), whereas the propagation described by means of the Ray-Tracing approach is reduced to an ensemble of independent initial value problems, whose initial conditions give the starting wave vector and power density at each point of a reference surface of the propagating wave. The required initial conditions for Ray-Tracing are usually determined by a Full-Wave solution of the field radiated by a finite-sized antenna but generally these conditions are obtained by means of auxiliary models describing the antenna or waveguide launcher and its surroundings.
Moreover, not only the Full-Wave solution can be used to provide the Ray-Tracing technique with the necessary initial conditions, but also the combined Ray-Tracing/Full-Wave approach has been applied[10], [19] when tunneling, mode conversion or local power deposition due to damping phenomena are important. Specifically, whenever a wave reaches a confluence region a transmitted wave, a reflected wave and a mode converted wave are excited; at each encounter of a ray with the confluence layer, the full wave model predicts the power redistribution and a set of sub-rays is relaunched.
Furthermore, Ray-Tracing has been coupled to the Fokker-Planck equation; considering the modeling of LH waves in the JET tokamak[11], the LH waves propagation solved by means of Ray-Tracing has been combined to a solver of the Fokker-Planck equation for the evaluation of the electron distribution function self-consistently[18] with the wave absorption.
Thanks to this treatment of electromagnetic propagation, Ray-Tracing techniques offer a fresh and new approach to study the wave propagation phenomena and power deposition mechanisms involved in efficient Helicon low temperature plasma sources, which are gaining more and more interest in a broad range of applications.
The physics of wave propagation and absorption involved in these plasma sources has been studied in many theoretical and experimental works so far, but only for confinement magnetostatic field considered constant and uniform along the axis of the plasma cylinder. This last assumption can be completly removed by the Ray-Tracing approach, unveiling new physical insight in the power deposition mechanisms. Moreover, such a technique so used in Tokamak plasma reactors could bring some heating modeling concepts coming from plasma fusion scenarios and exploit them in low temperature plasma devices to design and build innovative plasma sources.

2. Plasma source physical model

2.1. Low temperature plasma source applications

Plasmas are currently of considerable interest for a large variety of applications including semiconductor manufacturing, spacecraft propulsion, material surface modification, material processing, basic and applied laboratory research. Low-pressure discharges, whose ionized mechanism is related to helicon waves, are usually called Helicon plasma sources and have been studied for their possible use in industrial etching or surface treatment due to their high density and their unusually high ionization efficiency: the achieved densities of plasma (up to 10^{19} m^{-3}) are almost an order of magnitude higher than in other discharges (at comparable pressures and input powers). As a matter of fact, Helicon plasma sources have been recognized to be much more efficient rather than capacitive and inductive sources in generating dense plasmas [12],[13],[14],[15] for low temperature industrial application; their plasma potential is as low as in Electron Cyclotron Resonant (ECR) sources but it needs a magnetic field much

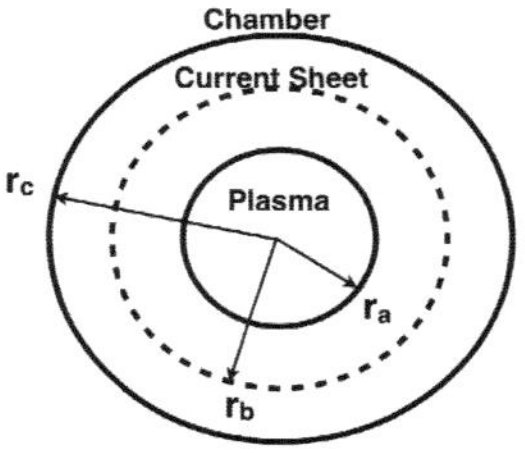

Figure 1. Geometry in the (r, θ) plane.

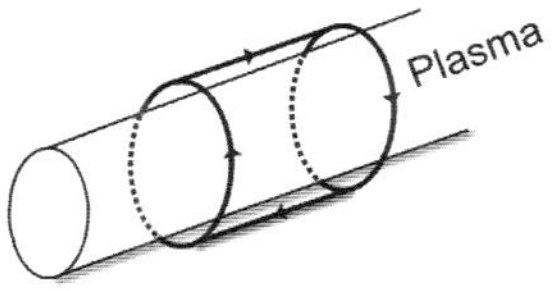

Figure 2. Nagoya Type III antenna.

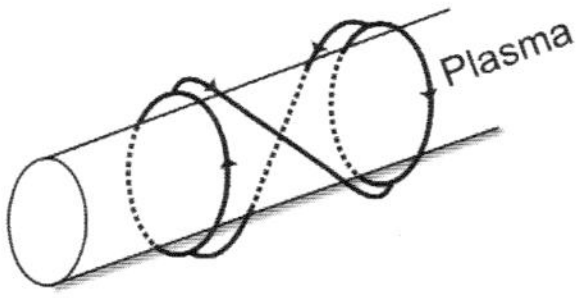

Figure 3. Fractional Helix antenna.

lower (below <0.1 T) than ECR and other high efficiency wave-heated industrial sources. A DC confinement magnetic field along the device axis is necessary for all these sources to be operative, increasing the cost and complexity unlike other plasma sources; however, electromagnets can be replaced by Permanent Magnets (PM), reducing cost and complexity issues, but this could lead to a magnetic field configuration that is no longer perfectly aligned with the axis inside the source region, deeply influencing the wave propagation and power deposition mechanisms. All these features become particularly severe for the development of a plasma source for an electric thruster for space application; these and many other aspects have been considered in the European project HPH.com[8] (Helicon plasma hydrazine combined micro) which aims to develop a space plasma thruster based on helicon plasma sources working in the radiofrequency regime at low-power (<100 W of RF power).

2.2. Helicon Plasma Model

As far as Helicon plasma sources are concerned, a plasma cylinder of radius r_a, surrounded by an antenna placed at radius r_b, and enclosed inside a conducting tube of radius r_c are to be considered; Fig. 1 shows an axial section of the geometry. Fields are forced by an RF oscillating current given by different antenna configurations; Fig. 2 and Fig. 3 show two common antennas used in Helicon experiments. A magnetostatic field $\boldsymbol{B_0}$ is directed along the z axis, $\boldsymbol{B_0} = B_0\hat{\boldsymbol{z}}$, perpendicular to the (r, θ) plane.

These sources are able to sustain a steady-state magnetized plasma at low pressure, by means of the propagation and absorption of radiofrequency electromagnetic waves in the range of $\omega_{ci} \ll \omega_{lh} \ll \omega \ll \omega_{ce}$, where ω_{ci} and ω_{ce} are the ion and electron cyclotron angular frequencies, and ω_{lh} is the lower hybrid frequency.

In this range of frequencies two coupled plasma waves can propagate inside the magnetized plasma column, namely the Helicon and Trivelpiece-Gould (TG) waves [16]. The first wave has a long transverse wavelength λ_1, the second a short one λ_2. The values of the two wavelengths $\lambda_{1,2}$ is expressed by the dispersion relation [16]:

$$\lambda_{\perp_{1,2}}\left(n_{\|}\right) = 2\pi \left[n_{\|}^2 \frac{\omega_{ce}^2}{4c^2} \left(1 \mp \sqrt{1 - 4\frac{\omega_{pe}^2}{\omega_{ce}^2}\frac{1}{n_{\|}^2}} \right)^2 - n_{\|}^2 \frac{\omega^2}{c^2} \right]^{-1/2}, \tag{1}$$

where $\lambda_{\perp_{1,2}}$ is the transverse wavelength of the Helicon and TG modes respectively, ω_{pe} is the plasma frequency, ω_{ce} is the electron cyclotron frequency, $\omega = 2\pi f$ is the antenna angular frequency, c is the speed of light, and $n_{\|}$ is the parallel wave number, defined as $n_{\|} = k_z/k_0$, where $k_0 = \omega/c$ is the vacuum wave number.

Once that source geometry (in terms of r_a, r_b and r_c), antenna type, working frequency f and plasma characteristics (in terms of gas specie, neutral pressure p_n, DC magnetic field B_0, plasma density n_0) are given, Eq. 1 gives all the wave modes that can propagate and deposite power inside the plasma column. The power deposition due to collisional or collisionless damping and for all the accessible propagative modes can be evaluated by the following equation:

$$P = \frac{1}{2}\int_{V_{plasma}} \boldsymbol{E^*} \cdot \boldsymbol{J_{plasma}} dV = \frac{|I_0^2|}{2}(R_p + iX_p) \, , \tag{2}$$

where P is the power coupled to the plasma by the antenna, $\boldsymbol{E^*}$ is the conjugate electric field inside the plasma, $\boldsymbol{J_{plasma}}$ is the plasma current density, I_0 is the antenna current, R_p and X_p are the electrical resistance and reactance, respectively. In the following, we show how the power deposition can be influenced by plasma parameters in a homogeneous Helicon source [9].

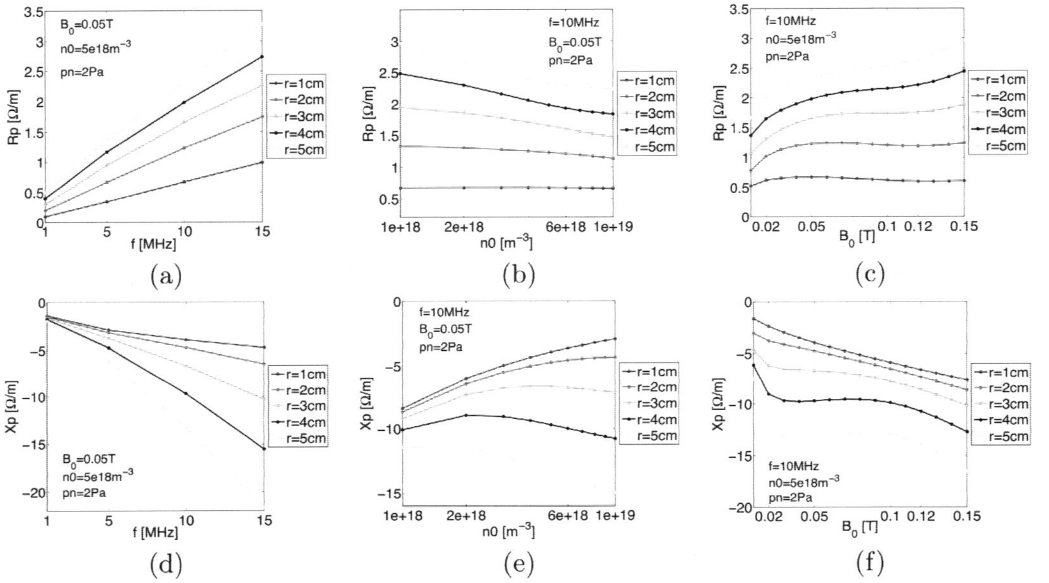

Figure 4. Electrical impedance as a function of the plasma parameters for five ($r_a = 1.0 \cdot 10^{-2}$ m, $r_a = 2.0 \cdot 10^{-2}$ m, $r_a = 3.0 \cdot 10^{-2}$ m, $r_a = 4.0 \cdot 10^{-2}$ m, $r_a = 5.0 \cdot 10^{-2}$ m) homogeneous magnetized plasma sources : (a),(d) $B_0 = 0.05$ T, $n_0 = 5 \cdot 10^{18}$ m^{-3}, $p_n = 2.0$ Pa, (b),(e) $B_0 = 0.05$ T, $f = 10$ MHz, $p_n = 2.0$ Pa, (c),(f) $f = 10$ MHz, $n_0 = 5 \cdot 10^{18}$ m^{-3}, $p_n = 2.0$ Pa. Parameters are: $m = 0$ antenna, $I_0 = 1.0$ A.

The power deposition mechanism can be influenced whenever a more realistic plasma density profile is considered, slightly changing the plasma electrical impedance; but still the behaviour - as a function of the working frequency, the DC confinement magnetic field and the plasma density - showed in Fig. 4 for a uniform source can be considered representative.

Once we remove the hypothesis that the DC magnetic field is aligned along the axial direction of the plasma cylinder, and a general confinement magnetic field is allowed, a completely new picture in the wave propagation and wave absorption phenomena arises. This analysis can be done by means of the Ray-Tracing technique, which can at the same time unveil new physical evidence in the power deposition mechanisms and exploit some heating modeling concepts coming from plasma fusion scenarios in order to design innovative and more efficient low temperature plasma sources.

Journal of Physics: Conference Series **401** (2012) 012015 doi:10.1088/1742-6596/401/1/012015

3. WKB methods for 3D Maxwell equation

The physics involved in the coupling, propagation and absorption of RF helicon waves (electronic whistler) in low temperature plasma sources is investigated by solving the 3D Maxwell-Vlasov model equations using a WKB asymptotic expansion in $\delta_0 = (\omega a)/c$ (expansion parameter), where a is the plasma radius of the plasma cylinder, and c is the speed of light. We solve numerically the reduced set of the WKB equations for the wave phase and for the power damping. In this way we deduce the propagation characteristics of the whistler wave (e.g. mode conversion on the Trivelpiece-Gould branch) and the absorption level of the wave (e.g. power deposition profiles), in the quasi-homogeneous plasma approximation.

The Ray-Tracing approach, which uses the asymptotic technique (WKB) for the solution of Maxwell wave equations, describes the wave propagation in a similar way to the geometric optics where light propagation is described in terms of rays. The reduced set of equations, after the WKB treatment at the lowest order in the expansion parameter, describes a bundle of trajectories in the phase space $(\boldsymbol{r}, \boldsymbol{k})$, and allows for the reconstruction of the wave front of the propagating wave. These equations are formally Hamiltonian.

In the past, this technique has been used extensively in the laboratory plasma physics to study the propagation and absorption of electromagnetic waves in magnetized plasmas[2]. The equations written in cylindrical geometry are:

$$
\begin{aligned}
&\frac{dx}{d\tau} = \delta_0^{-1}\frac{\partial \mathcal{H}}{\partial n_x}; \quad \frac{d\theta}{d\tau} = \delta_0^{-1}\frac{\partial \mathcal{H}}{\partial m}; \quad \frac{d\hat{z}}{d\tau} = \delta_0^{-1}\frac{\partial \mathcal{H}}{\partial n_z}; \\
&\frac{dn_x}{d\tau} = -\delta_0^{-1}\frac{\partial \mathcal{H}}{\partial x}; \quad \frac{dm}{d\tau} = -\frac{\partial \mathcal{H}}{\partial \theta}; \frac{dn_z}{d\tau} = -\delta_0^{-1}\frac{\partial \mathcal{H}}{\partial \hat{z}}; \\
&\frac{dS_0}{d\tau} = n_x\frac{\partial \mathcal{H}}{\partial n_x} + m\frac{\partial \mathcal{H}}{\partial m} + n_z\frac{\partial \mathcal{H}}{\partial n_z}; \\
&\frac{dt}{d\tau} = \frac{\partial \mathcal{H}}{\partial \omega}; \quad \frac{d\omega}{d\tau} = -\frac{\partial \mathcal{H}}{\partial t} = 0 \Rightarrow \omega = const; \\
&\frac{dP_{ower}}{dt} = -2\Gamma_{new}(\boldsymbol{r}, \boldsymbol{k}, \omega)\, P_{ower}
\end{aligned}
\tag{3}
$$

where $\boldsymbol{r}/a = (x, \theta, \hat{z})$ is the normalized space coordinate, and $\boldsymbol{k}c/\omega = (n_r, m, n_z)$ is the wave number. Here ω is the wave frequency, S_0 is the wave front surface, P_{ower} is the power carried by the wave along the trajectory, and Γ_{new} is the damping coefficient related to the anti-hermitian part of the Stix dielectric tensor and which accounts for Electron Landau Damping (ELD) and the collisional damping. Note that $\mathcal{H}$ is:

$$
\mathcal{H}(\boldsymbol{r}, n_{\parallel}, n_{\perp}) \equiv A(\boldsymbol{r}, n_{\parallel})\, n_{\perp}^4 + B(\boldsymbol{r}, n_{\parallel})\, n_{\perp}^2 + C(\boldsymbol{r}, n_{\parallel}), \tag{4}
$$

where

$$
\begin{aligned}
A(\boldsymbol{r}, n_{\parallel}) &= S \\
B(\boldsymbol{r}, n_{\parallel}) &= \left[\left(n_{\parallel}^2 - S\right)(P + S) + D^2\right] \\
C(\boldsymbol{r}, n_{\parallel}) &= P\left[\left(n_{\parallel}^2 - S\right)^2 - D^2\right]
\end{aligned}
\tag{5}
$$

and S, D, P are the usual Stixs elements of the dielectric tensor, with $n_{\parallel} = \hat{b_r} n_r + \hat{b_z} n_z$, $n_{\perp}^2 = n_r^2 + \left(\delta_0^{-1} m/x\right)^2 + n_z^2 - n_{\parallel}^2$. Eq. 4 gives the dispersion relation for the cold electromagnetic electron whistler wave when $\mathcal{H} = 0$. The structure of the external magnetic field $\hat{b_r} = B_r/B$, $\hat{b_z} = B_z/B$ can be accounted exactly in the definition of the parallel and perpendicular wave-number. This is important because this kind of high frequencies wave (at short wavelength)

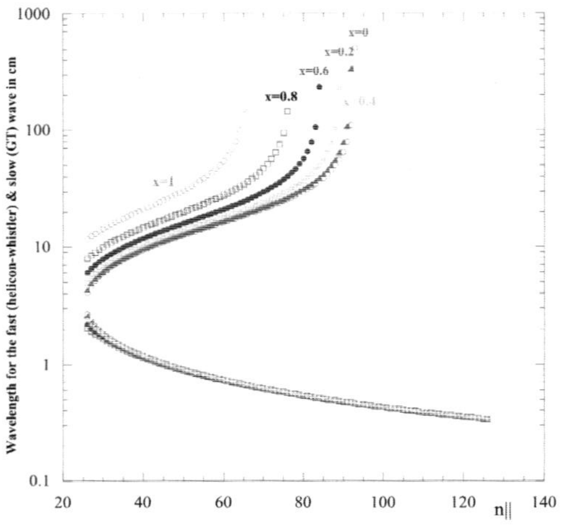

Figure 5. Branches of the dispersion relation for different radial positions.

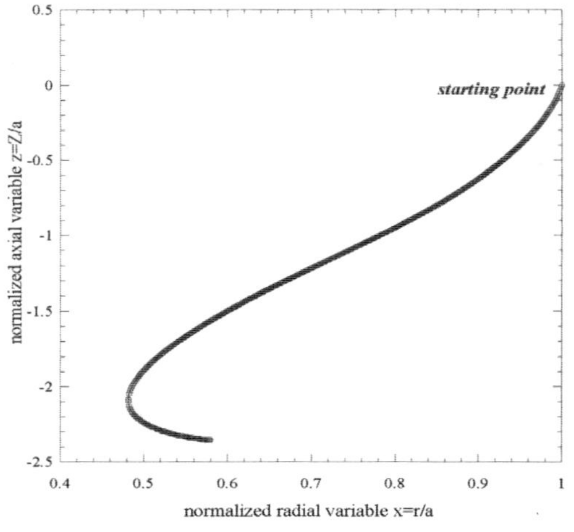

Figure 6. Surface wave front 2D trajectory in the $x - z$ plane.

are influenced by the structure of the medium (the density and the surrounding magnetic field) during the propagation and their absorption.

It is worth to recall that the validity of the WKB hypothesis relies on the assumption that: i) the wavelength of the propagating mode is much less of the characteristic scale length L: $\lambda \ll L = \left(\frac{1}{A_0}\frac{dA_0}{dr}\right)^{-1}$ where A_0 is the electric field amplitude, ii) the curvature radius of the wavefront is much greater than the wavelength $k \propto (\nabla \cdot \nabla S_0) \gg \lambda$ (absence of diffraction effects in the wave propagation).

We show now the numerical solution of the Ray-Tracing equation system in the case of whistler applied to the HPH.com facility[8] just with few examples, referring for a more exhaustive analysis of all the features of the propagation, conversion, and absorption of the wave to a subsequent paper[17]. The plasma parameters we have used in the simulation are: i) plasma radius $a = 10$ cm, length of the cylinder $L = 20$ cm, applied frequency $f = 13.56$ MHz, magnetic field on axis $B = 250$ Gauss, density at the plasma center $n_{center} = 10^{12}$ cm^{-3}, density at the edge $n_{edge} = 5 \times 10^{11}$ cm^{-3}, and 2D profile: $n_e(r,z) = Ce^{-(r/a)^2-(z/L)^2} = C^{-x^2-\hat{z}^2}$, central temperature $T_{central} = 10$ eV, edge temperature $T_{edge} = 1$ eV, argon plasma with neutral density $n_{neutral} = 10^{13}$ cm^{-3}, the equivalent neutral pressure is $p = 3$ mTorr, poloidal wave-number $m = 0$, and $30 < n_z < 90$. In Fig. 5 we have plotted the dispersion branches of the dispersion relation Eq. 4 for a typical situation with constant confinement magnetic field and the density profile $n_e(x) = Ce^{-x^2}$. Fig. 5 shows the perpendicular wavelength (in cm) vs the z-component of the wave number n_z of the launched wave at several radial positions from the edge ($x = 1$) to the center ($x = 0$). Notice that the perpendicular wavelength of the slow branch of the dispersion (Trivelpiece-Gould branch) is less than 1 cm while for the fast branch (Whistler-Helicon branch) the wavelength remains of the order of the device dimensions (10 cm) for a large fraction of the parallel refraction index $26 \leq n_z \leq 56$. At higher value of $n_z > 60$, the perpendicular wavelength of the fast branch increases rapidly with an asymptotic behavior for several radial position so that cut-off occurs for each of these radial positions; this behavior causes the breakdown of the WKB asymptotic expansion.

Notice in Fig. 6 the presence of the various component of the confinement magnetic field and the 2D character of the density profile, which induce a radial reflection around half radius ($x/a \sim 0.5$).

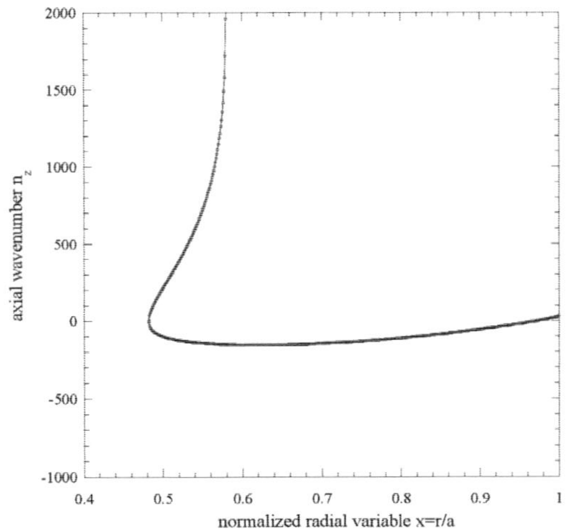

Figure 7. Axial wave-number n_z vs the radial normalized variable x.

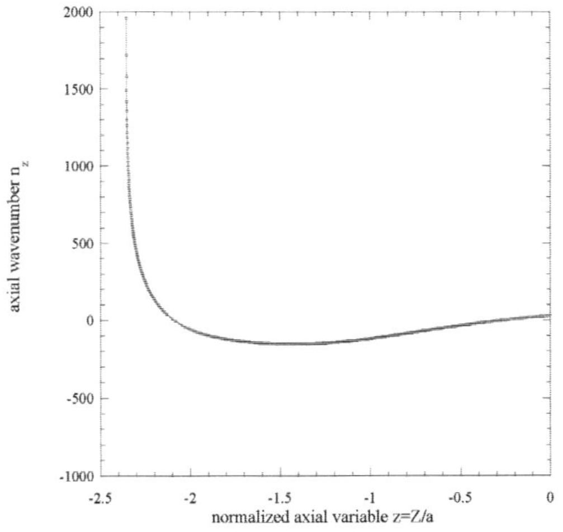

Figure 8. Axial wave-number n_z vs the normalized axial coordinate $\hat{z}$.

Moreover, Figs. 7-8 show the very strong variation of the axial wave-number and an axial resonance after a radial reflection, which occurs at $x \sim 0.5$ (half radius). It is important to recall that in a simple 1D geometry with a constant confinement magnetic field B_z and density profile depending only on the adimensional radial coordinate x, the axial wave-number remained constant inside the plasma. This feature of the wave propagation in a complex $2D$ or 3D geometry is accounted very well by using this asymptotic technique, without integrating explicitly the Maxwell-Vlasov or Maxwell-Newton system, which is a very expensive task still when considering a small portion of plasma. Obviously it is necessary to pay attention and monitor along the calculation that the WKB expansion remains satisfied. A detailed comparison of the WKB reduced analysis and full Maxwell-Vlasov integration is under consideration[17].

4. Conclusion

In this paper, we presented a brief overview of Ray-Tracing technique for the analysis and design of innovative and more efficient low temperaure plasma sources. This technique is based upon the WKB approximation and it has been extensively employed to provide new insight on wave propagation and absorption for simulation of LH, ECH, CD and ICRH to carry out heating modeling studies in fusion plasmas. More importantly, Ray-Tracing is still of high interest in heating scenarios even when mode conversion or local power deposition due to damping phenomena are important and a coupled solution with a Full-Wave method is necessary.

We used Ray-Tracing technique to study the physics involved in the coupling, propagation and absorption of RF helicon waves (electronic whistler) in low temperature plasma sources; specifically, we solved numerically the 3D Maxwell-Vlasov equations using a WKB asymptotic expansion for the propagation characteristics of the whistler wave, and the absorption level of the wave. Though the physics of wave propagation and absorption involved in these kind of sources has been studied by many theoretical and experimental works, the axial confinement magnetostatic field hypothesis has been completly removed by the Ray-Tracing approach, unveiling new physical insight in the power deposition mechanisms and giving the chance to use heating modeling concepts coming from plasma fusion scenarios and exploit them in low temperature plasma devices to design and build innovative plasma sources.

References

[1] Brambilla M 1986 *Comp. Phys. Rep.* **4** 71-93
[2] Brambilla M and Cardinali A 1982 *Plasma Physics* **24** 1187-1218
[3] Hongxiu L and Changlin Z 1986 *Inter. J. Infrared and Millimiter Waves* **7** 391-404
[4] Eester v D, Bhatnagar V P and Koch R 1985 *Plasma Phys. Control. Fusion* **27** 1015-1026
[5] Sazhin S S and Strangeways H J 1989 *Planet. Space Sci.* **37** 739-747
[6] Draganov A B, Inan U S, Sonwalkar V S and Bell T F 1992 *Geophys. Res. Lett.* **19** 233-236
[7] Shao X, Eliasson B, Sharma A S, Milikh G and Papadopoulos K 2012 *J. Geophys. Res.* **117** A04311
[8] Pavarin D , Ferri F, Manente M, Curreli D, Melazzi D, Rondini D and Cardinali A 2011 *International Electric Propulsion Conference (Wiesbaden)*
[9] Melazzi D, Curreli D, Manente M, Carlsson J and Pavarin D 2012 *Comp. Phys. Comm.* **183** 1182-1191
[10] Eester V D 1992 *Plasma Phys. Control. Fusion* **34** 1539
[11] Imbeaux F and Peysson Y 2005 *Plasma Phys. Control. Fusion* **47** 2041-2065
[12] Boswell R W 1970 *Physics Letters* **33A** 457-458
[13] Boswell R W 1984 *Plasma Phys. Control. Fusion* **26** 1147-1162
[14] Komori A, Shoii T, Miyamoto K, Kawai J and Kawai K 1991 *Physics Fluids B* **3** 893-898
[15] Chen F F 1991 *Plasma Phys. Control. Fusion* **33** 339-364
[16] Chen F F and Arnush D 1997 *Physics of Plasmas* **4** 34113421
[17] Cardinali A, Manente M, Melazzi D and Pavarin D 2012 *submitted for publication on Phys. of Plasmas*
[18] Cardinali A 2000 *Transworld Res. Net. Magazine* **1** 185-197
[19] Cardinali A 2007 *Physics of Plasmas* **11** 4199

A Parametric Analysis of Nonlinear Lower Hybrid Effects

F Napoli[1], **C Castaldo**[2], **R Cesario**[2], **G Schettini**[1]

[1] Università Roma Tre, Dipartimento di Elettronica Applicata, Roma, Italy

[2] Associazione EURATOM/ENEA sulla Fusione, Centro Ricerche Frascati, Italy

E-mail: francesco.napoli@uniroma3.it

Abstract. Based on a new modeling of the nonlinear behaviour of lower hybrid waves launched in tokamak plasmas triggered by the noisy background of the radiofrequency power generator, a parametric analysis of the role of the plasma parameters on the onset of the nonlinearity is here proposed. As a result, the nonlinear effects could be prevented by increasing the edge electron temperature, the relative ion-electron temperature, the edge density gradient and the ion mass.

1. Introduction

Nonlinear effects occurring at the plasma edge, namely parametric instabilities (PIs) [1], play an important role in determining the power absorption of the lower hybrid (LH) waves externally launched in tokamak plasmas. Parasitic nonlinear absorption at the edge due to strong PI was invoked to prevent [2] the power penetration into the plasma core during the early LH experiments aimed at heating the plasma ions [3,4]. LH experiments aimed at current drive [5] operate successfully only at relatively low plasma density (line averaged plasma density $< n > \leq 3\times10^{19}\text{m}^{-3}$) [6] due to the onset of strong PI. Following modeling results [2], a new method was developed on FTU (Frascati Tokamak Upgrade) which allowed producing LH current drive effects at central plasma densities even higher that in ITER ($n_0 \approx 5\times10^{20}\text{m}^{-3}$), provided that the operating edge temperature was kept relatively high to reduce the PI [7].

Based on previous works on the ponderomotive effects driving the PI phenomenon [8,9], we present here new results that provide a more complete interpretation of the nonlinear mechanism of the PI-produced spectral broadening. In a previous work [10] we proposed a new nonlinear LH modeling able to describe the nonlinear mode coupling of the LH wave (pump wave) launched in tokamak plasmas with LH waves (sideband waves) excited by the noisy background of the radiofrequency (RF) power generator. We showed that the nonlinear mode coupling is driven by low frequency (≤ 10 MHz) density fluctuations, produced by the beating of the pump and the sideband waves. The amplitude spectra in the frequency and $n_{\|}$ domains of the low frequency driving density fluctuations were calculated in the frame of the kinetic theory, based on the self-consistent solution of the Vlasov-Poisson equations, obtained by a perturbative approach. A system of nonlinearly coupled LH wave equations was then derived in the frame of the cold plasma approximation, taking into account the nonlinear density fluctuations in evaluating the current density perturbations. Thus, based on an iterative approach, we provided a numerical solution of the system of nonlinearly coupled LH wave equations.

In the present work, we discuss the role of the plasma parameters on the onset of the LH nonlinearity in front of the LH grill, based on the new nonlinear LH model proposed in [10].

2. Nonlinear LH Modeling

We shortly review here the new nonlinear LH modeling proposed in [10]. Based on previous works on the ponderomotive effects driving the PI phenomenon [8,9], the new modeling gives a more accurate description of the LH physics near the antenna-plasma interface, by utilizing a full wave approach that avoids using the eikonal approximation. Indeed, this approximation usually fails in the plasma region near the LH antenna mouth as the radial wavelength is often larger than the characteristic radial scale lengths of the spatial amplification and plasma inhomogeneities. In addition, the modeling allows accounting for both the spectral broadening and the pump depletion. However, it is not suitable to discuss the amplification of LH sidebands emerging from the thermal background of plasma density fluctuations, as done instead in the standard PI modeling [1]. Indeed, poloidal symmetry of low frequency (LF) modes has been assumed, i.e. only LH waves with poloidal wavenumber $k_y = 0$ are considered here. This is in agreement with our choice to evaluate the LH nonlinearity triggered by a noisy background produced by the RF power generator. Therefore, the spectral noise has the same polarization and (approximate) poloidal symmetry of the pump waves, i.e. we assume $E_y \approx 0$, $k_y \approx 0$.

The mode coupling among different LH wave spectral components is driven by LF density fluctuations that have angular frequency much lower than the ion-cyclotron angular frequency, $\omega' \ll \Omega_{ci}$. These fluctuations are excited by the beating of the pump wave, with operating angular frequency ω_0, and the high frequency (HF) spectral components that are intrinsic to the noise of RF power generators. The latter are shifted from the pump wave frequency by an amount of ω'. In turn, under suitable plasma conditions, the beating of the pump wave and low frequency perturbation produces the growth of LH sidebands with angular frequency $\omega = \omega_0 \pm \omega'$.

A simple slab geometry is assumed, where $\hat{\boldsymbol{x}}$, $\hat{\boldsymbol{y}}$ and $\hat{\boldsymbol{z}}$ are, respectively, the radial (inward), poloidal and toroidal directions of the outer plasma region of a tokamak, faced to the mouth of the LH launcher. Steady state conditions are considered and symmetry of the plasma parameter in poloidal and toroidal directions is assumed, with density and temperature gradients occurring only in the radial direction. The confinement magnetic field is assumed constant and directed in the toroidal direction. These conditions allow solving the integro-differential problem by means of a spectral method, i.e. by Fourier transforms in y and z coordinates and in the time t. The low frequency density fluctuations are assumed constant in the poloidal direction, as occur for the HF modes.

The amplitude spectra in the frequency and $k_{\parallel} (= k_z)$ domains of the low frequency driving modes is calculated in the frame of the kinetic theory, based on the self-consistent solution of the Vlasov-Poisson equations obtained by a perturbative approach. A new nonlinear wave equation for the LH waves is then obtained in the frame of the cold plasma approximation, taking into account the nonlinear density fluctuations in evaluating the HF current density perturbation.

Moreover, to solve the nonlinear wave equation, we consider the following spectral representation of the total toroidal electric field:

$$\tilde{E}_{z,k_{\parallel},\omega}(x) \equiv \tilde{E}^{\mathrm{P}}_{z,k_{\parallel},\omega}(x) + \tilde{\psi}_{k_{\parallel},\omega}(x) \tag{1}$$

assuming a spectral separation of the sideband from the pump:

$$\tilde{E}_{z,k_{\parallel},\omega}(x) = \tilde{E}^{\mathrm{P}}_{z,k_{\parallel},\omega}(x) \;\text{ if } k_{\parallel}, \omega \in I_{k_{\parallel},\omega} \tag{2}$$

$$\tilde{E}_{z,k_{\parallel},\omega}(x) = \tilde{\psi}_{k_{\parallel},\omega}(x) \quad \text{if } k_{\parallel}, \omega \notin I_{k_{\parallel},\omega} \tag{3}$$

where $I_{k_{\parallel},\omega}$ is the spectral range of the pump.

We assume that the pump electric field in the domain $I_{k_{\|},\omega}$ is represented by a double delta function as follows:

$$\tilde{E}^{\mathrm{p}}_{z,k_{\|},\omega}(x) \equiv \tilde{E}_{\mathrm{o}}(x)\delta(\omega-\omega_{\mathrm{o}})\delta(k_{\|}-k_{\|\mathrm{o}})+\tilde{E}^{*}_{\mathrm{o}}(x)\delta(\omega+\omega_{\mathrm{o}})\delta(k_{\|}+k_{\|\mathrm{o}}) \quad (4)$$

and the sideband electric field in the domain complement of $I_{k_{\|},\omega}$ as follows:

$$\tilde{\psi}_{k_{\|},\omega}(x) \equiv \sum\nolimits_{l,m}\left[\tilde{\psi}_{l,m}(x)\delta(\omega-\omega_{l})\delta(k_{\|}-k_{\|m})+\tilde{\psi}^{*}_{l,m}(x)\delta(\omega+\omega_{l})\delta(k_{\|}+k_{\|m})\right] \quad (5)$$

The governing nonlinear LH wave equation splits in two nonlinearly coupled wave equations, one for the complex pump amplitude and one for the complex sideband amplitude, for each spectral component considered:

$$\begin{cases} \partial^{2}_{x}\tilde{E}_{\mathrm{o}}(x)+\alpha(x,\omega_{\mathrm{o}},k_{\|\mathrm{o}})\cdot\partial_{x}\tilde{E}_{\mathrm{o}}(x)+\beta_{\mathrm{L}}(x,\omega_{\mathrm{o}},k_{\|\mathrm{o}})\cdot\tilde{E}_{\mathrm{o}}(x)+\gamma(x,\omega_{\mathrm{o}},k_{\|\mathrm{o}},\sum\nolimits_{l,m}|\tilde{\psi}_{l,m}(x)|^{2})\cdot\tilde{E}_{\mathrm{o}}(x)=0 \\ \partial^{2}_{x}\tilde{\psi}_{l,m}(x)+\alpha(x,\omega,k_{\|})\cdot\partial_{x}\tilde{\psi}_{l,m}(x)+\beta_{\mathrm{L}}(x,\omega,k_{\|})\cdot\tilde{\psi}_{l,m}(x)+\sigma(x,\omega,k_{\|},|\tilde{E}_{\mathrm{o}}(x)|^{2})\cdot\tilde{\psi}_{l,m}(x)=0 \end{cases} \quad (6)$$

where:

$$\alpha(x,\omega,k_{\|}) \equiv \frac{\partial_{x}S(x,\omega)}{S(x,\omega)}\frac{k_{\|}^{2}}{\left(k_{\|}^{2}-\dfrac{\omega^{2}}{c^{2}}S(x,\omega)\right)} \quad (7)$$

$$\beta_{\mathrm{L}}(x,\omega,k_{\|}) \equiv -\frac{P(x,\omega)}{S(x,\omega)}\left(k_{\|}^{2}-\frac{\omega^{2}}{c^{2}}S(x,\omega)\right) \quad (8)$$

where $S(x,\omega)$ and $P(x,\omega)$ are the elements of the Stix tensor [11]. The γ and σ terms, taking into account the nonlinear phenomena, are defined as follow:

$$\gamma \equiv -\frac{q_{\mathrm{e}}^{2}\omega_{\mathrm{pe}}^{2}(x)\left(k_{\|\mathrm{o}}^{2}-\dfrac{\omega_{\mathrm{o}}^{2}}{c^{2}}S(x,\omega_{\mathrm{o}})\right)}{\omega_{\mathrm{o}}^{2}S(x,\omega_{\mathrm{o}})\sqrt{2m_{\mathrm{e}}T_{\mathrm{e}}^{3}(x)}}\sum\nolimits_{l,m}\frac{\omega_{\mathrm{o}}|\tilde{\psi}_{l,m}(x)|^{2}}{\omega_{l}^{2}}\frac{F_{\mathrm{c}}^{*}\left((\omega_{l}-\omega_{\mathrm{o}})/(k_{\|m}-k_{\|\mathrm{o}})v_{\mathrm{th,e}}(x)\right)}{(k_{\|m}-k_{\|\mathrm{o}})} \quad (9)$$

$$\sigma \equiv -\frac{q_{\mathrm{e}}^{2}\omega_{\mathrm{pe}}^{2}(x)\left(k_{\|\mathrm{o}}^{2}-\dfrac{\omega_{\mathrm{o}}^{2}}{c^{2}}S(x,\omega)\right)}{\omega_{\mathrm{o}}^{2}S(x,\omega)\sqrt{2m_{\mathrm{e}}T_{\mathrm{e}}^{3}(x)}}\frac{|\tilde{E}_{\mathrm{o}}(x)|^{2}}{\omega_{\mathrm{o}}}\frac{F_{\mathrm{c}}\left((\omega-\omega_{\mathrm{o}})/(k_{\|}-k_{\|\mathrm{o}})v_{\mathrm{th,e}}(x)\right)}{(k_{\|}-k_{\|\mathrm{o}})} \quad (10)$$

Here we define a coupling function F_{c} as follows:

$$F_{\mathrm{c}}(u_{\alpha}) \equiv C_{1}(u_{\alpha})-C_{2}(u_{\alpha})\sum\nolimits_{\beta}\left(\frac{q_{\beta}}{q_{\alpha}}\frac{m_{\alpha}}{m_{\beta}}\frac{v_{\mathrm{th},\alpha}}{v_{\mathrm{th},\beta}}\frac{C_{1}(u_{\beta})}{k_{\|}^{2}\lambda_{\beta}^{2}}\right)\left(1+\sum\nolimits_{\gamma}\left(\frac{C_{2}(u_{\gamma})}{k_{\|}^{2}\lambda_{\gamma}^{2}}\right)\right)^{-1} \quad (11)$$

where $u_\alpha \equiv \omega / k_{\|} v_{\text{th},\alpha}$, $v_{\text{th},\alpha}$ is the thermal speed of the plasma species (ions and electrons), $C_1(u_\alpha) \equiv \left(2u_\alpha + 2u_\alpha^2 Z(u_\alpha) - Z(u_\alpha)\right)$, $C_2(u_\alpha) \equiv \left(1 + u_\alpha Z(u_\alpha)\right)$, $\lambda_\alpha \equiv \sqrt{T_\alpha / 4\pi n_\alpha^{(0)} q_\alpha^2}$ is the Debye length of the species and $Z(u_\alpha)$ is the plasma dispersion function.

The system of coupled equations (6) is numerically tractable and we solved it by an iterative method. As a specific case, we evaluated the spectral broadening and pump depletion of LH waves injected at 8 GHz in FTU. The coupled power density is 3 kW/cm^2, the peak parallel refractive index $n_{\|\text{o}} = k_{\|\text{o}} c / \omega_\text{o} = 1.85$. Realistic density and temperature profiles in front of the LH launcher and a white noise of the RF generator at -60 dB relative power are considered.

As a result, LH spectral broadening, both in frequency and in $n_{\|}$ is predicted and a pump depletion up to 30% (of the total LH coupled power) might occur [10].

3. Parametric Analysis

We analyzed the intensity of the nonlinear effects at the antenna layer for *i*) different ion mass species, *ii*) different ion-electron temperature ratio $\tau = T_\text{i} / T_\text{e}$, *iii*) different electron temperatures T_e, *iv*) different density gradients.

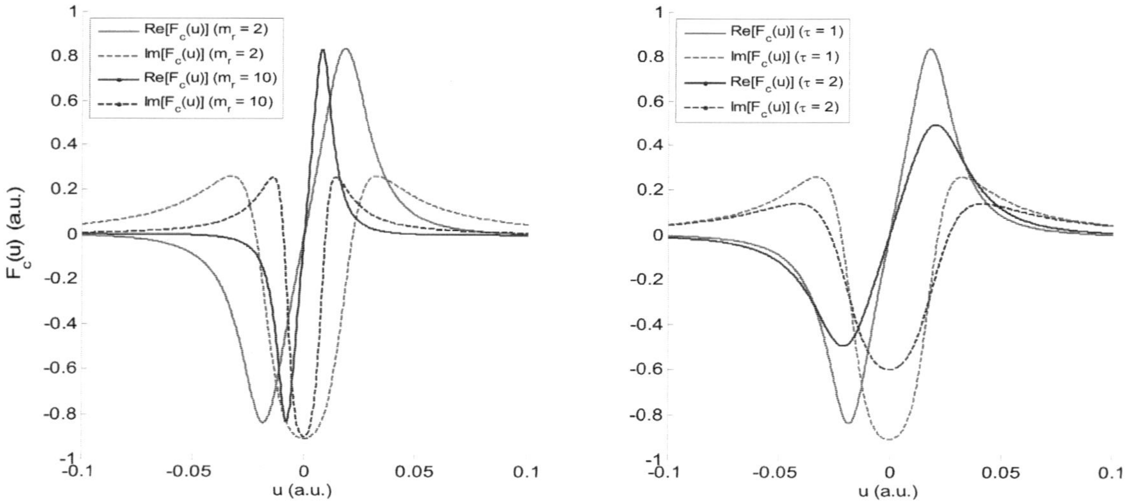

Figure 1. Analysis of the effects of the ion mass (left) and the ion temperature (right) on the coupling function.

In figure 1, on the left side, we compare the coupling function for a Deuterium plasma $(m_\text{r} \equiv m_\text{i}/m_\text{H} \approx 2)$ with single ionized Boron plasma $(m_\text{r} \approx 10)$; on the right side, we compare two cases with different ion temperatures $(\tau = 1, 2)$ for a Deuterium plasma. The comparison suggests that it can be very beneficial to operate with a hot heavy ion plasma in front of the LH antenna to reduce the nonlinear spectral broadening. For larger ion mass the coupling function is localized in a smaller spectral region (expressed in terms of u) thus a reduced spectral broadening is expected to occur, For larger ion/electron temperature ratio the magnitude of the coupling function is reduced. Therefore, the amplitude of the sidebands will be reduced, though the nonlinear interaction occur in the same spectral range. This result is not in contradiction with the experimental results of reduced spectral broadening obtained by increasing the electron temperature. It suggests that with $\tau > 1$ the amplitude of the sideband can be further reduced.

Numerical analysis suggests that the terms proportional to the gradient of the electric field do not have a leading role in determine the solution of the nonlinear system (6) in the range of plasma parameters considered here. Therefore, to evaluate the importance of nonlinear effect, we introduce a nonlinearity parameter ρ_{NL} defined as follows:

$$\rho_{\mathrm{NL}} \equiv \max\left(\left| \frac{\sigma(x,\omega,k_{\|})}{\beta_{\mathrm{L}}(x,\omega,k_{\|})} \right|_{x=0} \right) \quad \forall \omega, k_{\|} \tag{12}$$

We calculate the nonlinearity parameter at the antenna layer (x = 0) because the nonlinear effects are found most important in such low temperature plasma.

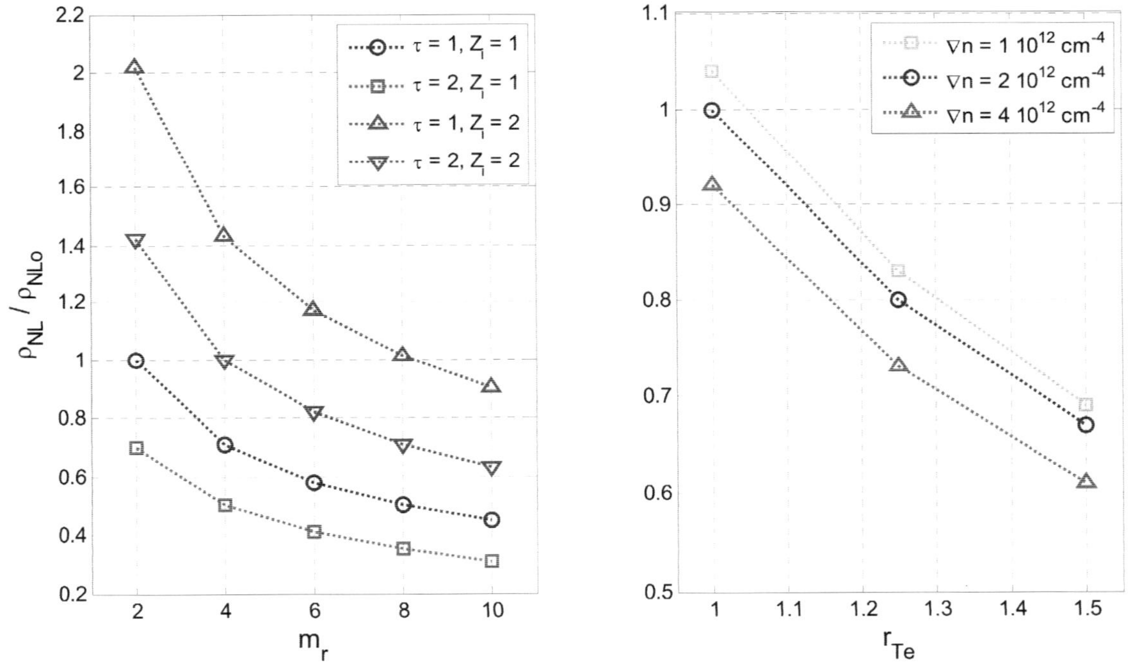

Figure 2. The ratio of the nonlinearity parameters $\rho_{\mathrm{NL}}/\rho_{\mathrm{NLo}}$ where ρ_{NLo} is calculated for a reference Deuterium plasma is plotted (left) versus the ion mass in units of Hydrogen mass ($m_{\mathrm{r}} \equiv m_{\mathrm{i}}/m_{\mathrm{H}}$), for different values of the ion-electron temperature ratio τ and ionization charge Z_{i}; on the right, $\rho_{\mathrm{NL}}/\rho_{\mathrm{NLo}}$ is plotted versus the ratio $r_{T_{\mathrm{e}}}$ of the electron temperature with the electron temperature of the reference plasma at the antenna layer, for different density gradients.

In figure 2 we plot the ratio $\rho_{\mathrm{NL}}/\rho_{\mathrm{NLo}}$, where ρ_{NLo} is the nonlinearity parameter for a reference Deuterium plasma, versus the ion mass (left side) and versus the electron temperature (right side), for different values of the ion-electron temperature ratio τ, ionization charge Z_{i} and density gradient. The

Journal of Physics: Conference Series **401** (2012) 012016 doi:10.1088/1742-6596/401/1/012016

reference Deuterium plasma parameters are $\tau = 1$, $n_e(x=0) = 2\times10^{12}\,\mathrm{cm}^{-3}$, $\nabla n_e(x=0) = 2\times10^{12}\,\mathrm{cm}^{-4}$, $T_e(x=0) = 10\,\mathrm{eV}$, $\nabla T_e(x=0) = 6\,\mathrm{eVcm}^{-1}$.

These parameters correspond to typical SOL profiles in FTU [12,13]. However, in order to show the effect of a strong density gradient, we increased the typical value observed in the experiments (with characteristic e-folding of the order of 1.5-2.0 cm) by a factor 2-4.

This analysis suggests that the nonlinearity can be strongly attenuated operating with hot(τ>1) heavy ions with single ionization and hot electrons along with a steeper density gradient at the antenna mouth.

The effect of higher ionization degree of impurity ions, as shown in figure 2, is to increase the nonlinear parameter. Therefore, the impurity seeding with the aim to reduce the spectral broadening is most effective at large atomic mass and low degree of ionization.

Numerical solutions of the system (6) suggest that the nonlinear effects are significant only if $\rho_{\mathrm{NL}} > 1$. Therefore this condition can be used to evaluate a threshold power density for the onset of the LH nonlinear behavior. For a monochromatic pump wave, the power flux threshold necessary for nonlinear spectral broadening, when $\omega_{\mathrm{pe}}^2/\omega_{\mathrm{o}}^2 \approx 2$ and $\tau = 1$ is of the order of $P_{\mathrm{th}}(\mathrm{kW/m^2}) \approx (n_{\|\mathrm{o}}^2 - 1)^{-1} \times |\Delta n_{\|}| \times T_e^{1.5}(\mathrm{eV}) \times f_o^2(\mathrm{GHz})$, where $|\Delta n_{\|}| \equiv |n_{\|} - n_{\|\mathrm{o}}|$ with $|\Delta\omega| \equiv |\omega - \omega_o| \approx (\omega_o/c)\times|\Delta n_{\|}|\times v_{\mathrm{th,i}}$. It is worth to note that a more general expression for sigma (not reported here) is necessary to take into account the LH pump power spectrum in frequency and wavenumber. In this case, the parameter sigma involves a sum over all the spectral components. For typical LH launched spectra, numerical analysis indicates that the power flux necessary for the onset of nonlinear behavior is increased at least by a factor 2 with respect to a pure monochromatic pump. In addition, we have neglected here the effect of the density gradient (as it is marginal with respect to the other relevant parameters). Thus the above threshold should be considered as a lower bound to the actual power flux threshold for the onset of nonlinear phenomena.

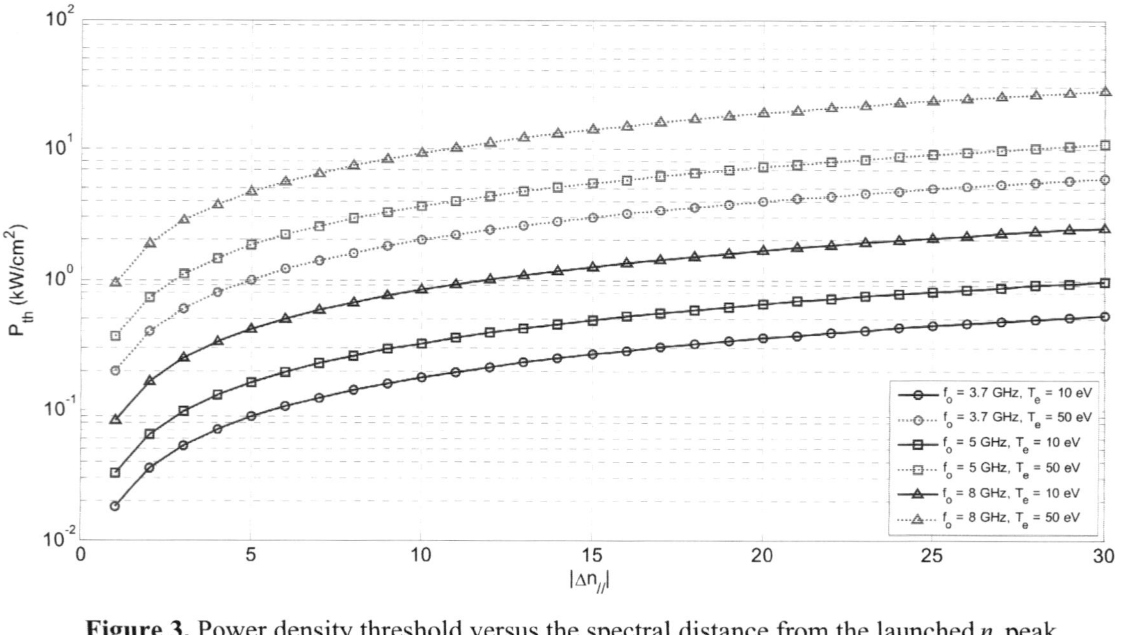

Figure 3. Power density threshold versus the spectral distance from the launched $n_{\|}$ peak.

In figure 3 the predicted power density threshold is plotted, for a launched $n_{\parallel}$ spectrum peaked in $n_{\parallel o} = 1.85$, versus the $n_{\parallel}$ distance from the pump, for three typical LH operating line frequencies (JET 3.7 GHz, ITER 5 GHz, FTU 8 GHz) and for two different electron temperatures (10 and 50 eV) of a reference Deuterium plasma with $\tau = 1$ and plasma density such that $\omega_{pe}^2 / \omega_o^2 \approx 2$.

It is worth to note that in a wide spectral range the power density threshold is lower than the typical value of 1 kW/cm^2 for LH wave launchers. A more accurate evaluation of the power density threshold can be obtained by the full wave solutions of the system (6). We do not discuss here the numerical solutions that we obtained, as they depend on specific experimental scenarios. The aim of the analysis proposed here is to discuss the general dependence of the LH nonlinear behavior on the plasma and RF parameters.

4. Conclusions

In [10] we derived a system of nonlinearly coupled LH wave equations describing the amplification of the LH sidebands and the depletion of the pump wave, considering the noisy background of the RF source. Full wave numerical solutions indicate that pump depletion and spectral broadening in frequency and wavenumber might significantly reduce the CD efficiency when operating at reactor grade high plasma density. In the present paper we analyzed the nonlinear LH behaviour by means of a parametric analysis. As a result, we found that the nonlinear effects could be prevented by increasing the edge electron temperature, the relative ion-electron temperature, the edge density gradient and the ion mass.

In particular, for ITER ($f_o = 5$ GHz, assuming $n_{\parallel o} = 1.85$ and 1 kW/cm^2 power density) if the electron temperature is ≥ 50 eV at the antenna mouth, the sidebands with spectral components $n_{\parallel} \geq 5$, emerging from the noisy background of the RF generator, cannot be amplified (in figure 3 it is shown that in this case the power density threshold to amplify sidebands with $|\Delta n_{\parallel}| \geq 3$ is greater than 1 kW/cm^2).

References

[1] Liu C S and Tripathi V K 1986 *Physics Reports* **130** 3 143
[2] Cesario R et al 2004 *Phys. Rev. Lett.* **92** 175002
Cesario R et al 2006 *Nucl. Fusion* **46** 462
[3] Stix T H 1965 *Phys. Rev. Lett.* **15** 878
Karney C F 1979 *Phys. Fluids* **22** 2188
[4] Porkolab et al 1977 *Phys. Rev. Lett.* **31** 230
[5] Fisch N J 1987 *Rev. Mod. Physics* **59** 175-234
[6] De Marco F 1985 *Course and Workshop on Applications of RF waves to Tokamak Plasmas* ed S Bernabei U Gasparino and E Sindoni Varenna (Italy) International School
[7] Cesario R et al 2010 *Nature Comms.* **1** 5 55
[8] Markov G A, Mironov V A and Sergeev A M 1979 *JETP Lett.* **29** 617
[9] Castaldo C et al 1997 *Phys. Lett.* A **230** 336
[10] Napoli F et al 2012 Modeling of the nonlinear mode coupling of lower hybrid waves triggered by a RF noisy background in tokamak plasmas, submitted to *Plasma Phys. Control. Fusion*
[11] Stix T H 1992 *Waves in Plasmas* (New York: Springer)
[12] Apicella M L et al 2005 *Nucl. Fusion* **45** 685–693
[13] Apicella M L et al 2007 *Journal of Nuclear Materials* 363–365 1346–1351

Evidence of Neoclassical Toroidal Viscosity on the Neoclassical Tearing Modes in TCV tokamak

S Nowak[1], E Lazzaro[1], O Sauter[2], G Canal[2], B Duval[2], L Federspiel[2], A N Karpushov[2], D Kim[2], H Reimerdes[2], J Rossel[2], D Wagner[2] and the TCV Team

[1]*Istituto di Fisica del Plasma IFP-CNR, Associazione EURATOM-CNR sulla Fusione, 20125 Milano, Italy*
[2]*Ecole Polytechnique Fédérale de Lausanne (EPFL), Centre de Recherches en Physique des Plasmas (CRPP), Association EURATOM-Confederation Suisse, 1015 Lausanne, Switzerland*

E-mail: nowak@ifp.cnr.it

Abstract. The interplay between the plasma toroidal rotation and the onset of magnetohydrodynamics instabilities, such as the Neoclassical Tearing Modes (NTMs), is an important issue for tokamak performance. An interesting mechanism characterizing this interaction is the breaking of axisymmetry due to the NTM helical structure, which is the source of a magnetic viscous drag parallel to the toroidal field. This effect, known as Neoclassical Toroidal Viscosity (NTV) depends on magnetic island width, and is responsible of the nearly global slowing down of the toroidal velocity across the profile. In the TCV tokamak the spontaneous plasma toroidal rotation profile, observed even in absence of other external momentum sources [1], can be modified by nearly central electron cyclotron heating (ECH) with a slight poloidal asymmetry and current drive (ECCD) [1,2,3]. The evidence of NTV effect on the plasma toroidal velocity profile of TCV is apparent as a pronounced flattening at the onset of m/n=3/2 and 2/1 tearing instabilities in the neoclassical regime in TCV discharges (I_p~150 kA, n_{e_av}~2 10^{19} m^{-3}, Te~3 keV) with 1.5 MW EC ramp up/down phases. Comparison of the measured and calculated toroidal plasma velocity is performed using the NTV formulation [4,5] applicable in the collisionless regimes. The different aspects of the NTM onset associated both with the ECH-coECCD effect on the current profile and with NTV observed in several TCV discharges are discussed, in the frame of classical and neoclassical tearing modes theory applied to 3/2 and 2/1 modes.

1. Evidence of NTV braking

The damping of the plasma toroidal rotation at the onset of magnetohydrodynamics instability, such as the NTMs, is an interesting effect expected when either or both resonant and nonresonant magnetic perturbations break the toroidal symmetry. The general fluid equations for electrons and ion species, include viscous force terms, given by the divergence of the viscous stress tensor that depends significantly on the collisionality regime and on the structure of the magnetic field. In the important special case of axisymmetric magnetic equilibria (e.g. tokamaks) there is *no* toroidal viscosity and the poloidal and perpendicular (radial) viscosities damp the plasma (ion) flow. Conversely, in non axisymmetric configurations there are both poloidal and toroidal viscous forces braking the flow in those directions. The perpendicular diffusion of the toroidal momentum is a transport effect depending on the collisional or turbulent regime, while the toroidal magnetic viscosity (NTV) depends both on collisionality and on the amplitude of the magnetic perturbation. New experimental evidence of the interplay between the plasma rotation and the presence of m/n=3/2 and 2/1 tearing instabilities in the neoclassical regime is observed in a series of highly reproducible TCV discharges in the same plasma conditions (I_p~150 kA, $n_{e,av}$~1.5 10^{19} m^{-3}, T_e~3 keV. T_i~0.25 keV, q_{95}~5.8) with slightly different 1.5 MW EC ramp up/down phases. The evolution of the 3/2 modes is characterized, initially, by a conventional growth rate driven by the ECH modifications of the plasma current profile, followed by a neoclassical behaviour during the EC power ramp, with subsequent onset of a 2/1 mode. Evidence of an associated neoclassical toroidal viscous torque is observed as a pronounced flattening and global braking of plasma rotation at the onset of these tearing instabilities 3/2 and 2/1. In Fig.1(left) the EC power evolution is shown for 2 sets of shots: #45225, #45226 with slow ramp up power along ~ 1 sec and #45246, #45247 with faster ramp up in ~ 0.2 sec and a slow ramp-down: the 3/2 onset is observed at 0.8 MW and the 2/1 onset at 1.25 MW in all these shots. Self-stabilization occurs at less than 0.4MW. In Fig.1(right) the spectrograms of #45226 and #45247 indicate the 3/2 (light lines around 11-12 kHz and the 2/1 (intense lines) at 4.5-5 kHz evolutions. In the latter one at the ramp down (~ 1.2 sec) the mode frequency slightly increases until the turn off of the power at β_p<0.5.

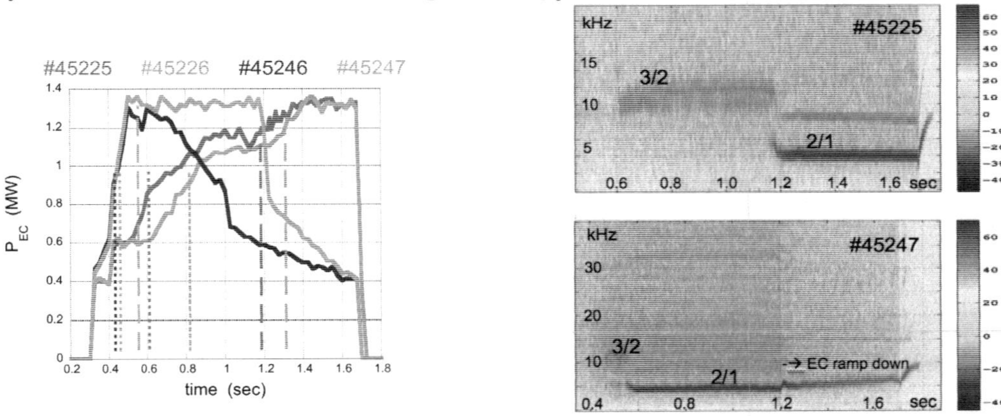

Figure 1. (left) EC power time traces and 3/2 (dotted lines), 2/1(dashed lines) onset for #45225, #45226, #45246, #45247. (right) Spectrogram of 3/2, 2/1 modes in #45226 and #45247 shots.

The time evolution of the toroidal v_φ profiles and of v_φ at different radial locations are shown in Figs.2-3 for #45225 and #45247, respectively. The natural intrinsic counter rotation in absence of external drive (t<0.3 s) changes when central EC heating and co-ECCD are added (t>0.3 s), as already observed in TCV [1]: the plasma is accelerated in the co-I_p direction and the toroidal rotation is reduced. At the onset time of 3/2 mode (t=0.605 s, t=0.555 s for #45225, #45247 respectively) the rotation is significantly modified and changes in sign. At this stage, the mode growth appears as a conventional tearing evolution driven by co-ECCD. When the mode appears, it creates a extremum near the mode surface (helping also to determine the mode location), which results in a local flattening. This interplay between the plasma toroidal rotation and the tearing onset is observed in the modification of the rotation profiles v_φ evolving in co-current direction, in particular outside the mode location. It is seen that the gradient inside the mode location reverses sign [1]. At the triggering of 2/1 mode (t=1.18 s, t=0.555 s for #45225, #45247) a global rotation braking towards the counter direction with the strongest deceleration at q=2 location is present as well. These kinds of braking affecting the whole v_ϕ profile could be related to the NTV effects. The aim of this paper is to attempt to describe the plasma rotation braking at the 3/2 and 2/1 positions as due to the NTV torque, expressed mainly through magnetic field resonant radial perturbations associated to the presence of magnetic island, in the TCV collisionless regime.

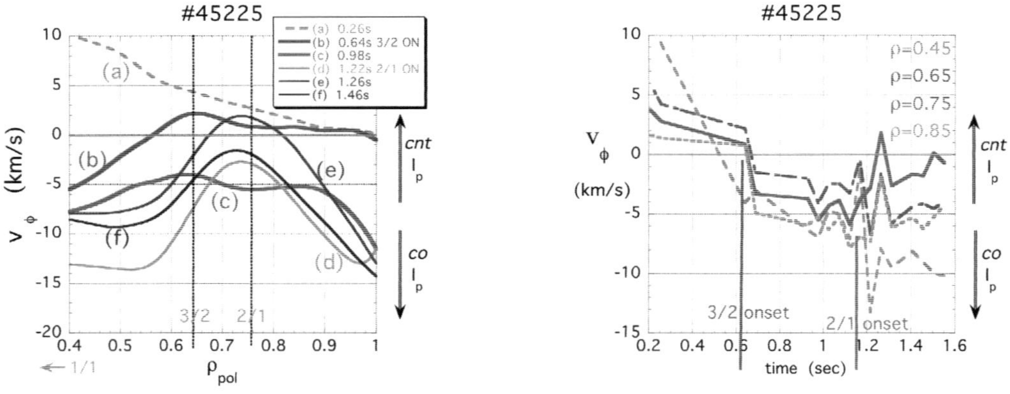

Figure 2. (left) v_ϕ profiles versus time for #45225. (right) v_ϕ evolution versus radii for #45225.

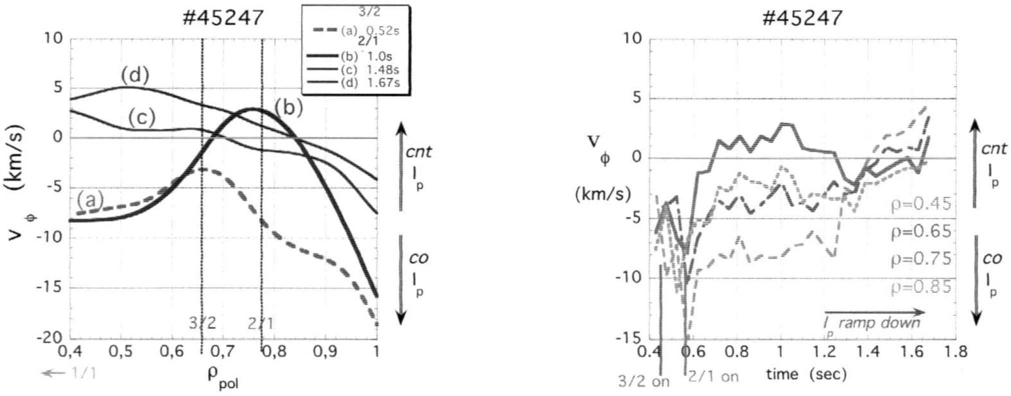

Figure 3. (left) As **Figure 2** (left) for #45247. (right) As **Figure 2** (right) for #45247.

2. Interpretation model of NTV effect

The observed damping of toroidal rotation in these TCV experiments can be mainly related to the action of *resonant* helical magnetic perturbations, with magnetic islands formation, with some contribution of *non resonant* perturbations. Such symmetry breaking magnetic perturbations 'modulate' the tokamak magnetic field, taking into account the tokamak magnetic field modulus. The magnetic field modulus, decomposed into helical harmonics, can be expressed in the Hamada coordinates by [7,8,9,10]:

$$\begin{aligned}|\mathbf{B}(\mathbf{r}+\underline{\delta \mathbf{r}})| &\approx |\mathbf{B}_0(\mathbf{r})| + \delta\mathbf{r}\cdot\nabla B_0 + \ldots \\ &\approx B_0\left(1-\varepsilon\cos\vartheta + \frac{1}{B_0}\sum_{m,n}[b_{m,nc}\cos(m\vartheta - n\varphi) + b_{m,ns}\sin(m\vartheta - n\varphi)]\right)\end{aligned} \quad (1)$$

where $\delta\mathbf{r}$ is the fluid displacement vector, $\varepsilon=r/R_0$, R_0 the major radius, $b_{m,nc}$ and $b_{m,ns}$ are the even and odd Fourier coefficients of the magnetic perturbations. The magnetic field toroidal non-uniformity $\partial B/\partial\phi \neq 0$ due to the perturbed field causes a non-ambipolar current in the minor radius r direction and a toroidal braking force proportional to the square of the fluid displacement $|\delta\mathbf{r}|$. In a fluid description the effect appears as a neoclassical *toroidal* viscous force (NTV) proportional to $\sim <B_\varphi\cdot\nabla\cdot\boldsymbol{\pi}_i>$ [4,8]. The contribution of the NTV effect to the evolution of the toroidal flow velocity v_φ is given by [6,7]:

$$\frac{\partial v_\phi}{\partial t} = \frac{\partial}{\partial t} < e_\phi \cdot v_\phi > = - < \frac{1}{\rho_M} e_\phi \cdot \nabla \cdot \pi_i > = v_{//}^{\lambda}\,(b^\lambda)^\alpha (v_\phi - v_{\phi,0}^{\lambda}) \quad (2)$$

where e_φ is unit vector in the toroidal direction, ρ_m the mass density, π_i the ion viscous stress tensor, being <…> the flux surface average. The term $v^\lambda_{//}$ is a frequency linked to the ion NTV viscosity and $v^\lambda_{\varphi,0}$ indicates the neoclassical toroidal flow velocity. This neoclassical rotation profile is comparable to the gradient of the ion temperature T_i of the ion species. The term b^λ represents the dimensionless effective magnetic field perturbation proportional to the $b_{m,n}$ in eq.(1). The terms $v^\lambda_{//}$, $v^\lambda_{\varphi,0}$ and b^λ refer to a given collisionality regime λ characterizing the plasma. The NTV force brakes the plasma momentum in the toroidal direction with a mechanism different from the traditional *perpendicular* diffusion. Non resonant helical perturbations (with kink-like displacement $|\delta\mathbf{r}|/R_0 \sim b_{m,n}/B_0$), without formation of magnetic islands, cause a *global* toroidal torque proportional to $(b^\lambda)^\alpha \propto (b_r/B_0)^2$ with $\alpha=2$, while resonant helical perturbations, with magnetic islands formed of full width w (where the displacement is $|\delta\mathbf{r}| \sim w$), provide a toroidal braking proportional to $(b^\lambda) \propto (w/R_0)^2$ with $\alpha=1$ [10], more localized near the q rational surfaces.

In the TCV experiments the relevant low collisionality appears to be in the banana regime $\lambda=\nu$ with frequencies in the range $\nu_{ii}/\varepsilon \leq q\omega_E$ [7], as shown in Figs. 4(a),(b),(c),(d), where ω_E is the ExB drift frequency, ν_{ii} the ion-ion collision frequency, ε the inverse local aspect ratio. In few very narrow time intervals the other banana regime $\lambda=1/\nu$ appears as well, with $\nu_{ii}/\varepsilon < q\omega_E < \omega_{ti}\,\varepsilon^{0.5}$, being $\omega_{ti} = v_{th,i}/(R_0 q) = (T_i/m_i)^{0.5}/(R_0 q)$ the ion transit frequency and $v_{th,i}$ the ion thermal speed.

In this regime, where the particles trapped in the banana orbits are collisionless and the ion neoclassical toroidal viscosity is dominant, the effective toroidal damping rate $v^\lambda_{//}$ in eq.(2) is given by $v^\nu_{//} = \nu_{ii}\,\omega^2_{ti} / \omega^2_E$, taking for the effective dimensionless amplitude b^ν of the field perturbation the expression:

$$(b^\nu)^\alpha \propto \sum_{m',n'} (b_{m',n'}(r)/B_0)^2 + \sum_{m.n}(b_{m,n}(r)/B_0) \quad (3)$$

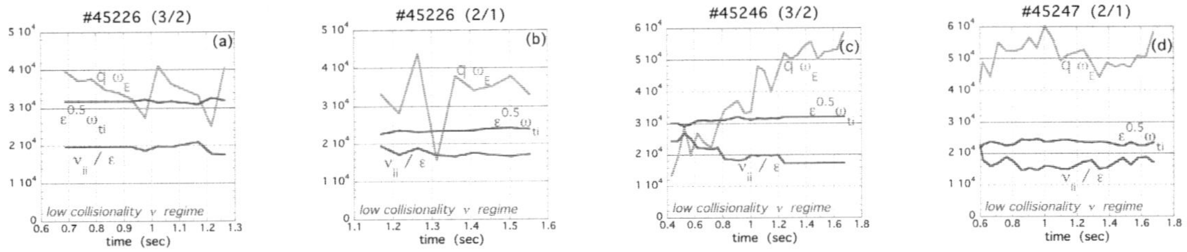

Figure 4. Evolution of frequencies characterizing the low-collisionality ν regime for #45226, #45246, #45247 at 3/2 and 2/1 mode locations. The collisionality 1/ν regime appears in very narrow time intervals, as shown in (b) and (c).

In eq.(3) m', n' are the poloidal and toroidal mode numbers of non resonant perturbations, while m, n, are those of the resonant perturbations associated to the full island width $w_{m,n} = 4 (b_{m,n})^{0.5} (R_0 L_q q /m)^{0.5}$ of an (m/n) mode. The $b^{\vee}$ term is calculated taking into account the experimental measured Mirnov signals and the experimental safety factor profile q. Since the temperature of the main deuterium ion species is considered equal to the temperature of the carbon ions, the radial profile of the carbon rotation is used for the evaluation of the experimental offset velocity $v^{\nu}_{\phi,0}$, comparable to the ion diamagnetic rotation speed v_d proportional to the ion temperature gradient through the constant $k_c = v^{\nu}_{\phi,0} / v_d$:

$$v^{\nu}{}_{\phi,0}(r) = k_c v_d(r) = k_c \frac{1}{Z_i e B_\theta} \frac{dT_i}{dr} \qquad (4)$$

with $Z_i = 1$ is the charge of the main deuterium species and B_θ the poloidal magnetic field. In this ν regime the theoretical limit of the value of k_c is 0.92 [7]. This offset velocity $v^{\nu}_{\phi,0}$ is comparable with the measured rotation near the zero torque dv_ϕ/dt. In particular, it should be noted that $v^{\nu}_{\phi,0}$ is experimentally observed to be also associated to the ion diamagnetic velocity $v_{d,p}$, expressed in terms of the ion temperature and the gradient of the ion pressure:

$$v^{\nu}{}_{\phi,0}(r) = v_{d,p}(r) = \frac{k_c\, m\, T_i}{Z_i e\, r_s B_0} \frac{\nabla p_i}{p_i} \frac{R_0}{n} \qquad (5)$$

with p_i the ion pressure and r_s the minor radius at the resonant q. Both these offset velocities $v^{\nu}_{\phi,0} = k_c v_d$ and $v^{\nu}_{\phi,0} = v_{d,p}$ have been considered in eq.(2) to model the plasma toroidal rotation evolution. Tests of local rate of decay of $v_\phi(r_s)$ according with the NTV model of eq.(2) are performed near the rational surfaces $r=r_s$.

3. *Simulations of NTV braking*

Magnetic braking of the NTV kind, associated to the appearance of few low order tearing modes, seems to affect significantly the plasma toroidal rotation profile in presence of tearing mode in these experimental TCV scenarios. In Figs.5-9 we show the satisfactory agreement between the toroidal velocity decay rate dv_ϕ/dt, using eq.(2) including the 2 expression (4)-(5) for $v^{\nu}_{\phi,0}$, and the experimental one. In our calculations the best simulations have been obtained with $|k_c|$=0.12-0.35 to be compared with the theoretical value $k_c \approx 0.92$ [7] in the ν regime limit. Our values fall below the neoclassical collisionality ν-regime limit confirming that our estimation is consistent with the neoclassical prediction. It should be noted that in these experiments the plasma toroidal velocity v_ϕ and the neoclassical toroidal flow velocity (offset rotation) $v^v_{\phi,0}$ are typically in the *co* plasma current direction (negative sign) in the plots. We found that the offset rotation (~ -2-4 km/s) is comparable to the ion diamagnetic speed, as predicted by the neoclassical theory. In Figs.5-6 the evolution rate at 2/1 resonant location are shown for #45225 and #45226, respectively. Both resonant helical magnetic perturbations from 2/1 mode and non resonant from 3/2 island contribute in eq.(2) to the NTV torque using eq.(3). Before 2/1 onset only the non resonant perturbation by 3/2 instability plays a role, while after this onset both resonant and non resonant perturbations can contribute. The measured $dv_\phi/dt=0$ gives an estimate of the equilibrium plasma rotation in terms of an offset toroidal rotation comparable to the ion diamagnetic speed, in agreement with [7].

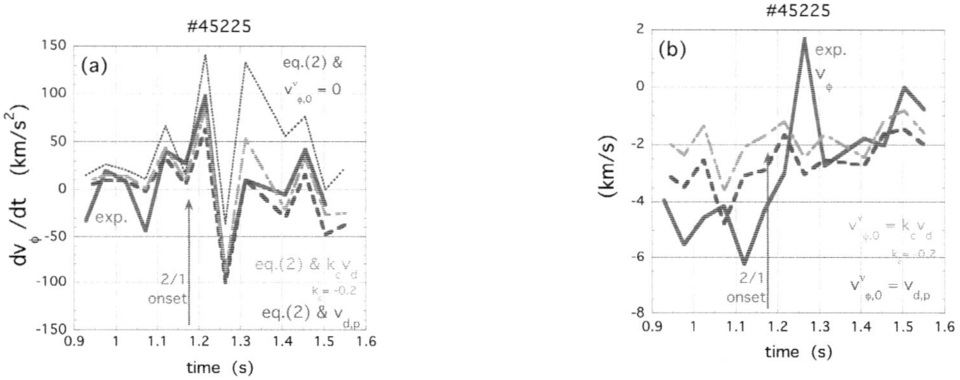

Figure 5. (a) Measured (solid line) and calculated by ion NTV drag (dashed, dotted lines) evolution rate of v_ϕ at q=2 for #45225. (b) Experimental time velocity evolution v_ϕ at q=2 and offset rotations by eqs.(3)-(4) for discharge #45225.

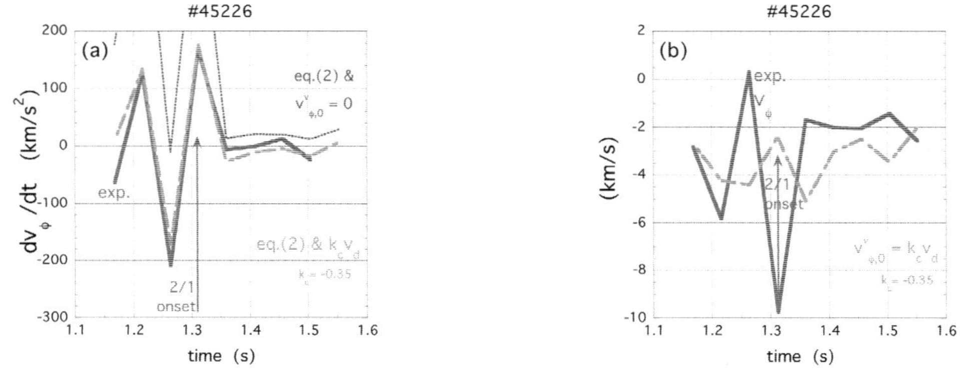

Figure 6. (a) As **Figure 5** (a) for discharge #45226. (b) As **Figure 5** (b) for discharge #45226.

In Fig.7 the growth/decay rate of the toroidal velocity is shown at 3/2 resonant position. The NTV model of eq.(2) reproduces well the measured evolution rate and the offset $v^v_{\phi,0}$ is still observed in *co* plasma current direction (negative sign). A very interesting feature appears modeling the v_ϕ rate for the #45246 (with 3/2) and #45247 (with 2/1) where the modes occur on the fast EC ramp up phase and evolve from the power plateau until the end of the ramp down phase. The mode dynamics is different from the evolution of instabilities in the 2 previous shots. Before the EC ramp down the toroidal rotation profiles are decelerated towards the counter direction up to a nearly zero rotation (#45246, #45247), while in the ramp down phase the rotations are pushed to other constant velocities, in counter direction for #45246 and co direction for the #45247, as shown in Figs. 8-9. So, different offsets seem to characterize the plasma in the EC ramp up and plateau phase and in the power ramp down. Best fits of the evolution rate of the toroidal velocity are obtained using both co- $v^{\vee}_{\phi,0}$ and $v^{\vee}_{\phi,0} = 0$ in the former phase and $v^{\vee}_{\phi,0} \neq 0$ with co/cntr $v^{\vee}_{\phi,0}$ in the latter one.

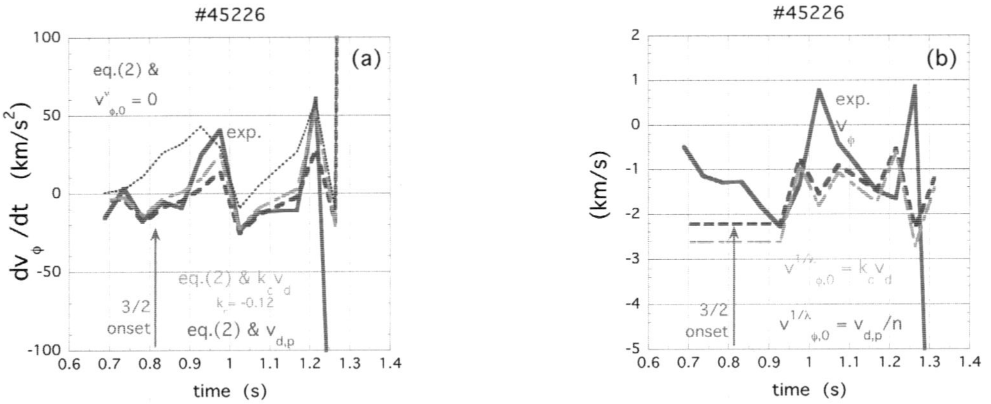

Figure 7. (a) As **Figure 5** (a) for discharge #45226 at q=3/2. (b) As **Figure 5** (b) for discharge #45226 at q=3/2.

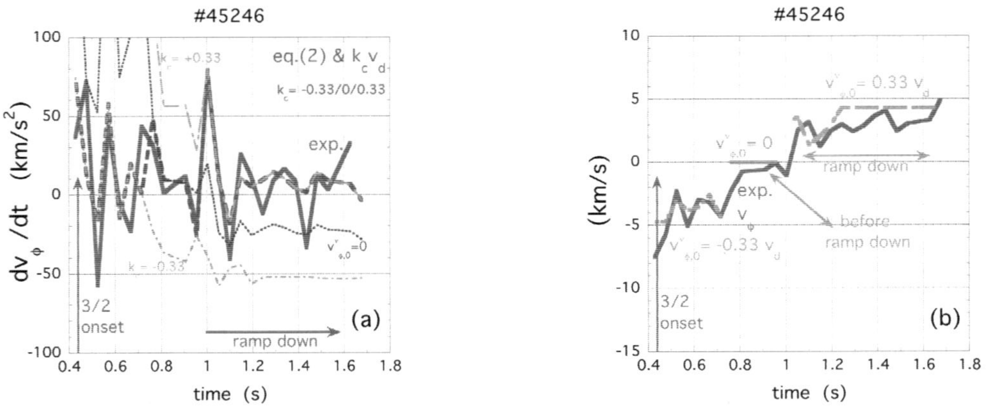

Figure 8. (a) As **Figure 7** (a) for discharge #45246 at q=3/2. Effects of $v_{\phi,0}$ =0 (dotted line) and $v_{\phi,0}$ #0 (dashed-dotted line). Best fit (dashed line) with $v_{\phi,0}$ =0 before the EC ramp down. (b) As of **Figure 7** (b) for discharge #45246 at q=3/2. The different offsets are shown:co $v_{\phi,0}$ and $v_{\phi,0}$ =0 before ramp down and $v_{\phi,0}$ #0 in cntr direction during the ramp down.

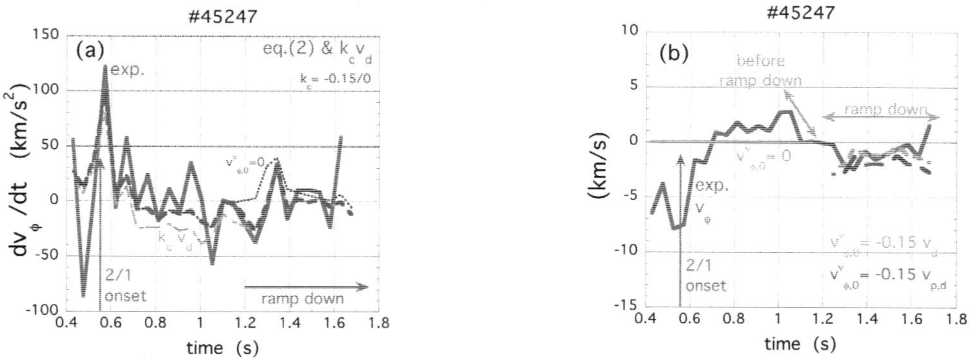

Figure 9. (a) As **Figure 8** (a) for discharge #45247 at q=2/1. (b) As **Figure 8** (b) for #45247 at q=2/1 with $v_{\phi,0}$ #0 in co direction in ramp down.

4. Conclusions

In low collisionality TCV discharges with just tearing and neoclassical tearing instabilities, but no external fields, it has been found that within the uncertainty of the measurements (± 2 km/s) the evolution of the toroidal rotation profile is consistent with the effect of a NTV torque proportional to the difference between the toroidal velocity and the neoclassical toroidal flow velocity $v^{*}_{\phi,0} \neq 0$ (offset speed), evaluated in steady state conditions $dv_{\phi}/dt \sim 0$. In some cases, when the mode is triggered on the fast EC power ramp up, the toroidal offset rotation $v^{*}_{\phi,0} = 0$ is used to better simulate the experimental evolution rate of the toroidal rotation velocity before the EC ramp down. It should be noted that in our simulations we best fit the experimental v_{ϕ} evolution taking into account an offset in co- plasma current direction (as well shown in the plots for evolution), despite the usual counter offset from the neoclassical theory. This can be perhaps due to the presence of resonant magnetic perturbations changing the toroidal rotation sign and further experimental investigations should be performed.

Acknowledgments

We are grateful to the CRPP for the efficient operation of the TCV tokamak. This work, performed under the European Fusion Development Agreement, was supported by the European Communities and Istituto di Fisica del Plasma CNR and the Swiss National Science Fundation.

References

[1] Sauter O *et al.,* 2010 *Proc. of the 23th IAEA FEC* (Daejon, Korea) Paper EXS/P2-17
[2] Reimerdes H, Sauter O *et al.*, 2002 *Phys. Rev. Lett.* **88** 105005
[3] Martinell J J and Gutiérrez-Tapia C 2001 *Physics of Plasmas* **8** 2808
[4] Shaing K C 2003 *Physics of Plasmas* **10** 1443
[5] Lazzaro E, Zanca P 2003 *Physics of Plasmas* **10** 2399
[6] Zhu W, Sabbagh S A, Bell R E *et al.,* 2006 *Phys. Rev. Lett.* **96** 225002
[7] Cole A J, Hegna C C and Callen J D 2008 *Physics of Plasmas* **15** 056102
[8] Smolyakov A I, Hirose A, Lazzaro E *et al.*, 1995 *Physics of Plasmas* **2** 1581
[9] Shaing K C 2002 *Physics of Plasmas* **9** 3470
[10] Shaing K C 2001 *Phys. Rev. Lett.* **87** 245003

Free-boundary perturbed MHD equilibria

C Nührenberg

Max-Planck-Institut für Plasmaphysik, IPP-Euratom Assocation, Teilinstitut Greifswald, Wendelsteinstraße 1, 17491 Greifswald, Germany

E-mail: carolin.nuehrenberg@ipp.mpg.de

Abstract. The concept of perturbed ideal MHD equilibria [Boozer A H and Nührenberg C 2006 *Phys. Plasmas* **13** 102501] is employed to study the influence of external error-fields and of small plasma-pressure changes on toroidal plasma equilibria. In tokamak and stellarator free-boundary calculations, benchmarks were successful of the perturbed-equilibrium version of the CAS3D stability code [Nührenberg C *et al.* 2009 *Phys. Rev. Lett.* **102** 235001] with the ideal MHD equilibrium code NEMEC [Hirshman S P *et al.* 1986 *Comput. Phys. Commun.* **43** 143].

1. Introduction

In the overlap between stellarator and tokamak research, the 3D modeling of the plasma edge for divertor studies is in need of efficient methods treating plasma equilibria under the influence of external error fields. In tokamaks, e.g. the coil ripple due to the finite number of toroidal field coils and, possibly, error fields from ferritic in-vessel components or, in the control of edge-localized modes (ELMs), the applied resonant magnetic perturbations (RMPs) destroy the axisymmetry of the ideal configuration, break up magnetic surfaces, and corrugate the plasma surface. Another example is the island divertor of the Wendelstein 7-X stellarator, where the plasma edge structure is exploited and use is made of the natural island chains conforming with the boundary rotational transform. As a complement to vacuum-field studies, it is important to know how a finite plasma pressure influences the load imbalances on the divertor plates caused by magnetic field perturbations.

Though important, such changes of the magnetic field are small and, therefore, may be treated as perturbations of an equilibrium state. Here, the concept of perturbed ideal MHD equilibria [1] is employed, which was computationally realized in the CAS3D ideal MHD stability code [2] allowing, in its latest version, the study of plasma edge distortions under the influence of external error-fields and small plasma-pressure changes.

For perturbed tokamaks, the IPEC code [3] exists which follows a similar concept, but incorporates a control surface with surface currents to account for external error-fields.

In Secs. 2 and 4 of this paper, stellarator and tokamak applications are shown together with their benchmarks to free-boundary NEMEC [4] results. In Sec. 3 of this paper, a generalization of the vacuum part of the CAS3D code will be described which is necessary for the correct treatment of perturbations which preserve the periodicity of the unperturbed equilibrium.

2. Stellarator-type plasma

This section describes the influence of (i) an external error-field, (ii) a small plasma-pressure changes, and (iii) the combination of both, on a zero-β stellarator.

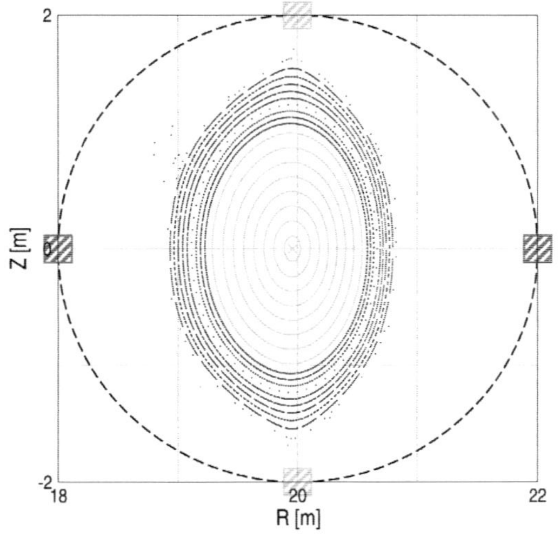

Figure 1. Cross-section at the beginning of a field period for a turning-ellipse $\ell = 2$, zero-plasma-β stellarator equilibrium with large aspect ratio, $A = R/a \approx 23.4$, a low-shear rotational transform, $\iota_{\text{axis}} = 0.23$, and $\iota_{\text{edge}} = 0.244$, and five field periods. NEMEC free-boundary calculation for the plasma domain (green); GOURDON field-line tracing in vacuum field (black). The two stellarator coils are indicated, too.

For the free-boundary benchmarks of the perturbed-equilibrium method with the MHD equilibrium code NEMEC [4], a geometrically simple case has been chosen: a turning-ellipse $\ell = 2$, zero-plasma-β stellarator equilibrium, with large aspect ratio, $A = R/a \approx 23.4$, a low-shear rotational transform, $\iota_{\text{axis}} = 0.23$, and $\iota_{\text{edge}} = 0.244$, and five field periods. The vacuum field for the free-boundary equilibrium calculation is the superposition of the purely toroidal, axisymmetric field of a line-current on the z-axis and equal-magnitude, opposite-sign currents in two continuous stellarator coils each of which makes two toroidal transits before closing on itself after one poloidal transit. This basic setup is shown in Fig. 1.

A deformation of the plasma boundary may be described in terms of the ideal MHD displacement vector, $\boldsymbol{\xi}$, as

$$\delta \boldsymbol{r} = \boldsymbol{r}_{\text{perturbed}} - \boldsymbol{r}_{\text{unperturbed}} = \boldsymbol{\xi}_{\text{n}} = (\boldsymbol{\xi} \cdot \boldsymbol{n})\, \boldsymbol{n} \quad . \tag{1}$$

In Eq. (1), the outer unit normal $\boldsymbol{n}$ points from the plasma boundary into the unbounded vacuum region. The normalized toroidal flux, s, is used as magnetic-surface label, $s = 1$ on the plasma boundary and $F_{\text{T}}(s) = s\, F_{\text{T}}(1)$ the toroidal flux enclosed by a magnetic surface. The normal displacement is $\xi^s = \boldsymbol{\xi} \cdot \nabla s = |\nabla s|\ \boldsymbol{\xi} \cdot \boldsymbol{n}$. In the perturbed-equilibrium calculations, magnetic coordinates, (s, θ, ϕ), are used [2], in which the magnetic field lines are straight. In one poloidal transit, θ increases by unity. In a toroidal transit of one of the N_p field-periods, ϕ increases by unity. For stellarator-symmetric configurations including up-down symmetric tokamaks, the Fourier decomposition of an even-parity normal displacement is $\xi^s(s, \theta, \phi) = \sum_j \widehat{\xi^s}_{m_j n_j}(s) \cos 2\pi(m_j\, \theta + n_j\, \phi/N_p)$.

The superposition of a constant purely vertical external field, $\delta B_z^{\text{ext}} = 0.0004$ T, leads to an inward shift of the plasma column of e.g. 0.023 m at the outside of the vertical ellipse and of 0.04 m at the outside of the horizontal ellipse (Fig. 2, bottom right). The normal component of the error-field on the plasma boundary is the boundary condition of the CAS3D calculation. The corresponding ξ^s harmonics are shown in the top left frame of Fig. 2. The perturbation preserves the periodicity of the configuration, i.e. belongs to the $N = 0$ mode-family. The dominant harmonics are shown and their poloidal (m) and toroidal (n) Fourier indices are indicated. Here, the results of NEMEC and CAS3D differ by 0.0005 m at the outside of the vertical ellipse and by 0.0015 m at the outside of the horizontal ellipse.

The introduction of a small plasma pressure in the $\beta = 0$ configuration, leading to an average plasma-β of $\langle \beta \rangle \approx 0.00016$, results in an outward shift of the plasma column of approximately

Journal of Physics: Conference Series **401** (2012) 012018 doi:10.1088/1742-6596/401/1/012018

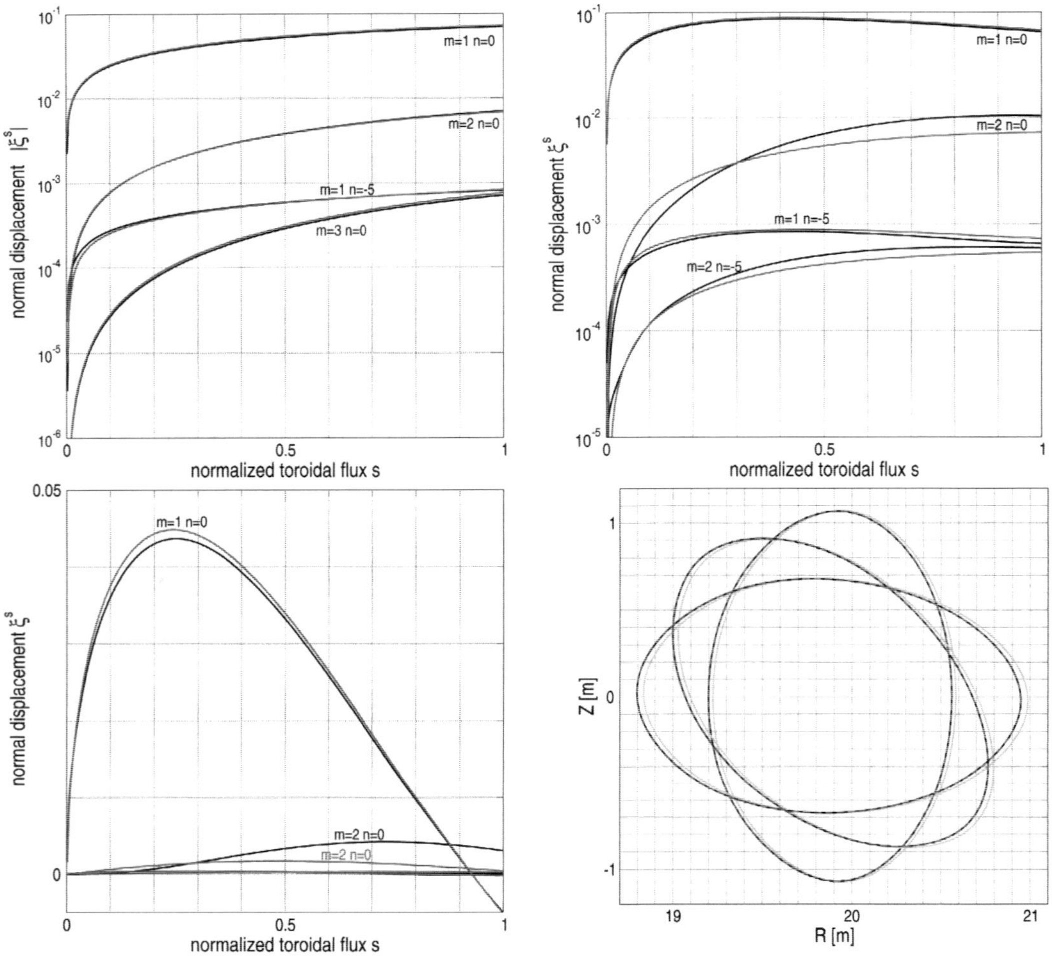

Figure 2. Results for the configuration of Fig. 1: Normal displacement harmonics versus normalized toroidal flux s, describing the influence of an external, inward-shifting $\delta B_z^{\mathrm{ext}}/B_0(\mathrm{axis}) = 0.0004$ (top left, log-scale for the ξ^s-axis), of $\langle\beta\rangle = 0.00016$ (top right, log-scale for the ξ^s-axis), and of the combination of both, external field perturbation and plasma pressure (bottom left, linear scale for the ξ^s-axis). The perturbation belongs to the $N = 0$ mode family, the Fourier indices of the dominant harmonics as indicated. Black: NEMEC code; red: CAS3D code; 301 radial points; 4 of 10 perturbation harmonics are shown. The effect of the $\delta B_z^{\mathrm{ext}}$ perturbation (top left) is also shown in the bottom right frame: Unperturbed and perturbed boundaries at the beginning, quarter, and middle of a field period. Green: unperturbed; black: perturbed, NEMEC; red: perturbed, CAS3D.

the same amount, as may be seen in the top right frame of Fig. 2. Consequently, applying this error-field and this plasma-pressure at the same time leaves the plasma boundary nearly unaffected as can be seen from the nearly vanishing ξ^s-harmonics in the bottom right frame of Fig. 2. The results of the NEMEC code using a non-linear minimization method and of the linear stability CAS3D stability code compare very well when applied for small perturbations.

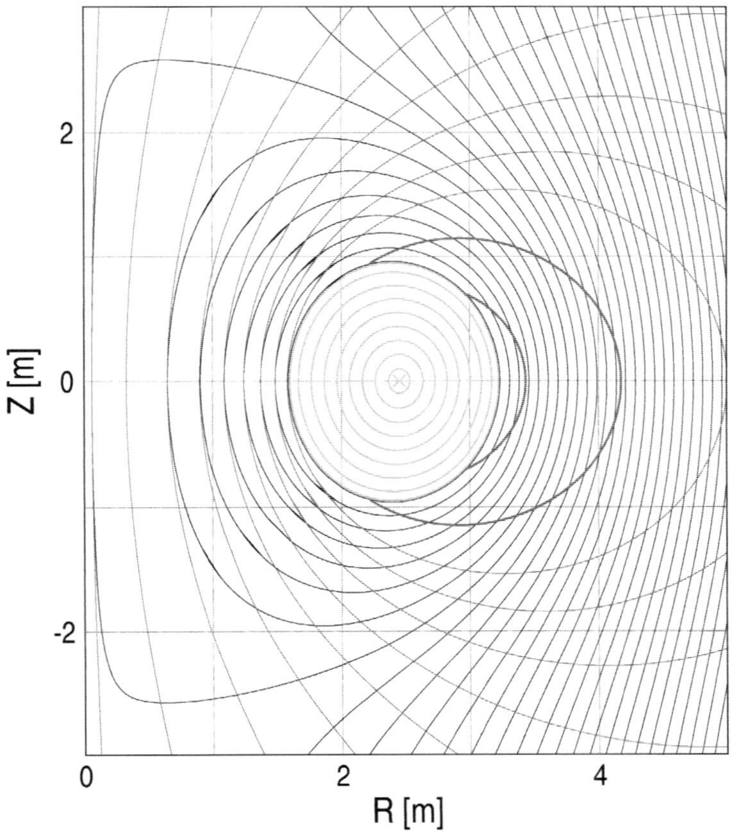

Figure 3. Slightly elongated, force-free, axisymmetric equilibrium, aspect ratio $A = R/a \approx 2.7$, inverse safety factor $\iota_{\text{axis}} = 0.5$ and $\iota_{\text{edge}} = 0.3$. NEMEC free-boundary calculation for the plasma domain (green); GOURDON field-line tracing: plasma-current field in the vacuum domain (EXTENDER, red); poloidal component of equilibrium field (black).

3. Construction of a boundary-conforming vacuum magnetic field

Free-boundary ideal MHD stability studies and perturbed-equilibrium calculations employ the perturbed magnetic field in the vacuum region surrounding the toroidal plasma domain. In contrast to the vector potential, the scalar potential representing the perturbed vacuum magnetic field may be multi-valued because of the non-simply connected vacuum region.

In Ref.[5], Lüst and Martensen describe how to construct the perturbed vacuum magnetic field from the gradient of a single-valued scalar potential fulfilling the boundary condition and a, possibly multi-valued, vector field tangential to the unperturbed plasma boundary. Since it is defined in the vacuum region, the latter vector field must be divergence- and curl-free, and, for uniqueness, normalized to unity net current in the plasma domain. For convenience, some results of Ref. [5] are repeated in Appendix B.

For perturbations which do not preserve the periodicity of the plasma, the multi-valued part does not contribute to the perturbed vacuum energy, and in the case of periodicity-preserving perturbations it acts stabilizing [5, 6], so that it may be omitted from the free-boundary stability calculation.

In the tokamak application of Sec. 4, the poloidal component of the equilibrium magnetic field, $\boldsymbol{B}_0^{\text{pol}}$, is used as *Lüst-Martensen* field. The plasma-current contribution to this field may be calculated using Biot-Savart's law for a surface current density on the plasma boundary, $\mu_0^{-1}\boldsymbol{B}_0 \times \boldsymbol{n}$, with $\boldsymbol{n}$ the outer unit normal pointing into the unbounded vacuum region and μ_0 the permeability of free space. For this procedure, also known as *virtual casing* principle [7], implementations are available in the MFBE [8] and EXTENDER [9] codes, the latter being used

Journal of Physics: Conference Series **401** (2012) 012018 doi:10.1088/1742-6596/401/1/012018

here. In Fig. 3 an example is shown, a slightly elongated, force-free, axisymmetric case with aspect ratio $A = R/a \approx 2.7$, inverse safety factor $\iota_{\mathrm{axis}} = 0.5$, and $\iota_{\mathrm{edge}} = 0.3$.

Alternatively, or in plasma configurations without a net toroidal current, for the construction of the *Lüst-Martensen* field, a line current may be used positioned on the unperturbed magnetic axis. Its magnetic field will have, in general, a finite normal component on the unperturbed plasma boundary. For componsation, a vacuum magnetic field is needed with the normal component equal in magnitude, but opposite in sign. The single-valued scalar potential of this magnetic field can be obtained from the solution of Laplace's equation in the vacuum region with its normal derivative given on the plasma boundary. The NESTOR code implementation of this *exterior Neumann* problem yields the values of the scalar potential on the plasma boundary which is sufficient for the use in 3d ideal MHD equilibrium [4] and stability [10] calculations. As explained in Appendix B, Eq. (B.4), the perturbed-equilibrium application, however, needs to know the scalar potential in a part of the vacuum region. The implementation of Green's representation formula for the exterior Neumann problem will be part of future work.

4. Tokamak-type plasma

As in Sec. 2 for the stellarator case, the perturbed-equilibrium method is used for a force-free axisymmetric case to evaluate the influence of (i) an external error-field, (ii) a small plasma-pressure change, and (iii) the combination of both.

As before, a geometrically simple case has been chosen for the benchmarks with the NEMEC code. The axisymmetric toroidal field of a line-current on the z-axis superposed with a constant field purely in z-direction are used as vacuum field for the free-boundary equilibrium calculation.

The perturbations which are used for this case preserve the axisymmetry of the plasma, as external error-field a small change of the equilibrium B_z-field is used. Therefore, the MHD displacement is axisymmetric, too.

In all CAS3D perturbed-equilibrium computations of this section, the normalized poloidal part of the equilibrium field, i.e. the plasma-current and the coil-current fields, is used as the *Lüst-Martensen* field $\boldsymbol{Y}$ of Eqs. (B.1) and (B.4). Only with the implementation of the $L\,\gamma^2$-term of Eq. (B.5) in the perturbed-vacuum energy does the perturbed-equilibrium calculation agree with the respective non-linear equilibrium result.

In Fig. 4, the normal displacements are shown resulting from the perturbed-equilibrium calculations for the axisymmetric benchmark case of Fig. 3 for the application of an external error-field (top left sub-frame, $\delta B_z^{\mathrm{ext}} = 0.00055$ T), of a small plasma-pressure change (top right, to $\langle\beta\rangle = 0.0003$), and of the combination of both (bottom left). The benchmark results from the NEMEC (black) and CAS3D (red) codes compare well. The change of the vertical field by 1% together with an increase of the initially zero plasma-β to an average of $\langle\beta\rangle = 0.0003$ results in an outward shift of the plasma column of ≈ 0.022 m at the high-field side, and of ≈ 0.033 m at the low-field side of the plasma region. For these shifts, the values obtained by NEMEC and CAS3D differ by ≈ 0.0005 m at the high-field side, and ≈ 0.0027 m at the low-field side. The very good agreement between the corresponding perturbed plasma boundaries from the NEMEC non-linear equilibrium calculation and the CAS3D linear perturbed-equilibrium calculation is shown in the bottom-right subframe of Fig. 4.

5. Summary and outlook

In stellarator and tokamak benchmark cases, the non-linear ideal MHD equilibrium code NEMEC and the linear ideal MHD stability code CAS3D in its perturbed-equilibrium version yield results in the study of small plasma-pressure changes and of small external error-fields that are in good agreement. Therefore, the perturbed-equilibrium method appears to be a safe tool for the study of scenarios which are expensive for equilibrium codes.

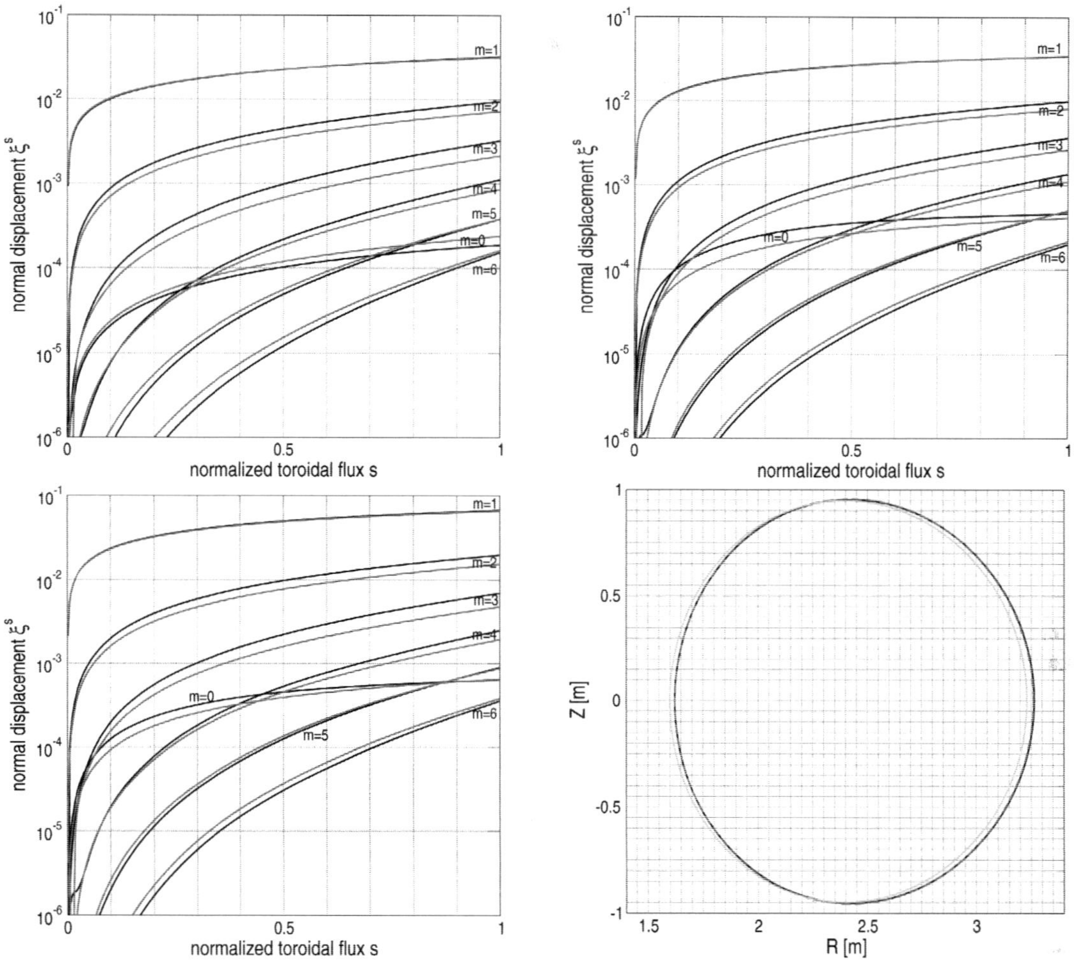

Figure 4. Results for the axisymmetric plasma configuration of Fig. 3: Dominant normal displacement harmonics versus normalized toroidal flux s, describing the influence of an external $\delta B_z^{\mathrm{ext}} = 0.00055$ T (top left), of $\langle\beta\rangle = 0.0003$ (top right), and of the combination of both, external field perturbation and plasma pressure (bottom left). In each of these sub-frames, a log-scale ξ^s-axis is used. The perturbation is purely $n = 0$. Black: NEMEC code; red: CAS3D code; 301 radial points, poloidal perturbation harmonics: $0 \leq m \leq 14$. Unperturbed and perturbed boundaries are shown for the combined perturbation (bottom right); green: unperturbed; black: perturbed, NEMEC; red: perturbed, CAS3D.

The vacuum contribution to the potential energy of the energy principle has been complemented with the multi-valued parts in the perturbed vacuum magnetic field. The free-boundary tokamak calculations show that the inclusion of these *Lüst-Martensen* terms is indeed essential for the correct treatment of symmetry-preserving perturbations with the perturbed-equilibrium method.

It is planned to generalize the implementation of the Lüst-Martensen terms to the stellarator case.

Journal of Physics: Conference Series **401** (2012) 012018 doi:10.1088/1742-6596/401/1/012018

Appendix A. MHD energy principle and perturbed equilibria

In this section, the basic equations of the perturbed-equilibrium method [2] are summarized. The inclusion of an external error-field is described, too.

The determination of perturbed equilibria needs the variation of the ideal MHD energy through second order, so

$$\delta W = \delta^1 W + \delta^2 W = \iiint (\nabla p - \boldsymbol{j} \times \boldsymbol{B}) \cdot \boldsymbol{\xi} \, \mathrm{d}^3 r - \frac{1}{2} \iiint \boldsymbol{\xi} \cdot \mathcal{F}[\boldsymbol{\xi}] \, \mathrm{d}^3 r \quad . \tag{A.1}$$

In the second order term, $\mathcal{F}$ is the ideal MHD force operator acting on the displacement vector, $\boldsymbol{\xi}$. The integrand of the first order term can be rewritten to show the dependence on a change of the plasma pressure, p, and on an external error-field, $\boldsymbol{B}_1^{\text{ext}}$,

$$(\nabla p - \boldsymbol{j} \times \boldsymbol{B}) \cdot \boldsymbol{\xi} = (p'_{\text{perturbed}} - p'_{\text{unperturbed}})\, \xi^s + (\boldsymbol{j}_0 \times \boldsymbol{B}_1^{\text{ext}}) \cdot \boldsymbol{\xi} \quad . \tag{A.2}$$

Equation (A.2) is valid if a perfect MHD equilibrium is perturbed. Primes denote the derivative of a scalar function constant on magnetic surfaces. The unperturbed current density is $\boldsymbol{j}_0$. Stationarity of δW leads to the perturbed equilibrium given by $\mathcal{F}[\boldsymbol{\xi}] = \boldsymbol{g}$. The left-hand side is the center-piece of ideal MHD stability calculations which minimize $\delta^2 W$. The right-hand side, $\boldsymbol{g}$, derives from the first-order term, $\delta^1 W$, which is linear in the displacement. For a free-boundary perturbed equilibrium, no further boundary conditions are used.

Appendix B. Perturbed vacuum magnetic field

In ideal MHD stability, the plasma is treated as a toroidal domain, P, with boundary ∂P, being surrounded by a vacuum region, V. In contrast to standard tokamak stability studies, in the perturbed-equilibrium-method implementation so far no enclosing wall is present, so $V = \mathbb{R}^3 \setminus (P \cup \partial P)$.

In Ref. [5], a method is described and, for convenience, summarized here how to find the, possibly multi-valued, scalar magnetic potential, Φ, for the first-order perturbed magnetic field in the vacuum region,

$$\boldsymbol{B}_{1\mathrm{V}} = \boldsymbol{\nabla}\Phi = \boldsymbol{\nabla}\Phi^* + \gamma \boldsymbol{Y} \quad . \tag{B.1}$$

The split-up into a sum of the gradient of a single-valued scalar potential, Φ^*, and the multiple of a vector field, $\boldsymbol{Y}$, can always be accomplished and may be used as the defining equation for Φ^*. Like $\boldsymbol{B}_{1\mathrm{V}}$, the vector field $\boldsymbol{Y}$ must be divergence- and curl-free, $\boldsymbol{\nabla} \cdot \boldsymbol{Y} = 0$ and $\boldsymbol{\nabla} \times \boldsymbol{Y} = \boldsymbol{0}$ in V. Furthermore, it must be tangential to the plasma boundary, $\boldsymbol{Y} \cdot \boldsymbol{n} = 0$ on ∂P, and normalized to unity net toroidal current, $\oint \boldsymbol{Y} \cdot \mathbf{d}l = 1$, integrated along any poloidally closed curve, $\mathcal{C}$, created by the intersection of the plasma boundary and a toroidal cut of the plasma domain. With these prerequisites, γ is the first-order perturbed net toroidal current flowing in the plasma. $\mu_0^{-1}\gamma$ has the unit of current. Expressed in terms of the displacement vector, this so-called *period* is the line integral

$$\gamma = \oint_{\mathcal{C}} \boldsymbol{B}_{1P} \cdot \mathbf{d}l = \oint_{\mathcal{C}} \mathbf{d}l \cdot \boldsymbol{\nabla} \times (\boldsymbol{\xi} \times \boldsymbol{B}_0) \quad , \tag{B.2}$$

integrated along any poloidally closed curve, $\mathcal{C}$, on the plasma boundary, encircling the magnetic axis once. The net perturbed toroidal flux in the vacuum region can be expressed by the constant in the normal displacement taken on the plasma boundary, ξ_{00},

$$L\, \gamma_{00} = F'_{\mathrm{T}}\, \xi_{00} \quad . \tag{B.3}$$

In the flux constraint, Eq. (B.3), the flux linkage is used to relate the net perturbed toroidal flux to net perturbed toroidal current, Eq. (B.2). With D the part of the $z = 0$ plane enclosed by the inner hole of the toroidal domain, the *inductance* L is defined by

$$L = \iiint_V |\boldsymbol{Y}|^2 \,\mathrm{d}^3 r = \iint_D Y_z \,\mathrm{d}^2 r \quad . \tag{B.4}$$

It depends only on the geometry of the toroidal plasma domain and has the unit of length. The change of the vacuum energy expressed in terms of the ideal MHD displacement vector $\boldsymbol{\xi}$ is then given by

$$\begin{aligned} \delta W_{\mathrm{V}} - L\,\gamma^2 &= \frac{1}{2}\iiint_{\mathrm{V}} |\boldsymbol{\nabla}\Phi^*|^2 \,\mathrm{d}^3 r \\ &= -\frac{1}{2}\iint_{\partial P} \Phi^* \,\boldsymbol{n}\cdot\boldsymbol{\nabla}\Phi^* \,\mathrm{d}^2 r \\ &= -\frac{1}{2}\iint_{\partial P} \Phi^* \,\boldsymbol{n}\cdot\boldsymbol{\nabla}\times(\boldsymbol{\xi}\times\boldsymbol{B}_0)\,\mathrm{d}^2 r \quad . \end{aligned} \tag{B.5}$$

References

[1] Boozer A H and Nührenberg C 2006 *Phys. Plasmas* **13** 102501
[2] Nührenberg C, Boozer A H and Hudson S R 2009 *Phys. Rev. Lett.* **102** 235001
[3] Park J k, Boozer A and Glasser A 2007 *Phys. Plasmas* **14** 052110
[4] Hirshman S P, van Rij W I and Merkel P 1986 *Comput. Phys. Commun.* **43** 143
[5] Lüst R and Martensen E 1960 *Z. Naturforsch.* **15a** 706–13
[6] Grimm R C, Greene J M and Johnson J L 1976 *Methods in Computational Physics* (*Controlled Fusion* vol 16) ed Killeen J (New York: Academic Press) pp 253–280
[7] Shafranov V D and Zakharov L E 1972 *Nucl. Fusion* **12** 599
[8] Strumberger E 1997 *Nucl. Fusion* **37** 19
[9] Drevlak M, Monticello D and Reiman A 2005 *Nucl. Fusion* **45** 731–740
[10] Merkel P, Nührenberg C and Cooper W A 1996 *Theory of Fusion Plasmas Varenna* (Bologna: Società Italiana di Fisica) p 233

Inertial magnetic reconnection effects on the electron dynamics in a three-dimensional configuration

A Perona[1], **D Borgogno**[2], **D Grasso**[3,2]

[1] Centre for Fusion, Space and Astrophysics, University of Warwick, United Kingdom
[2] Dipartimento di Energia, Politecnico di Torino, Italy
[3] Istituto Sistemi Complessi - CNR, Roma, Italy

Abstract. The three-dimensional response of a population of thermal electrons to a collisionless magnetic reconnection process is investigated through a relativistic Hamiltonian guiding centre formulation of the particle dynamics. The resulting electron equations have been implemented in a test-particle code that evolves the spatial position and the velocity of the electrons according to the fields calculated by the numerical solutions of a two-fluid reconnection model. In particular, we consider here reconnection events triggered by single helicity perturbations. By following the evolution of the electron distribution function we can quantify the number of suprathermal electrons generated during the nonlinear phase of reconnection and observe how the increase in the electron energy deforms the moments of the electron distribution function with respect to the corresponding fluid quantities.

1. Introduction

One of the most characteristic features of magnetic reconnection is the conversion of part of the stored magnetic energy into heating of the plasma and particle acceleration. Energetic electrons have been observed in the diffusion region of the Earth's magnetotail [1, 2, 3, 4], as well as during solar flares by *in situ* space measurements [5, 6, 7, 8, 9] and indirectly by the γ-ray lines produced in the chromosphere and in the photosphere. Electron acceleration also occurs following the reconnection of the magnetic field during sawtooth oscillations or disruptions in laboratory Tokamak experiments [10, 11], where the electrons can acquire such energies as to pose a potential danger to surrounding surfaces when they are finally lost. Understanding what conditions control the particle acceleration efficiency during the reconnection process is therefore important. Depending on the plasma parameters, on the magnetic field topology and on the type of reconnection, in fact, the energy distribution rate from magnetic field energy to the plasma particle energy can vary significantly.

In the past, test particles investigations of the effects produced on the thermal electrons populating the reconnection region have been carried out by adopting either analytically prescribed fields [12, 13, 14, 15], or the output of self-consistent magnetohydrodynamic (MHD) simulations [16, 17, 18]. Some of these studies, in particular, have explored particle acceleration in the solar corona [19, 20], and in astrophysical plasmas [21] considering three-dimensional (3D) resistive MHD fields.

As these works have pointed out, the precise structure of the reconnection electric field is the determining factor of the energization process in the reconnection region. In our work, we have

reconstructed the behaviour of a population of thermal electrons during a 3D inertial magnetic reconnection process characterized by a strong longitudinal ('guide') magnetic field. Such a process is described by a two-fluid model, considering a rarefied, high temperature plasma where the finite electron mass, and not the collisions between ions and electrons, can account for the violation of the ideal conductivity [22]. We have expressed the electron dynamics in the presence of the reconnection fields through a set of equations, based on a relativistic Hamiltonian approach, that represents the extension to the 3D geometry of a previous model aimed at investigating the electron behaviour in two-dimensional (2D) configurations [18].
Differently from the 2D case, 3D magnetic reconnection effects introduce regions of magnetic field line stochasticity produced by the nonlinear coupling of initial unstable perturbations with different helicities [23]. The investigation of the particle dynamics in these chaotic regions will be possible by means of the numerical code where the equations of the electron model have been implemented. In this paper, however, we present the numerical results related to a single-helicity case, introduced in Section 2. Although depending on all three spatial coordinates, the single-helicity case is equivalent to a 2D mode after a proper coordinate rotation. This allows us to perform a first 3D numerical investigation with our newly developed tool and to compare the results with those provided by the 2D study described in Ref. [18]. The numerical results describing the electron response in the presence of the reconnection fields are presented in Section 3. Finally, in Section 4 these results are discussed and conclusions are drawn.

2. The two-fluid reconnection model

We consider a dissipationless model of magnetic reconnection in a plasma immersed in a strong and uniform magnetic field [22]. This model, where the electron inertia is the term responsible for the breaking and reconnection of the magnetic field lines, is particularly apt at describing magnetic reconnection events in laboratory and space plasmas, where the density and the temperature are such as to make the contribution of collisions negligible [24]. The three-dimensional evolution of the magnetic flux and of the stream function is described by two-fluid, quasineutral ion and electron equations characterized by two characteristic scale lengths in the form of constant parameters. The electrons skin depth, $d_e = \sqrt{m/(\mu_0 n e^2)}$, is associated with the electron inertia, while the sonic Larmor radius, $\rho_s = \sqrt{T_e M/(eB^2)}$, is defined in terms of the ion inertia, M, and of the electron temperature, T_e. Ion temperature effects are neglected. The magnetic field, $\mathbf{B} = B_0\mathbf{e}_z + \mathbf{e}_z \times \nabla\psi$, is expressed through the magnetic flux, $\psi = \psi(x, y, t)$, and has a strong constant guide field component, B_0, along z. The resulting equations, normalized to the macroscopic scale length, L, and to the Alfvén time, $\tau_A = \sqrt{nM}L_x/B_{y0}$, with B_{y0} the characteristic magnitude of the equilibrium magnetic field, are

$$\frac{\partial F}{\partial t} + [\phi, F] - \rho_s^2[U, \psi] = \frac{\partial \phi}{\partial z} - \rho_s^2 \frac{\partial U}{\partial z}, \tag{1}$$

$$\frac{\partial U}{\partial t} + [\phi, U] + [\psi, J] = -\frac{\partial J}{\partial z} \tag{2}$$

where $F = \psi + d_e^2 J$ is the generalized flux function, $U = \nabla^2\phi$ is the vorticity, $J \equiv -\nabla^2\psi$ is the current density and ϕ is the stream function, which in turn is related to the electrostatic potential Φ by the relation $\phi = (c/B_0)\Phi$.
The evolution of these fields is represented in a 3D slab geometry where magnetic curvature effects are not taken into account and periodic boundary conditions are imposed along the y and z directions. Vanishing boundary conditions and a non-uniform mesh with an increasing density of points in the reconnection region are assumed along x. The reconnection process is triggered by small amplitude, unstable reconnecting modes that are superimposed to the "Harris pinch" equilibrium, $\psi_{eq} = -A\ln(\cosh(x))$, where the constant $A = 1$ represents the amplitude

of the equilibrium.
This 3D model can be traced back to the 2D limit when considering only one single-helicity (oblique) perturbation $\delta J(x,y,z;t) = -\nabla^2\delta\psi(x,y,z;t) = \hat{J}(x)exp(ik_y y + ik_z z)$, where $k_y = \pi m/L_y$ and $k_z = \pi n/L_z$ ($2L_y = 8\pi$ and $2L_z = 32\pi$ are the y- and z-sizes of the 3D box and (m,n) are the mode wave numbers along y and z). This single-helicity case, in fact, is equivalent to a two-dimensional mode after a proper coordinate rotation, $y_* = y + (k_z/k_y)z$. However, in the 3D representation of the single-helicity mode the equilibrium effective shear field is not symmetric with respect to the mode resonant surface. This causes a drift of the magnetic island with a velocity close to the fluid velocity at the magnetic X-point [23], while in the 2D case the magnetic island evolves on a rational surface coinciding with the position of the null line of the equilibrium sheared magnetic field. In our investigation we consider a collisionless reconnection process in the limit of the so-called large Δ' regime ($\Delta' d_e > min(1,(d_e/\rho_s))^{1/3}$ [25]), where Δ' is the standard stability parameter of reconnection modes [26], recently extended to oblique modes [27]. This parameter depends on the analytical function describing the equilibrium magnetic flux. In particular, for the most unstable mode and for the (normalized) value of $d_e = 0.1682$ chosen for our study, we obtain $\Delta' d_e \approx 1.26$.
The electric field parallel to the magnetic field lines can be expressed through the fields ψ and ϕ by remembering the relation that links the stream function to the electrostatic potential. As shown in Fig. 1, the parallel electric field exhibits a monopole structure peaked at the X-point of the magnetic island throughout the process, while it has an opposite sign along the separatrix of the island. In particular, during the nonlinear phase it is more negative along one arm of the separatrix than along the other, as an effect of the island asymmetry.

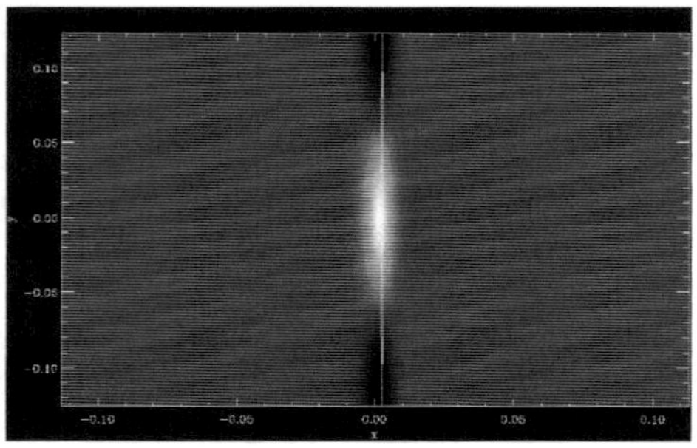

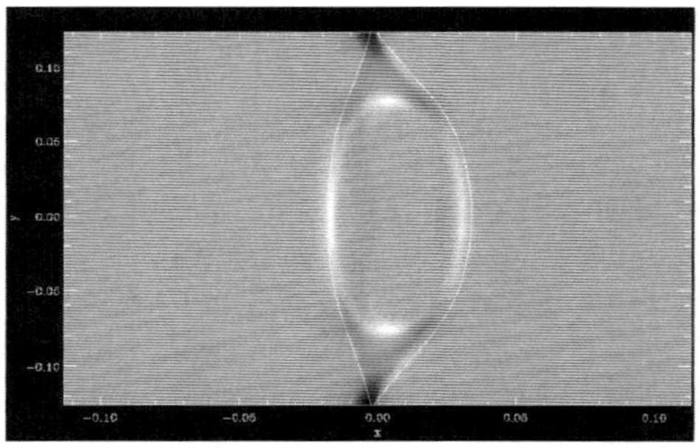

Figure 1. Contour levels in the (x,y) plane of the parallel electric field during the linear phase (left) and the nonlinear phase (right) of the reconnection process.

3. The electron response to the reconnection fields

In order to reconstruct the behaviour of the electrons in the presence of the reconnection fields we have derived a set of equations, based on a relativistic Hamiltonian formulation of the particle dynamics [28], which describes the evolution in time of the spatial position and of the parallel velocity of the electron guiding centres. These equations have been implemented in a numerical code that reads at each time step the reconnection fields calculated by a fluid code solving the magnetic reconnection equations introduced in Sec. 2. In order the electron dynamics to be coherent with the collisionless reconnection model considered in this paper, collisions have not been taken into account in the electron code. However, we have provided the code with a collisional operator reproducing their effect, which can be switched on when considering resistive models of magnetic reconnection characterized by longer time-scales.
After evaluating the reconnection fields in the whole reconnection region through a cubic spline

interpolation, the test electron code integrates the electron equations using a fourth-order Runge-Kutta method with adaptive stepsize control.
The evolution of the equilibrium Maxwellian electron distribution function and of its moments can be reconstructed at the end of each time step by considering the velocity and the spatial location of the particles at this step. However, this "*full-f*" approach produces effective results when the amplitude of the perturbation is large enough to overcome the background noise. For this reason, a δf technique has also been adopted in the code [29], which allows the small changes in the equilibrium quantities occurring during the linear phase of the process to be observed.
The fluid reconnection simulation considered in this work has been performed using a grid of $801\times128\times128$ points and chosing the values of the two (normalized) parameters d_e and ρ_s in such a way as to correspond to those of typical fusion devices. Coherently, the test electron equilibrium density and temperature are initially taken as constant and uniform in the real space. Thus, simulations have been performed for $d_e = 0.1682\cdot10^{-2}$ m and $\rho_s = 0.3245\cdot10^{-2}$ m, corresponding to an equilibrium density $n_0 = 10^{19}\mathrm{m}^{-3}$ and an electron thermal temperature, $T_e \approx 1$ keV, respectively. As described in Fig. 2, after the initial transient, the reconnection process has a linear exponential behaviour between 60 τ_A and 180 τ_A, followed by a superexponential growth and a saturated state [30]. The results of the electron simulations have been obtained by reconstructing the dynamics of 10^7 particles in the reconnection region up to the end of the nonlinear phase.

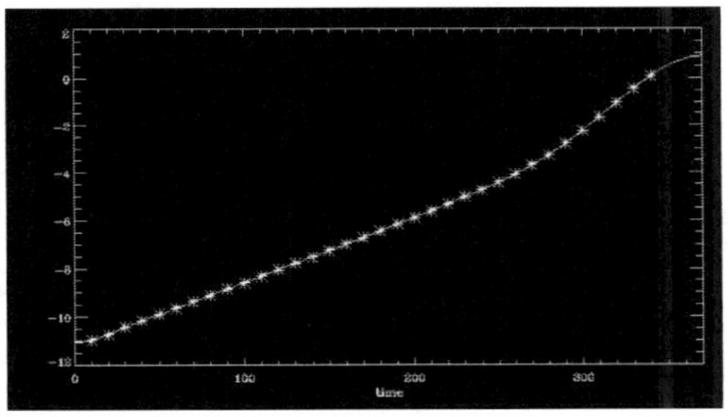

Figure 2. $\ln|\tilde{\psi}|$ evaluated at the X-point, vs the normalized time t/τ_A. The fluid fields are saved each 2 Alfvén times. This choice represents the minimum sampling required in order to properly describe the electron dynamics. The stars represent up to which stage of the process the electron response has been reconstructed.

A good agreement characterizes the evolution of the moments provided by the electron simulations and the corresponding fluid quantities during the linear phase of the process. During the nonlinear phase, the parallel velocity of the electrons passing through the X-point increases, so that these particles move from the thermal bulk towards the tails of the electron distribution function. This acceleration affects the *parallel* electron temperature, $T_{\|e} \equiv p_{\|e}/n$, where n is the electron density and $p_{\|e}$ represents the longitudinal kinetic energy of the electrons.

Contrary to what expected by the isothermal closure of the fluid reconnection model, at this stage the temperature becomes significantly higher along the magnetic island separatrix with respect to the surrounding area and reaches the highest value in the denser regions, as shown in Fig. 3. In the following steps of the process, the number of the energized electrons becomes increasingly relevant, wherefrom a very significant distortion of the electron distribution function arises. The change with respect to the equilibrium Maxwellian distribution function at the end of the nonlinear phase is drawn in Fig. 4, taken at t=340 τ_A. At this time we can see that about 1/3 of the thermal electrons has run away from the original thermal bulk and that their velocity distribution generate a second Maxwellian centered at $v_\|/c = -0.15$.
In this mixed fluid-kinetic approach the electron response is not fed back to the fluid reconnection fields. A further development of the work that would allow us to reconstruct the process beyond the point at which the electron response and the fluid description do not agree any more

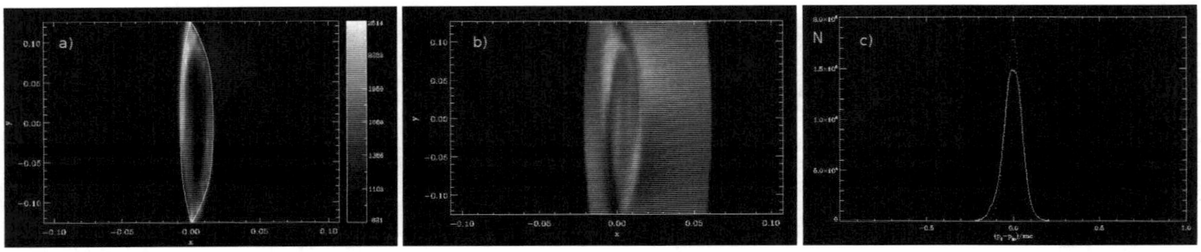

Figure 3. Contour levels in the (x, y) plane of the *parallel* electron temperature (a) and of the electron density (b) at t=320 τ_A. The white line represents the separatrix of the magnetic island. The distortion of the electron distribution function (continuous line) with respect to the equilibrium Maxwellian distribution function (dotted line) at this time is represented in Fig. (3c).

would consist in providing the fluid fields at each time step with the electron feedback, thus making the description of the process fully self-consistent.

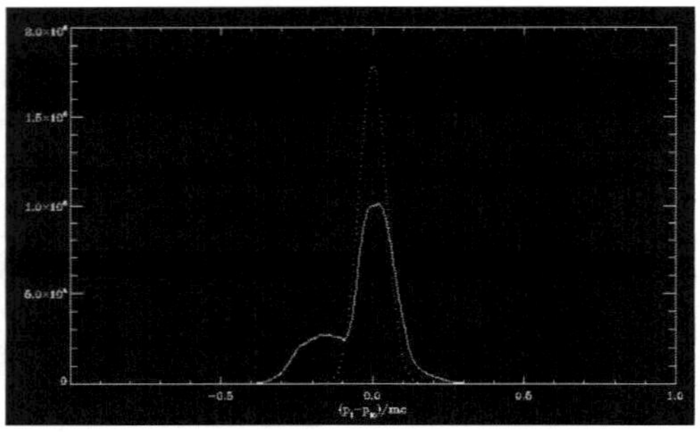

Figure 4. Distortion of the electron distribution function (continuous line) with respect to the equilibrium Maxwellian distribution function (dotted line) at the end of the nonlinear phase (t=340 τ_A). The shift of the non-thermal Maxwellian towards negative values of the velocity is due to the choice of a negative guide field component in the fluid simulations.

4. Discussion and conclusions

We have numerically investigated the effects of the fields characterizing an inertial magnetic reconnection process through a recently developed test particle code, which reproduces the dynamics of the electron guiding centres in a 3D cartesian geometry. As a first step, a single-helicity perturbation has been considered. Although this case is equivalent to a 2D mode, it still depends on all three spatial coordinates and is therefore useful for the validation of the correct behaviour of the 3D electron tool. Moreover, the fact that the single-helicity mode can be traced back to a 2D perturbation gives us the opportunity to compare the electron response with the one obtained in a previous investigation where the 2D version of the electron model had been used [18]. In that case, a fluid reconnection event described by the 2D version of Eqs. (1)-(2) had been studied with a different equilibrium function and a smaller equilibrium amplitude, $\psi_{eq} = 0.1/\cosh(x)^2$. Although the structure of the parallel electric field was still peaked at the X-point of the magnetic island throughout the process, similarly to the 3D electric field, no significant electron energization was observed and the electron response did not yield neither energetic nor relativistic tails in the distribution function. However, by the relation expressing the magnetic field configuration, $\mathbf{B} = B_0\mathbf{e}_z + \mathbf{e}_z \times \nabla\psi$, we know that the perpendicular component of the equilibrium magnetic field, $b_y = -\partial\psi_{eq}/\partial x$, depends on the equilibrium amplitude, A. The strong acceleration that we obtain in the 3D case can therefore be explained by the large amount of magnetic energy that characterizes the reconnection process when $A = 1$. The strength of the perpendicular magnetic field naturally depends on the physical environment where magnetic reconnection occurs. In a Tokamak, for instance, the equilibrium amplitude that has been taken into account in this paper should be reduced in order to give a more realistic representation of the ratio of the toroidal to the poloidal component of the magnetic field also at the edge of the reconnection region. Then, it will be interesting to observe whether this choice still leads to a significant production of the highly energetic electrons experimentally observed during reconnection events in fusion devices. This aspect of the problem will be explored in the next step of our work.

References

[1] Oieroset M, Lin R P, Phan T D, Larson D E and Bale S D, *Phys. Rev. Lett.*, **89**, 195001 (2002).
[2] Wang R-S, Lu Q-M, Guo J and Wang S, *Chin. Phys. Lett.*, **25**, 3083 (2008).
[3] Vaivads A, Retinò A, Khotyaintsev Y V and André M, *Ann. Geophys.*, **29**, 1917 (2011).
[4] Chen L-J, Bessho N, Lefebvre B, Vaith H, Asnes A, Santolik O, Fazakerley A, Puhl-Quinn P, Bhattacharjee A, Khotyaintsev Y, Daly P and Torbert R, *Phys. Plasmas*, **16**, 1056501 (2009).
[5] He J-S, Zong Q-G, Deng X-H, Tu C-Y, Xiao C-J, Wang X-G, Ma Z-W, Pu Z-Y, Lucek E, Pedersen A, Fazakerley A, Cornilleau-Wehrlin N, Dunlop M W, Tian H, Yao S, Tan B, Fu S-Y, Glassmeier K-H, Reme H, Dandouras I and Escoubet C P, *Geophys. Res. Lett.*, **35**, L14104 (2008).
[6] Lin R P, Krucker S, Hurford G J, Smith D M, Hudson H S, Holman G D, Schwartz R A, Dennis B R, Share G H, Murphy R J, Emslie A G, Johns-Krull C and Vilmer N, *The Ap. J.*, **595**, L69 (2003).
[7] Holman G D, Linhui S, Schwartz R A and Emslie A G, *Astrophys. J.*, **595**, L97 (2003).
[8] Mann G and Warmuth A, *A&A*, **528**, A104 (2011).
[9] Vilmer N, *Phil. Trans. R. Soc. A*, **370**, 3241 (2012).
[10] Savrukhin P V, *Phys. Rev. Lett.*, **86**, 3036 (2001).
[11] Klimanov I, Fasoli A, Goodman T P and the TCV team, *Plasma Phys. Control. Fusion*, **49**, L1 (2007).
[12] Heerikhuisen J, Litvinenko Y E and Craig I J D, *ApJ*, **566**, 512 (2002).
[13] Craig I J and Litvinenko Y E, *ApJ*, **570**, 387 (2002).
[14] Dalla S and Browning P K, *A&A*, **289**, 491 (2008).
[15] Sakai J, *Sol. Phys.*, **140**, 99 (2002).
[16] Karlický M and Bárta M, *ApJ*, **647**, 1472 (2006).
[17] Liu W J, Chen P F, Ding M D and Fang C, *ApJ*, **690**, 1633 (2009).
[18] Perona A, Eriksson L-G and Grasso D, *Phys. Plasmas*, **17**, 042104 (2010).
[19] Turkmani R, Cargill P J, Galsgaard K, Vlahos L and Isliker H, *A&A*, **449**, 749 (2006).
[20] Guo J-N, Büchner J, Otto A, Santos J, Marsch E and Gan W-Q, *A&A*, **513**, A73 (2010).
[21] Schopper R, Birk G T and Lesch H, *Phys. Plasmas*, **6**, 4318 (1999).
[22] Schep T J, Pegoraro F and Kuvshinov B N, *Phys. Plasmas*, **1**, 2843 (1994).
[23] Borgogno D, Grasso D, Porcelli F, Califano F, Pegoraro F and Farina D, *Phys. Plasmas*, **12**, 032309 (2005).
[24] Porcelli F, Borgogno D, Califano F, Grasso D, Ottaviani M and Pegoraro F, *Plasma Phys. Control. Fusion*, **44**, B389 (2002).
[25] Porcelli F, *Phys. Rev. Lett.*, **66**, 425 (1991).
[26] Furth H P, Killeen J and Rosenbluth M N, *Phys. Fluids*, **6**, 459 (1963).
[27] Daughton W, Roytershteyn V, Karimabadi H, Yin L, Albright B J, Bergen B and Bowers K J, *Nat. Phys.*, **7**, 539 (2011).
[28] Brizard A J and Chan A A, *Phys. Plasmas* **6** , 4548 (1999).
[29] Candy J, *J. Comput. Phys.* **129**, 160 (1996).
[30] Grasso D, Califano F, Pegoraro F and Porcelli F, *Phys. Rev. Lett.*, **86**, 5051 (2001).

Exploitation of a general-coordinate guiding centre code for the redistribution of fast ions in deformed hybrid tokamak equilibria.

D Pfefferlé[1], J P Graves[1], W A Cooper[1]

[1] Centre de Recherches en Physique des Plasmas (CRPP), EURATOM/CH Association, EPFL, 1015 Lausanne, Switzerland

E-mail: david.pfefferle@epfl.ch

Abstract. Self-consistent fast ion distributions are usually obtained using a code that solves the guiding-centre equations, with an appropriate fast ion source (e.g. NBI pinis) and sink (e.g. collision operators). Straight field-line coordinate systems, such as Boozer coordinates, are ordinarily convenient due to the simple separation of longitudinal and cross-field motion, and the simple expression of magnetic differential operators. However, these coordinates are found to be near-singular at the boundary of the internal helical region associated with an $n = m = 1$ infernal mode. These important configurations are associated with many tokamak phenomena, including snakes and long-lived modes [1] in spherical or more conventional devices. Such internal helical states occur when there is a radially extended region where the safety factor is close to unity. Recent calculations predict the possibility of helical equilibria in ITER hybrid scenarios [2]. The ANIMEC code [3] conveniently produces an equilibrium helical state despite choosing for example an axisymmetric fixed boundary. The corresponding magnetic field in these coordinates can now be fed to the newly devised Particle-In-Cell (PIC) code VENUS-LEVIS, which has been upgraded with phase-space Lagrangian guiding-centre orbit equations [4], embodying full 3D anisotropic electromagnetic fields and a formulation that is independent of coordinate choice, despite retaining intrinsic Hamiltonian properties. The simulations are applied to MAST experiments where the presence of a long-lived mode can effect confinement of neutral beam ions, potentially affecting NBI heating and current drive [1]. Neighbouring equilibria from ANIMEC, one helical in the core and the other axisymmetic, permits a precise means of identifying the effect of 3D geometry on the simulated confinement properties of MAST's neutral beam fast ion population. In agreement with the compared experimental data from MAST neutron camera, a significant fraction of particles are pushed out of the helical core region affecting both the measured radial neutron distribution and the heating and current drive properties of the neutral beam population.

1. Motivation

Modern fusion devices are nourished by powerful heating systems and operate at very energetic regimes. Large populations of non-thermal particles are produced, either via ICRH, NBI and/or α-emission. Confinement of such fast particles is key to maintain plasma control and to avoid disruption.

Tokamak instabilities form non-trivial electromagnetic structure on top of which the motion of particles becomes complex and sometimes chaotic. Some instabilities will saturate at large amplitude and form stationary configurations. Despite the absence of electric fields, particle orbits are again complex because of the 3D magnetic geometry. This is usually associated with loss of density, decrease of temperature and degradation of the conditions for fusion.

The study of fast ion transport is an important starting point for understanding the interplay between the fields and particles. Analytic progress has been made in simple cases, like axisymmetric geometry, linearized electromagnetic perturbations, simple fast ion distributions, etc., but it is necessary to treat more realistic situations, like helical tokamak core, Stellarator geometry, full-field perturbations (especially relevant for high beta plasmas), etc. For such investigations, numerical work becomes a necessity.

The $n = m = 1$ infernal mode consists of a helical displacement of the inner core region and is related to many Tokamak phenomena including snakes, long-lived modes in spherical or more conventional devices.

There is experimental evidence that such helical states strongly affect fast particles. This evidence is highlighted by neutron emissivity, which is directly related to ionization processes between fast ions and background deuterium/tritium. Observing MAST discharge with Long-lived mode (LLM)[1], it is found that the LLM develops as q_{min} approaches unity and correspondingly the neutron signal drops at a location assumed to be close to the magnetic axis. Then, when q_{min} crosses unity and the LLM fades away (equivalently the helical displacement shrinks), fast ion confinement is restored and the neutron signal returns to its initial intensity. It appears that the helical geometry is fully responsible for the expulsion of the hot particles out of the core region. Such interpretation begs for theoretical and numerical confirmation.

2. Model & Tools

A qualitative and quantitative estimate of the redistribution of NBI induced fast ions during a helical internal state is achieved by numerically solving the orbits about this particular magnetic configuration. The model assumes that *a*) the helical core is described as a static MHD equilibrium, and is unaffected by the fast ion distribution throughout the simulation[1]; *b*) spatial variation of the magnetic field is small with respect to Larmor radius of ions; *c*) plasma rotation is much slower than particle bounce motion but much faster than diagnostic and slowing-down time, i.e. equations of motion do not take into account centrifugal effects but particle deposition is *randomized* toroidally.

2.1. Guiding-centre drift equations

For the purposes of fast ion transport, it is arguably sufficient to consider the lowest order expansion in the guiding-centre approximation - the higher orders only being relevant for gyro-kinetics and the study of turbulence. Hence, the guiding-centre drift equations (GCDE) are derived by neglecting the particle's fast and minute gyro-motion against the slow drift motion of the guiding-centre across the field-lines. It is equivalent to considering the first order magnetic moment $\mu = mv_\perp^2\gamma^2/qB$ as being an adiabatic constant (and relativitic factor γ is defined below).

There are many approaches to derive the GCDE. The non-canonical phase-space lagrangian technique given by Littlejohn [4, 5] is especially suited for a general and coordinate independent formulation. Following his approach, we start with the relativistic guiding-centre phase-space Lagrangian

$$\mathcal{L}_{GC}(\mathbf{X}, \rho_{||}, t) = (\mathbf{A} + \rho_{||}\mathbf{H}) \cdot \dot{\mathbf{X}} - (\tfrac{m}{q}c^2\gamma + \Phi) \tag{1}$$

[1] The effect of energetic particles on the evolution towards a saturated helical state is not treated. While linear growth of the mode might be affected by fast ions, it is reasonable that the final saturated amplitute of the kink is not sensitive to the NBI population. In other words, the linear stage of the internal kink growth would be sensitive to FLR effects from the thermal ions and kinetic effects from the fast ions but the 3D equilibrium is not, especially for the cases considered here for which the pressure ratio $P_{hot}/P_{th} \sim 5 - 10\%$.

Journal of Physics: Conference Series **401** (2012) 012020 doi:10.1088/1742-6596/401/1/012020

where $\mathbf{X}$ is the position of the guiding-centre, $\rho_{||}$ is the parallel gyroradius (change of variable from $v_{||} = \rho_{||} qH/m\gamma$), $\mathbf{A}$ is the vector potential providing $\mathbf{B} = \boldsymbol{\nabla} \wedge \mathbf{A}$, $\mathbf{H} = \sigma \mathbf{B}$ is the magnetic field intensity, permeability σ accounts for pressure anisotropy [6], Φ is the electrostatic potential providing $\mathbf{E} = -\boldsymbol{\nabla}\Phi + \partial_t \mathbf{A}$ and γ is the relativistic factor carrying kinetic energy

$$\gamma = \sqrt{1 + \frac{2\mu B}{mc^2} + \frac{\rho_{||}^2 H^2}{(m/q)^2 c^2}}$$

The definition of parallel gyroradius was altered to include anisotropy explicitly. Since the phase-space Lagrangian does not contain a $\dot{\rho_{||}}$ term, playing with its definition is transparent. This variable can be seen as a *pivot* variable to load physics in the GCDE ($A_{||}$ perturbations, plasma rotation,...). From (1), the Euler-Lagrange equations are derived and inverted to yield the following system of four differential equations as GCDE:

$$\dot{\rho_{||}} = -\partial\mathcal{H} \cdot \mathbf{B}^* / \mathbf{H} \cdot \mathbf{B}^* \quad (2)$$

$$\dot{\mathbf{X}} = \left(v_{||} H \mathbf{B}^* + \mathbf{H} \wedge \partial\mathcal{H} \right) / \mathbf{H} \cdot \mathbf{B}^* \quad (3)$$

where $\partial\mathcal{H} = \frac{\mu}{q\gamma}\nabla B + v_{||}\rho_{||}\nabla H - \mathbf{E} + \rho_{||}\partial_t \mathbf{H}$ represents essentially the space-time gradient of the Hamiltonian.

It can be verified that these equations share the same conservation properties as those derived in the Hamiltonian canonical formalism, The underlying symplectic structure entails conservation of energy in time-independent cases, conservation of toroidal momentum in axisymmetric geometry, Liouville equation for phase-space volume, etc. These are essential for stable and consistent numerical implementation of guiding-centre motion. The vector equations of motion for $\mathbf{X}$ highlights the freedom of choice in space coordinates and field representation.

2.2. `VENUS-LEVIS` orbit solver & PIC code

`VENUS-LEVIS` is an offspring of the `VENUS` code [7] rewritten in `FORTRAN 90` to match performance, modularity and clarity criteria. Its primary function is to evolve particles solving the above GCDE (2), using a *RK4* integrator scheme and fast linear interpolation of the field quantities over a tight 3D mesh. The choice of coordinate system and field representation is established via an independent *equilibrium module* loaded at compile time. Modules for straight field-line Boozer, `ANIMEC` [8] and `MINERVA` [9] coordinates are currently available.

Presently, `VENUS-LEVIS` is used as a particle-in-cell (PIC) code following an ensemble of weighted markers representing a slowing-down distribution of fast ions. An NBI module has been devised to provide the source of markers and a collection of Monte-Carlo operators, mimicking Coulomb collisions with background plasma, act as sinks [10]. The fields are not affected by the hot distribution and mutual interaction between fast particles is not taken into account.

The outcome of the code is a self-consistent estimation of the saturated fast ion distribution from NBI and the associated flux-averaged moments (fast ion density, pressure, current, etc.).

It is noted here that particular numerical challenges arise around the magnetic axis and near the transition region bounding the helical core. In particular, the magnetic axis is commonly accompanied by singularities in toroidal flux coordinate systems. A pseudo-cartesian patch combined with an analytic prescription of the fields partially overcomes this difficulty. In helical core equilibria this issue is enhanced by the geometrical spread of the magnetic axis and the large radial excursions of particles.

2.3. `ANIMEC` equilibrium solver

`ANIMEC` [8] is an equilibrium solver based on the premise that $n = m = 1$ instabilities eventually saturate into stationary internal helical core configurations satisfying the force balance equation.

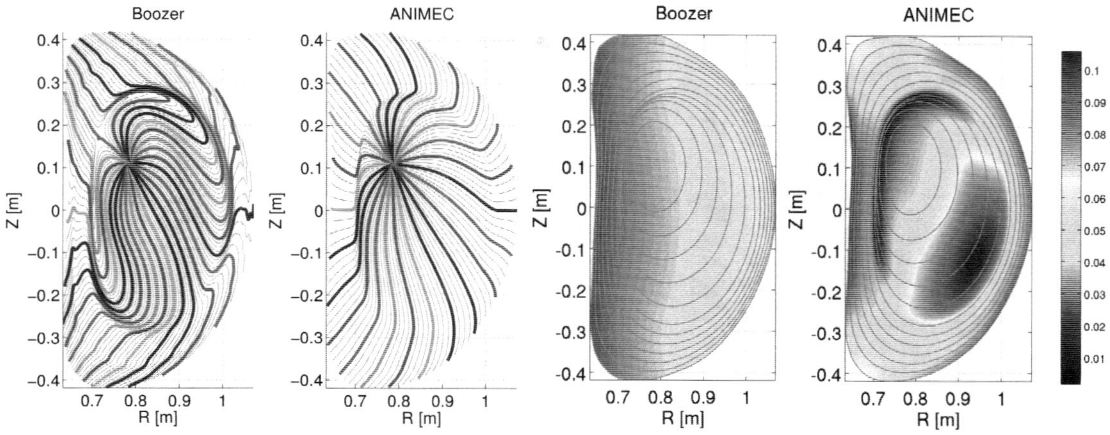

(a) Lines of constant poloidal angle (b) Flux surfaces and Jacobian

Figure 1. Coordinate grid comparison between Boozer and ANIMEC coordinate systems for helical core in TCV. Poloidal cut at toroidal angle $\varphi = -23°$

These states are typically found for reversed-shear q-profiles with $q_{min} \sim 1$. The boundary, the pressure profile and the initial guess for the magnetic axis are prescribed to ANIMEC, which then uses an MHD-energy minimization principle to work out the 3D geometry of the flux surfaces and the components of the magnetic field. As is typical with root-finders, the bifurcation between axisymmetric and helical solution is very sensitive.

ANIMEC uses its own flux coordinate system - neither straight field-line nor orthogonal - resulting from the minimization of the Fourier spectrum. TERPSICHORE can be used to translate them into Boozer-like coordinates. Although straight field-line coordinates are more convenient to express field components (become flux quantities), perturbations and magnetic differential operators, the translation process is computationally heavy, as it requires a massive spectrum of modes. This is because Boozer coordinates end up being very contorted, if not singular, in the transition region between the helical core and axisymmetric mantel (see Fig.1(a)).

On the other hand, the Jacobian of ANIMEC coordinates is close to zero near the transition region between helical core and axisymmetric mantel due to the packing of flux surfaces (see Fig. 1(b)). This leads to the spiking of some components of the field (currents) which is equivalently problematic.

3. Results

The tools described above were used to investigate numerically fast ion redistribution in MAST device. Based on experimental background profiles and plasma boundary measurements from MAST, ANIMEC was successful at recreating an axisymmetric and helical equilibrium. Markers, deposited in accordance with NBI pinis in MAST, were evolved within VENUS-LEVIS and multiple *snapshots* of the resulting slowing-down distribution were taken to provide the statistics. A comparison between axisymmetric and helical situations is discussed below.

As a first result, it was observed that, on top of their usual radial drift, fast ions undergo large radial excursions from the inner helical core to beyond the transition region (see Fig.2(b)). This implies that hot particles visit larger regions of the plasma and contribute to flatten the density profile. It is also noticed that many particles pass through the magnetic axis. This is a numerical difficulty because of the singular behaviour of flux coordinates. To avoid problems, it is possible to express the differential equations from polar (radial flux label s and poloidal

Journal of Physics: Conference Series **401** (2012) 012020 doi:10.1088/1742-6596/401/1/012020

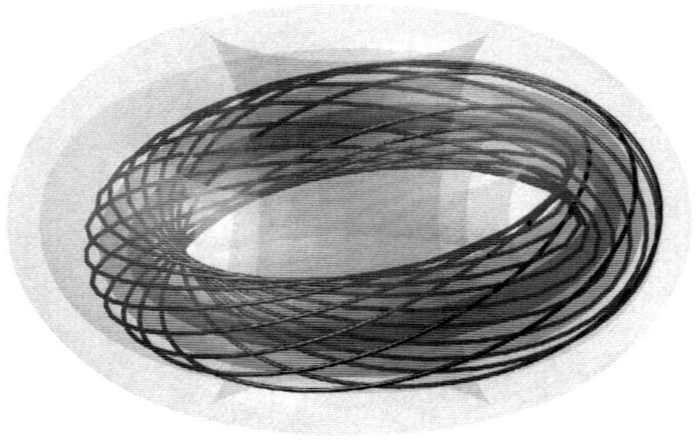

(a) Characteristic 3D orbit of passing particle (majority of NBI) in MAST helical core (flux surfaces). Initial conditions: deuterium at $E_{kin} = 6.5$[keV], $v_{||}/v = 0.9$

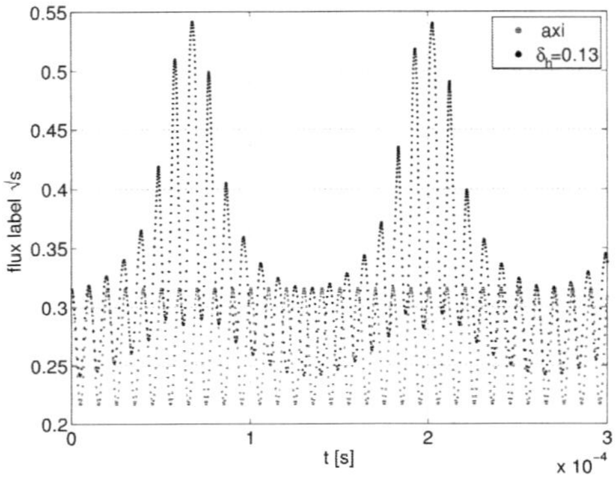

(b) Radial excursion of passing particle from Fig.2(a). Comparison between axisymmetric and helical cases.

Figure 2. In addition to their ordinary radial drift, passing particles undergo slower and larger radial excursion in helical geometry. This contributes to flattening the fast ion density profile near core region.

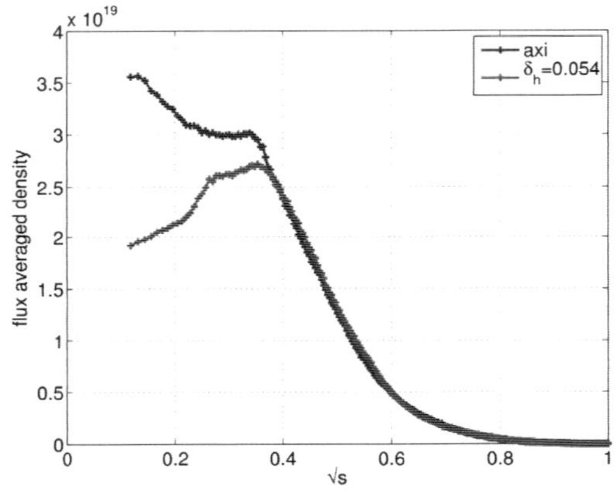

(a) Flux-averaged fast ion density as a function flux label $\sim$ minor radius.

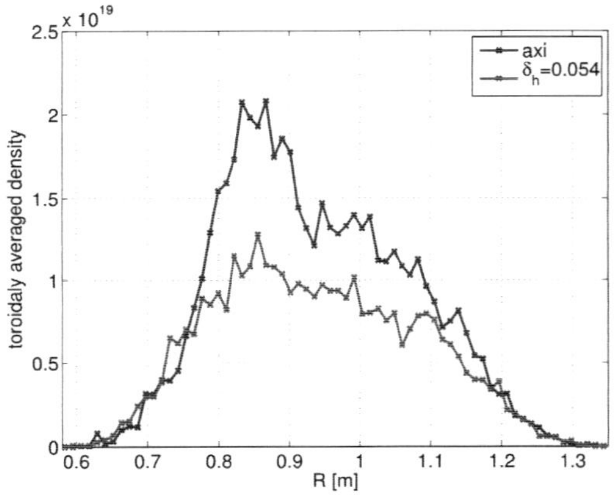

(b) Toroidally averaged fast ion density as at $Z = 0.2$[m] as a function of major radius.

Figure 3. Fast ion density is reduced in helical core region.

angle θ) to *pseudo*-cartesian variables $X = s\cos\theta$, $Y = s\sin\theta$. That way singular components for example $\partial B/\partial s \propto 1/\sqrt{s}$ entering in $\dot{\theta}$, are regularized by s. It is also essential that the interpolation scheme preserves the properties of the *magnetic field*[2], so high-order interpolation or even analytic prescription are found useful.

Secondly, it is observed that the fast ion density profile drops proportionally to the helical displacement in the region around the magnetic axis (see Fig.3(a)). The toroidally averaged density decreases as well, but also flattens and broadens (see Fig.3(b)). This is a combined effect of the magnetic axis twisting around the torus and the orbits laying on larger portions of the equilibrium. These observations relate to the experimental data at a qualitative level. More simulations will be commissioned to answer more quantitative questions, especially for direct

[2] In the sense that $\nabla \cdot \mathbf{B} = 0$, $\nabla \wedge \mathbf{B} = \mathbf{K}$, etc.

comparison with MAST neutron emissivity data.

4. Conclusion

`ANIMEC` equilbrium solver is used to obtain $n = m = 1$ internal mode configuration, which is relevant for experimental plasmas and ITER hybrid scenario. Helical equilibria obtained from MAST configuration are fed to **`VENUS-LEVIS`** drift-kinetic solver to simulate energetic particle dynamics. Starting with an NBI source, the slowing-down distribution recovers the experimentally observed expulsion of fast ions from the central region where the helical mode is located. The derived guiding-centre drift equations provide a formulation of the problem that is independent of the choice of coordinates. This advantage is used to follow particles in the `ANIMEC` coordinate system, which is established to be smoother at the transition region between helical and axisymmetric mantel than the Boozer coordinate system. It is shown that hot particles have larger radial excursion in the helical core than in the axisymmetric configuration. Despite choosing a helical amplitude much smaller than that measured in MAST, the fast ion density in the core is significantly reduced. These ions are redistributed to the axisymmetric boundary, or lost from plasma entirely.

Acknowledgments

The authors would like to thank I.T.Chapman and M.Turnyanskiy for close collaboration, as well as L.Villard and D.Brunetti for useful discussions. This work has been partially supported by Swiss National Science Foundation. Simulations have been performed on BlueGene cluster at EPFL, Lausanne (CH), on ROSA XT-5 cluster at CSCS, Manno (CH) and on Helios at CSC, Rokkasho (Japan).

References

[1] Chapman I T 2010 *Nucl. Fusion* **50** 045007
[2] Cooper W A *et al.* 2011 *Plasma Phys. Control. Fusion* **53** 024002
[3] Cooper W A *et al.* 2010 *Phys. Rev. Lett.* **105** 035003
[4] Littlejohn R G 1983 *Journal of Plasma Physics* **29** 111–125
[5] Grebogi C 1985 *Phys. Rev. A* **31**(2) 914–920
[6] Cooper G A *et al.* 2007 *Physics of Plasmas* **14** 102506 (pages 7)
[7] Fischer O *et al.* 2002 *Nuclear Fusion* **42** 817
[8] Cooper W A *et al.* 2009 *Computer Physics Communications* **180** 1524–1533
[9] Aiba N *et al.* 2009 *Computer Physics Communications* **180** 1282 – 1304 ISSN 0010-4655
[10] Albergante M 2011 *Interaction Between Fast Ions and Microturbulence in Thermonuclear Devices* Ph.D. thesis EPFL Lausanne

Gyrokinetic TEM stability calculations for quasi-isodynamic stellarators

J H E Proll[1], P Helander[1], P Xanthopoulos[1] and J W Connor[2,3]

[1] Max-Planck-Institut für Plasmaphysik, EURATOM Association, Teilinstitut Greifswald, Wendelsteinstraße 1, 17491 Greifswald, Germany

[2] Culham Centre for Fusion Energy, Abingdon OX14 3DB, United Kingdom

[3] Imperial College of Science, Technology and Medicine, London SW7 2BZ, United Kingdom

E-mail: josefine.proll@ipp.mpg.de

Abstract. Recent theoretical findings suggest that in perfectly quasi-isodynamic stellarators, the trapped-particle instability as well as the ordinary electron-density-gradient-driven trapped-electron mode are stable in the electrostatic and collisionless approximation. In these configurations contours of constant magnetic field strength B are poloidally closed, the second adiabatic invariant J is constant on flux surfaces and peaks in the centre. It follows that the diamagnetic drift frequency ω_{*a} and the bounce-averaged magnetic drift frequency $\overline{\omega_{da}}$ are in opposite directions, $\omega_{*a} \cdot \overline{\omega_{da}} < 0$, everywhere on the flux surface. This is the signature of average "good curvature" for trapped particles and, thanks to this property, particles that bounce faster than the frequency of any unstable mode must draw energy from it near marginal stability. Consequently, the point of marginal stability cannot exist for the collisionless trapped-particle mode, and hence this mode will be absent. By a similar argument, the ordinary trapped-electron mode is also stable. Because perfect quasi-isodynamicity can never be reached, it is necessary to test configurations approaching quasi-isodynamicity numerically in order to probe their resilience against trapped-particle modes. Progress has been made to extend gyrokinetic simulations to stellarator geometries, enabling us to perform the first linear gyrokinetic simulations of density-gradient-driven modes in stellarator configurations approaching quasi-isodynamicity, such as Wendelstein 7-X and more recently found configurations. These simulations appear to confirm the analytical predictions.

1. Introduction

Plasma confinement in a stellarator is achieved through a three-dimensionionally shaped magnetic field, rather than an axisymmetric field as in a tokamak. Due to this property stellarators usually suffer from higher levels of collisional (so-called neoclassical) transport than tokamaks. However, in recent numerically optimised stellarator configurations [1], the neoclassical transport is on par with axisymmetric devices (unless the collisionality is very low). An important class of these optimised stellarators are the quasi-isodynamic ones. It was shown recently [2] that in these stellarator configurations, the collisionless trapped-particle instability as well as the ordinary electron-density-gradient-driven trapped-electron mode (TEM) are stable, provided the temperature gradients are small enough compared with the density gradient. Since these instabilities - in combination with ion and electron-temperature gradient (ITG and ETG) modes - are believed to be the cause for much of the turbulence in tokamaks, quasi-isodynamic stellarators could also enjoy reduced turbulent transport.

A quasi-isodynamic field is characterised by poloidally, but not toroidally, closed contours of maximum and minimum magnetic field strength $B = |\mathbf{B}|$ and confinement of all collisionless orbits

[3, 4]. The second adiabatic invariant,

$$J = \int_{l_1}^{l_2} m v_\| dl,$$

where the integral is taken between two bounce points, is then constant on flux surfaces. If the magnetic field is given by $\mathbf{B} = \nabla\psi \times \nabla\alpha$, where ψ is the toroidal flux and α denotes the field line label, we can define the magnetic drift frequency $\omega_{da} = \mathbf{k}_\perp \cdot \mathbf{v}_{da}$ and the drift wave frequency $\omega_{*a} = (T_a k_\alpha / e_a) d \ln n_a / d\psi$ for each particle species a in the usual way, where the density n_a and temperature T_a are constant on flux-surfaces, and $\mathbf{v}_{da} = \hat{\mathbf{b}} \times ((v_\perp^2/2)\nabla \ln B + v_\|^2 \boldsymbol{\kappa})/\Omega_a$ denotes the drift velocity, with $\hat{\mathbf{b}} = \mathbf{B}/|\mathbf{B}|$ the unit tangent vector and $\boldsymbol{\kappa} = \hat{\mathbf{b}} \cdot \nabla\hat{\mathbf{b}}$ the curvature vector of the magnetic field. The wave vector perpendicular to the magnetic field is given by $\mathbf{k}_\perp = k_\alpha \nabla\alpha + k_\psi \nabla\psi$. In quasi-isodynamic configurations, the precessional drift frequency of the particles $\overline{\omega_{da}}$, where an overbar denotes the bounce average, is usually in the opposite direction to the diamagnetic drift frequency ω_{*a} everywhere on the flux surface. Trapped particles therefore experience an average "good curvature" and, if their bounce frequency ω_{ba} exceeds the frequency of any unstable mode, must draw energy from it near marginal stability. It is thus concluded that quasi-isodynamic stellarators are resilient against the ordinary TEM and collisionless trapped-particle mode [2]. Here we embark on the numerical validation of this finding by performing linear gyrokinetic simulations of these modes in stellarator geometries. This is important since perfect quasi-isodynamicity can never be realised and it is interesting to investigate how well these analytical findings apply to actual configurations such as those conceived in the last few years [1, 4, 5] and also in Wendelstein 7-X. In Section 2, we present a condensed version of the analytical calculation that lead to the result above. In Section 3, the simulation setup is explained, the results and mode analysis are presented in Section 4, and are summarised in the final Section 5.

2. Analytical considerations

We now briefly revisit the calculation why the non-resonance of the two characteristic frequencies of the system

$$\omega_{*a} \cdot \overline{\omega_{da}} < 0. \tag{1}$$

leads to resilience against trapped-particle modes. For the detailed calculation the reader is referred to [2]. The calculation is based upon the gyrokinetic equation in ballooning space,

$$i v_\| \nabla_\| g_a + (\omega - \omega_{da}) g_a = \frac{e_a \phi}{T_a} J_0(k_\perp v_\perp / \Omega_a) \left(\omega - \omega_{*a}^T\right) f_{a0}, \tag{2}$$

together with the quasi-neutrality condition,

$$\sum_a \frac{n_a e_a^2}{T_a} \phi = \sum_a e_a \int g_a J_0 \mathrm{d}^3 v. \tag{3}$$

Here ϕ is the electrostatic potential, J_0 is the zeroth order Bessel function of the first kind, $g_a = f_{a1} + \frac{e_a \phi}{T_a} f_{a0}$ denotes the non-adiabatic part of the perturbed distribution function, and the equilibrium distribution function f_{a0} is Maxwellian. The ratio between the temperature and density gradients is denoted by $\eta_a = d \ln T_a / d \ln n_a$, and we have written $\omega_{*a}^T = \omega_{*a}[1 + \eta_a(x^2 - 3/2)]$, with $x^2 = m_a v^2 / 2T_a$. From the gyrokinetic equation we can obtain a quantity

$$P_a = e_a \mathrm{Im} \left\{ (i v_\| \nabla_\| g_a - \omega_{da} g_a) \phi^* J_0 \right\}, \tag{4}$$

which, following [6], can be understood as the gyrokinetic energy transfer from the electrostatic fluctuations to species a. Here we use the notation

$$\{\cdots\} = \int_{-\infty}^{\infty} \frac{\mathrm{d}l}{B} \int (\cdots) d^3 v.$$

When P_a is directly obtained from the gyrokinetic equation and we sum over all species, we find for a complex mode frequency $\omega = \omega_r + i\gamma$ a relation for the energy budget of the fluctuations,

$$-\gamma \sum_a \frac{n_a e_a^2}{T_a} \int \frac{\mathrm{d}l}{B}(1-\Gamma_0)|\phi|^2 = \sum_a P_a, \tag{5}$$

where $b = k_\perp^2 T_a / m_a \Omega_a^2$ and

$$\Gamma_0(b) = n_a^{-1} \int J_0^2 f_{a0} d^3v < 1.$$

For a positive growth rate we thus require the sum on the right-hand side to be negative. For particles that bounce faster than the frequency of the mode, $\omega \ll \omega_{ba}$, which allows the ordering $\omega \sim \omega_{*a} \ll k_\parallel (T_a/m_a)^{1/2}$, we can find a solution for the distribution function g_a where we can neglect the passing particle contribution. We therefore obtain

$$P_a = \frac{e_a^2}{T_a} \mathrm{Im} \left\{ (\omega - \omega_{*a}^T) \left(\overline{|J_0\phi|^2} - \frac{\omega |\overline{J_0\phi}|^2}{\omega - \overline{\omega}_{da}} \right) f_{a0} \right\}. \tag{6}$$

If we now approach the point of marginal stability, $\gamma \to 0+$, we find

$$P_a = \frac{\pi e_a^2}{T_a} \left\{ \delta(\omega - \overline{\omega}_{da}) \overline{\omega}_{da} (\overline{\omega}_{da} - \omega_{*a}^T) |\overline{J_0\phi}|^2 f_{a0} \right\}. \tag{7}$$

Let us now limit ourselves to scenarios where the temperature gradient is smaller than the density gradient, more specifically $0 < \eta_a < 2/3$, which ensures that ω_{*a} and ω_{*a}^T have the same sign for all energies. For quasi-isodynamic configurations, we know in addition that ω_{*a}^T and $\overline{\omega}_{da}$ are of opposite signs, consequently $P_a > 0$ and energy is transferred from the electrostatic fluctuations to the particles of plasma species a. For modes whose frequency is well below the bounce frequency of all plasma constituents $\omega \ll \omega_{ba}$ for all a - as for the collisionless trapped-particle mode - we therefore obtain $\sum_a P_a > 0$. However, according to Eq. (5) we expect $\sum_a P_a = 0$ at the point of marginal stability. Consequently, this point cannot exist, and neither can the corresponding mode. The special case where the real part of the mode frequency vanishes, $\omega_r = 0$, needs to be treated separately, but this mode can be ruled out as well. We have therefore shown the resilience of quasi-isodynamic configurations against the collisionless trapped-particle mode. In addition, by explicitly solving the gyrokinetic equation in general geometry, we can demonstrate that if there are any modes whose frequency is below the electron bounce frequency but comparable to or larger than the ion bounce frequency, $\omega_{bi} \simeq \omega \ll \omega_{be}$, in a quasi-isodynamic configuration, it must propagate in the ion diamagnetic direction and draw its energy from the ions rather than the electrons at marginal stability. The ordinary TEM is thus expected to be stable as well. Since it is known that perfect quasi-isodynamicity cannot be achieved, simulating trapped-particle modes in realistic configurations is essential. Even if most, but not all, orbits satisfy $\omega_{*e} \cdot \overline{\omega_{de}} < 0$ one would expect a reduced growth rate compared with a tokamak or a non-optimised stellarator.

3. The simulation setup

The simulations are performed with the gyrokinetic code GENE [7, 8]. For computational efficiency, field-aligned coordinates are used. Here, x corresponds to the flux label, z the coordinate along the field line and y the binormal coordinate. Since we are interested in collisionless, electrostatic instabilities, we neglect both collisions and magnetic fluctuations. (It was shown in tokamaks that the effect of the plasma β on TEMs is very small [9, 10].) We will, however, keep the full magnetic drift when using a high-β equilibrium, since our modes are expected to be highly sensitive to the sign of ω_{da}. Both the electrons and the ions are treated kinetically, and the mass ratio is $m_e/m_i = 5.4466 \cdot 10^{-4}$, which corresponds to a hydrogen plasma. The simulations are performed in a flux tube, therefore only a small fraction of the surface is covered. While definite statements about the stability of the entire plasma cannot be made without the knowledge of the stability properties of the entire flux surface, flux tube simulations can give an estimate of the stability of a given magnetic geometry.

We investigate the stability of the six-periodic stellarator of Subbotin *et al.* - in a configuration very similar to that presented in [1] and with major and minor radii of $R_0 = 11.85\mathrm{m}$ and $a = 0.95\mathrm{m}$ respectively - and compare the performance of the vacuum case with a high - β case with $\beta = 6\%$. As we shall see, the configuration with $\beta = 6\%$ is nearly quasi-isodynamic whilst the vacuum field is not. In order to investigate the influence on microinstabilities, we will compare these in the $\alpha = 0$ flux tube on the surface at half radius of each configuration. In Fig. 1 and 2 the bounce averaged drift frequency $\overline{\omega_{de}}$ is shown for both cases, with the closed black lines denoting the zero contour and the open black line the magnetic field line with $\alpha = 0$.

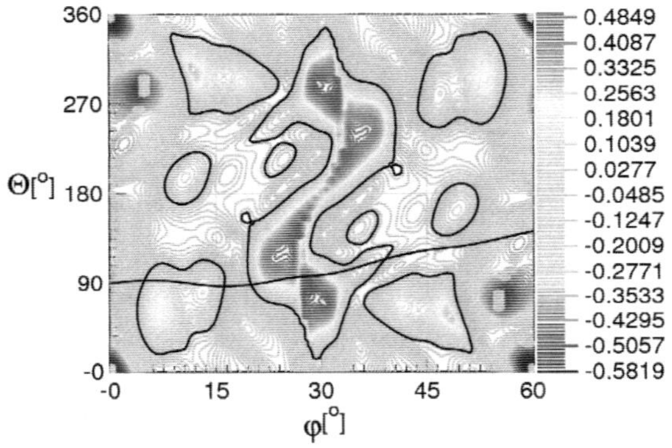

Figure 1. $\overline{\omega_{de}}$ for the vacuum configuration, shown as a function of the poloidal and toroidal angles at the bounce points

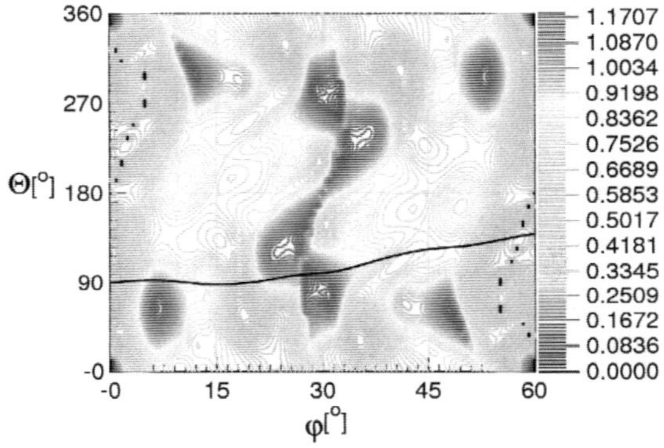

Figure 2. $\overline{\omega_{de}}$ for the high β configuration, shown as a function of the poloidal and toroidal angles at the bounce points

Note that in Fig. 2 there is no zero contour, and since ω_{*e} has been defined to be negative for conventional density gradients our stability criterion $\omega_{*a} \cdot \overline{\omega_{da}} < 0$ holds everywhere on the flux surface. It must be noted, however, that even though the $\beta = 6\%$ case comes very close to quasi-isodynamicity, the surfaces of constant J do not quite coincide with flux surfaces.

For both cases the shear $\hat{s}$ is very small, and in order to decrease the computational cost the simulations were limited to only one poloidal revolution. The modes were observed to decay sufficiently fast towards the edges of the flux tube, and did not change if the computational domain was extended to cover several turns around the torus.

4. Simulation results

4.1. Comparison of stability properties

For the simulations the normalised density gradient $a/L_n = -a\ \mathrm{d}\ln n/\mathrm{d}r$ with the minor radius a was varied simultaneously for both species between values of 0.0 and 3.0 In addition the normalised electron temperature gradient $a/L_{T_e} = -a\ \mathrm{d}\ln T_e/\mathrm{d}r$ was also varied between 0.0 and 3.0. The ion temperature profile was deliberately kept flat $a/L_{T_i} = 0$ to avoid the onset of ion-temperature-gradient-driven modes. For each pair of gradients we scanned over the binormal wave vector $k_y\rho = [0.1 \ldots 0.9]$ to find the most unstable mode (excluding the ETG mode, which has higher k_y). For comparison we also calculated the equivalent stability diagram for the Cyclone base case tokamak [11]. The stability diagram of this tokamak is displayed in Fig. 3. It can be seen that increasing the temperature gradient has a destabilising influence and that the highest growth rates slightly exceed $\gamma = 0.7c_s/a$, where c_s is the sound speed.

Journal of Physics: Conference Series **401** (2012) 012021 doi:10.1088/1742-6596/401/1/012021

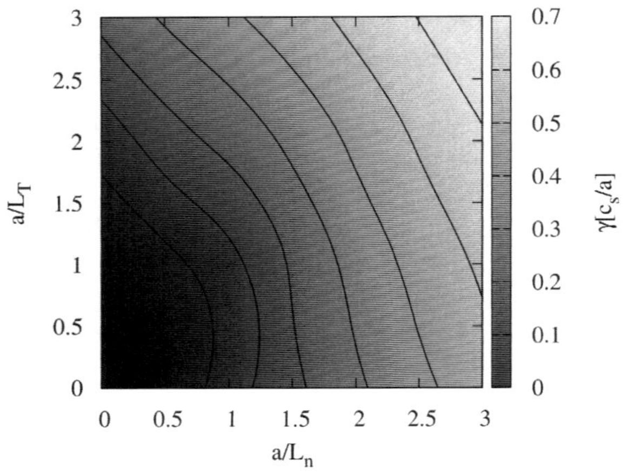

Figure 3. Stability diagram showing the growth rates of each most unstable low-frequency mode for the Cyclone base case tokamak. The lines denote the contours of $\gamma = 0.1c_s/a, \ldots, 0.7c_s/a$.

From the stability diagrams of the six-periodic stellarator, we see that the highest growth rates of the vacuum case (Fig. 4) are comparable to those in the tokamak. However, the electron temperature gradient is stabilising for the stellarator. In contrast, the high-β-case (Fig. 5) is hardly unstable at all. The fastest growing mode is found at $a/L_n = 3.0$ with a flat temperature profile and reaches a growth rate of mere $\gamma = 0.12\ c_s/a$.

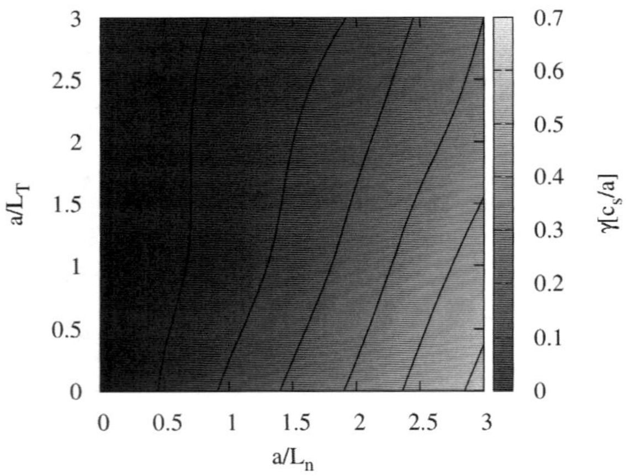

Figure 4. Stability diagram showing the growth rates of each most unstable mode for the vacuum configuration. The lines denote contours of $\gamma = 0.1c_s/a, \ldots, 0.6c_s/a$.

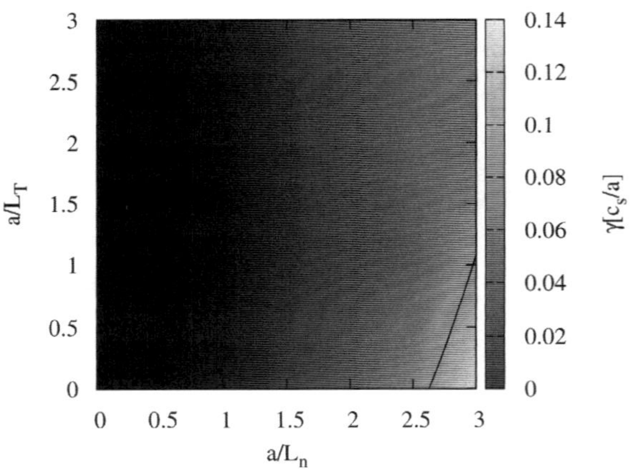

Figure 5. Stability diagram showing the growth rates of each most unstable mode for the high β configuration. The line denotes the $\gamma = 0.1c_s/a$ - contour.

The improved stability of the high-β-case becomes even more evident when one juxtaposes the density gradient scans at different temperature gradients for the two configurations. To display the influence of the density gradient more clearly, only one $k_y\rho$ is chosen for each temperature gradient.

From Fig. 6 we conclude that the temperature gradient acts stabilising unless the density gradient is very small. We also see that there is no critical density gradient for the vacuum configuration. The high-β-case on the other hand clearly has a critical density gradient at $a/L_n = 0.5$, but even above this

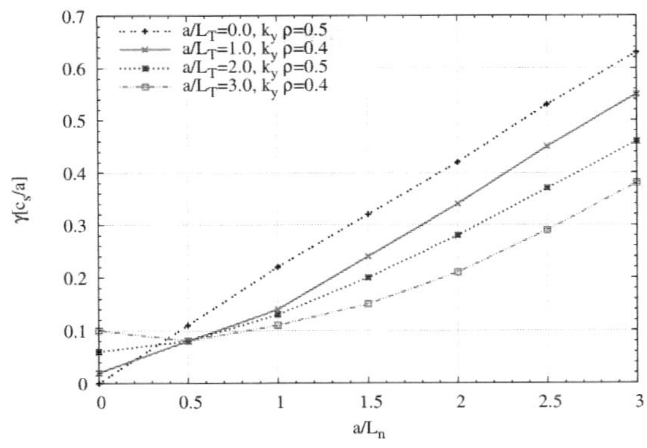

Figure 6. Density gradient scans for different temperature gradients for the vacuum configuration

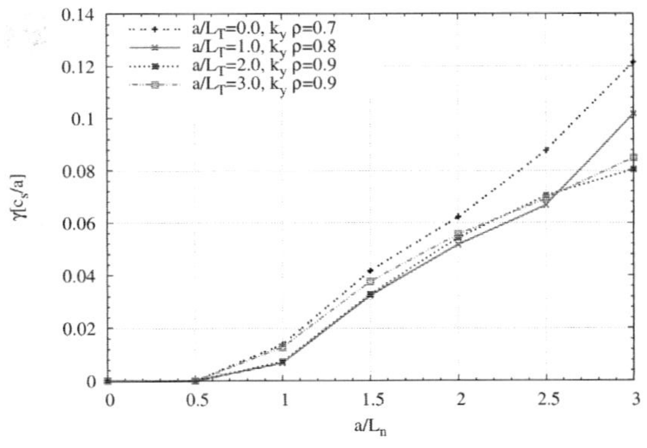

Figure 7. Density gradient scans for different temperature gradients for the high β configuration

gradient do the growth rates remain far below the vacuum case.

4.2. Mode analysis

The structure of the electrostatic potential of most of the modes observed (Fig. 8), where the peaks coincide with magnetic wells, clearly hints at a trapped-particle mode.

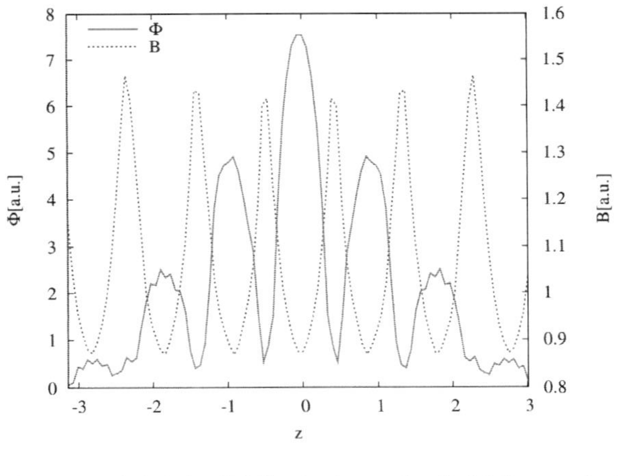

Figure 8. The electrostatic potential ϕ (solid line) along the field line and the magnetic field strength B (dashed line) along the field line, displaying a characteristic trapped-particle mode. Modes of this structure were observed almost everywhere in the parameter space.

In order to gauge the nature of the instabilities observed in our simulations, it is useful to compare the mode frequency ω with the bounce frequency of both the electrons and the ions. For the quasi-isodynamic high-β-case, the most unstable mode observed at high density gradient and with a high $k_y\rho$ had a ratio of $\omega/\omega_{be} \approx (0.22 - 0.27)$, while the majority of modes observed had lower frequencies. The frequencies are thus in general higher than the ion bounce frequency ω_{bi} but below the electron bounce frequency ω_{be},

$$\omega_{bi} < \omega < \omega_{be}.$$

Journal of Physics: Conference Series **401** (2012) 012021 doi:10.1088/1742-6596/401/1/012021

As already mentioned, for instabilities of this kind it can be shown analytically that at the point of marginal stability only modes travelling in the ion direction exist, and they would need to be driven by the ions rather than the electrons. The modes we found in the quasi-isodynamic high-β-configuration indeed had a positive frequency, i.e. they were propagating in the ion diamagnetic direction. Now we pick two modes very close to the point of marginal stability - for the vacuum case $a/L_{T_e} = 1.0$, $a/L_n = 1.5$ and $k_y\rho = 0.8$, for the high-β case the same gradients but $k_y\rho = 0.3$ - and analyse the energy transfer from the particles to the mode with an energy diagnostic built into GENE by A. Bañón Navarro [12]. While for the vacuum case (Fig. 9) both electrons and ions drive the instability, it is only the ions in the high-β-case (Fig. 10) that pump energy into the mode while the electrons - as predicted analytically - draw energy from the instability. It is thus not surprising that the quasi-isodynamic high-β-case is more stable than the vacuum case.

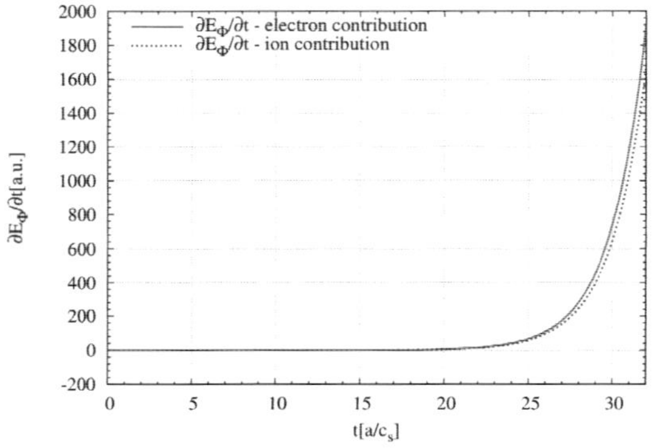

Figure 9. Change of electrostatic energy of the mode due to the electrons (solid line) and the ions (dashed line) in the vacuum configuration.

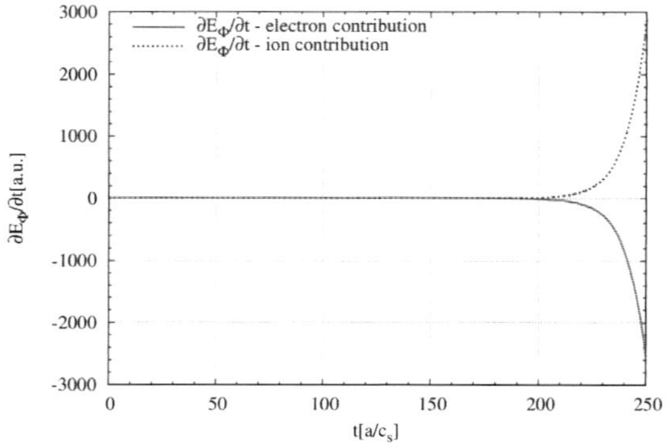

Figure 10. Change of electrostatic energy of the mode due to the electrons (solid line) and the ions (dashed line) in the high-β configuration.

5. Conclusions and Outlook

In this paper we have presented electrostatic trapped-particle mode simulations in an almost perfectly quasi-isodynamic stellarator with $\beta = 6\%$ and the corresponding non-quasi-isodynamic vacuum configuration. Extensive scans over both electron temperature gradient and density gradient revealed a stabilising influence of the temperature gradient for both cases and significantly reduced growth rates for the quasi-isodynamic configuration. While no critical density gradient could be found for the vacuum case, there was a very clear critical gradient at $a/L_n = 0.5$ in the high-β-case, and the growth rates were much smaller. An analysis of the modes found in the high-β-case showed pronounced peaks of the mode structure in the magnetic wells, thus pointing at a trapped-particle mode. In the energy transfer diagnostic it was revealed that for the vacuum case both ions and electrons were driving the mode close to marginal stability while for the quasi-isodynamic configuration only the ions were providing the drive (through their density gradient) whereas the electrons were drawing energy from the mode. This is consistent with the analytical calculations discussed in Sec. 2.

It must be kept in mind that here only one particular flux tube was simulated. Further simulations using other flux tubes are thus required to strengthen the results. Ultimately simulations of an entire flux surface, or the entire plasma volume, might be required. Other configurations should be tested as well, such as Wendelstein 7-X, which is also approaching quasi-isodynamicity, or LHD and NCSX.

Acknowledgments
The authors wish to thank Frank Jenko, Thomas M Bird, Daniel Told, Tobias Görler and Gabriel G Plunk for the fruitful discussions as well as Yuriy Turkin for his help with the visualisation. We would also like to thank Mikhail Mikhailov for providing us with the equilibrium files and Alejandro Bañón Navarro for providing the energy diagnostic. Some of the GENE calculations were performed on the Helios supercomputer (Japan).

References

[1] A A Subbotin, M I Mikhailov, V D Shafranov, M Yu Isaev, J Nührenberg, C Nührenberg, R Zille, V V Nemov, S V Kasilov, V M Kalyuzhnyj and W A Cooper 2006 *Nucl. Fusion* **46** 921
[2] J H E Proll, P Helander, J W Connor and G G Plunk 2012 *Phys Rev. Lett.* **108** 245002
[3] P Helander and J Nührenberg 2009 *Plasma Phys. Control. Fusion* **51** 055004
[4] J Nührenberg 2010 *Plasma Phys. Control. Fusion* **52** 124003
[5] M I Mikhailov, V D Shafranov and J Nührenberg 2009 *Plasma Phys. Rep.* **35** 529
[6] A A Schekochihin, S C Cowley, W Dorland, G W Hammett, G G Howes, E Quataert and T Tatsuno 2009 *Ap. J. (Suppl. Series)* **182** 310
[7] F Jenko, W Dorland, M Kotschenreuther and B N Rogers 2000 *Phys. Plasmas* **7** 1904
[8] T Dannert and F Jenko 2005 *Phys. Plasmas* **12** 072309
[9] M J Püschel, M Kammerer and F Jenko 2008 *Phys.Plasmas* **15** 102310
[10] M J Püschel and F Jenko 2012 *Phys. Plasmas* **17** 062307
[11] A M Dimits, G Bateman, M A Beer, B I Cohen, W Dorland, G W Hammett, C Kim, J E Kinsey, M Kothschenreuther, A H Kritz, L L Lao, J Mandrekas, W M Nevins, S E Parker, A J Redd and D E Shumaker 2000 *Phys. Plasmas* **7** 969
[12] A Bañón Navarro, P Morel, M Albrecht-Marc, D Carati, F Merz, T Görler and F Jenko 2011 *Phys. Rev. Lett.* **106** 055001

Joint Varenna-Lausanne International Workshop 2012 IOP Publishing
Journal of Physics: Conference Series **401** (2012) 012022 doi:10.1088/1742-6596/401/1/012022

Nonlinear alfvénic fast particle transport and losses

M Schneller[1], Ph Lauber[1], M García-Muñoz[1], M Brüdgam[1] and S Günter[1]

[1] Max Planck Institute for Plasma Physics, EURATOM Association, Boltzmannstr. 2, 85748 Garching, Germany

E-mail: `mirjam.schneller@ipp.mpg.de`

Abstract. Magnetohydrodynamic instabilities like Toroidal Alfvén Eigenmodes or core-localized modes such as Beta Induced Alfvén Eigenmodes and Reversed Shear Alfvén Eigenmodes driven by fast particles can lead to significant redistribution and losses in fusion devices. This is observed in many ASDEX Upgrade discharges. The present work aims to understand the underlying resonance mechanisms, especially in the presence of multiple modes with different frequencies. Resonant mode coupling mechanisms are investigated using the drift kinetic HAGIS code [Pinches 1998].

Simulations were performed for different plasma equilibria, in particular for different q profiles, employing the availability of improved experimental data. A study was carried out, investigating double-resonant mode coupling with respect to various overlapping scenarios. It was found that, depending on the radial mode distance, double-resonance is able to enhance growth rates as well as mode amplitudes significantly. Small radial mode distances, however can also lead to strong nonlinear mode stabilization of a linear dominant mode.

With the extended version of HAGIS, losses were simulated and directly compared with experimental loss measurements. The losses' phase space distribution as well as their ejection signal is consistent with experimental data. Furthermore, it allowed to characterize them as prompt, resonant or stochastic. It was found that especially in multiple mode scenarios (with different mode frequencies), abundant incoherent losses occur in the lower energy range, due to a broad phase-space stochastization. The incoherent higher energetic losses are "prompt", i.e. their initial energy is too large for confined orbits.

1. Introduction

Fusion devices contain fast particle populations due to external plasma heating and (eventually) fusion-borne α-particles. Fast particle populations can interact with global electromagnetic waves, leading to the growth of MHD-like and kinetic instabilities – e.g., Toroidicity Induced Eigenmodes (TAE) [1], Reversed Shear Alfvén Eigenmodes (RSAE) [2] or Beta Induced Alfvén Eigenmodes (BAE) [3]. In this work, drift-kinetic fast particle simulations performed with the HAGIS code [4] are carried out to obtain a deeper understanding of the dynamics of wave-fast particle interaction in scenarios with two modes of different frequencies: what is the mode coupling mechanism, and how does it dependent on radial mode distance? In a second study, the extended version of HAGIS [5] is used to simulate fast particle losses and to compare them with experimental loss measurements [6] in ASDEX Upgrade (AUG). The losses' phase space distribution as well as their ejection signal allows to characterize them as prompt, resonant or stochastic.

2. Simulation Tool and Experimental Reference Case

The numerical investigations are performed with the HAGIS Code [4], a nonlinear, drift-kinetic, perturbative Particle-in-Cell code, that models the interaction between a distribution of energetic particles and a set of Alfvén Eigenmodes. The plasma equilibrium for HAGIS is based on the CLISTE [7] and HELENA [8] codes. The data for the MHD equilibria originate from the ICRH minority heated ASDEX Upgrade discharge #23824, at times $t = 1.16$ s and $t = 1.51$ s. At the earlier time point, the q profile is slightly inverted ($q_0 = -1.55$, $q_{\text{min}} = -1.43$) at the later time point, it is monotonic, with lower absolute values ($q_0 = q_{\text{min}} = -0.97$)[1]. This particular reference scenarios were chosen due to the availability of detailed experimental data concerning fast particle-wave interaction and losses. The comparison between these data and numerical results is subject of section 4.

3. Numerical Study on Double-Resonance

Theoretical Picture Theory (e.g. Ref. [9]) predicts that conversion of free energy to wave energy is enhanced in a multiple-mode scenario, i.e., the interaction of multiple modes produces energy conversion rates higher than that which would be achieved with each mode acting independently. This effect is called "double-resonance" and can be partially explained by the principle of gradient driven mode growth – according to[2]: $\gamma \propto \nabla f(s)$ [10] – which can be extended to multiple modes [9, 11]. This picture of ***gradient driven double-resonance*** is based on the precondition that modes share resonances in the same phase space area. Through the resulting redistribution by each mode, a steeper gradient is produced at the other mode's position, enhancing its drive. The radial overlapping of modes leads then to a much larger conversion of free energy to wave energy.
However, this mechanism can only work if the modes not only share resonance regions in phase space, but if there is also *spatial* mode overlap in the radial direction. In Ref. [5] simulations were carried out, finding a double-resonant effect also without this precondition. Furthermore, a superimposed oscillation on the modes' amplitudes was observed, clearly indicating mode-mode interaction. The modes without radial overlap are then coupled radially through the particles' trajectories: A population of particles that shares resonances in phase space with both modes and passes both modes' location at once, can transfer energy from one mode to the other [5]. Due to the particle orbits' width, it is not necessary that both modes have a radial overlap. In the following, this mechanism is called ***inter-mode energy transfer***.

Simulation Conditions As the first question, the understanding of double mode resonance is very fundamental, the simulations were performed under quite simple physical conditions: radial particle distribution $f(\psi)$ with constant gradient (to avoid different mode drive at different radial mode positions only due to a steeper gradient) and a slowing down function as energy distribution function. The volume averaged fast particle beta is β_f=1%. The particles are distributed isotropically in pitch angle (as e.g., fusion borne α-particles would be).

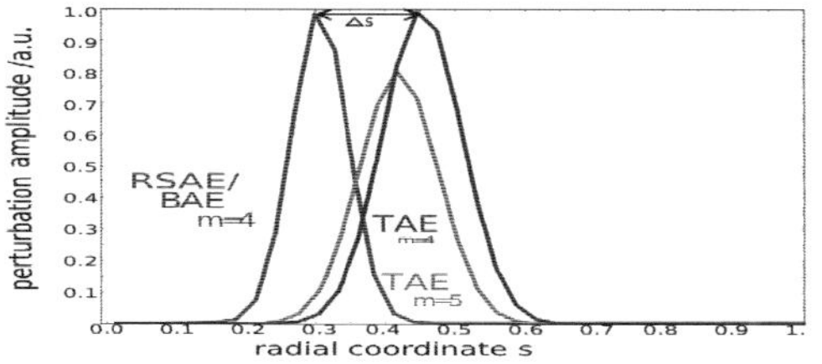

(a) analytical perturbation

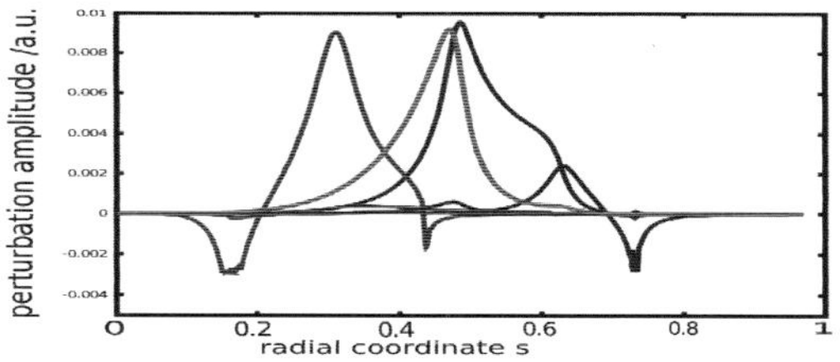

(b) numerical perturbation

Figure 1: *Two Alfvénic modes: (a) gauss-shaped perturbation as used in the following simulations, (b) for comparison: perturbations occurring in AUG discharge #23824, as calculated numerically with* LIGKA *[12].*

[1] Note that the q profile is negative here, due to the AUG current direction.
[2] s refers to the radial coordinate as the square root of the normalized poloidal flux: $s = (\psi_{\text{pol}}/\psi_{\text{pol,edge}})^{1/2}$.

As MHD perturbations, analytic, gauss shaped functions are used (see figure 1), without background damping. The mode frequencies are chosen to match experimental data: one high frequency mode 120 kHz (TAE) and one lower frequency mode at 55 kHz (possible RSAE or BAE).

Simulation Results A scan over the radial mode distance reveals an effective double-resonance as shown in figure 2: depicted are the ratios of the linear growth rates (a) and the amplitudes (b) in the double mode case vs. the single mode case over the radial mode distance Δs. The amplitude level was compared after ≈ 300 TAE periods ($= 2.5$ ms) of simulation time. This time is sufficient for the single amplitudes to saturate, but still significantly below energy slowing down time. One can see that the **growth rates** of both the TAE and the RSAE are enhanced in all double mode cases compared to the single mode ones. However, the growth rate of the outer TAE is enhanced most strongly and independently of the radial mode distance – i.e. gradient driven double-resonance works even if there is no radial mode overlap. In contrast, the enhancement of the inner and weaker RSAE decreases with the radial mode distance for small Δs. Then it increases again for $\Delta s > 0.15$. These larger mode distances match the double-resonant particle orbits and therefore enable inter-mode energy exchange, driving the weak mode. For higher Δs, the larger, i.e. higher energetic orbits fit the mode distance and lead to even more energy exchange. Furthermore, with larger radial mode distances, the modes are able to tap energy from a wider gradient region. The **amplitude** ratios, however, are even lowered in the double-resonant case compared to the respective single mode levels, if the radial mode distance is small. This happens due to the mutual gradient depletion at the other mode's radial position. If the modes are at a larger radial distance ($\Delta s > 0.15$), the double mode scenario amplitudes are much higher compared to the single mode amplitudes, both for the TAE and the RSAE. Both modes drive each other – most for a radial distance of about $\Delta s \approx 0.25$.

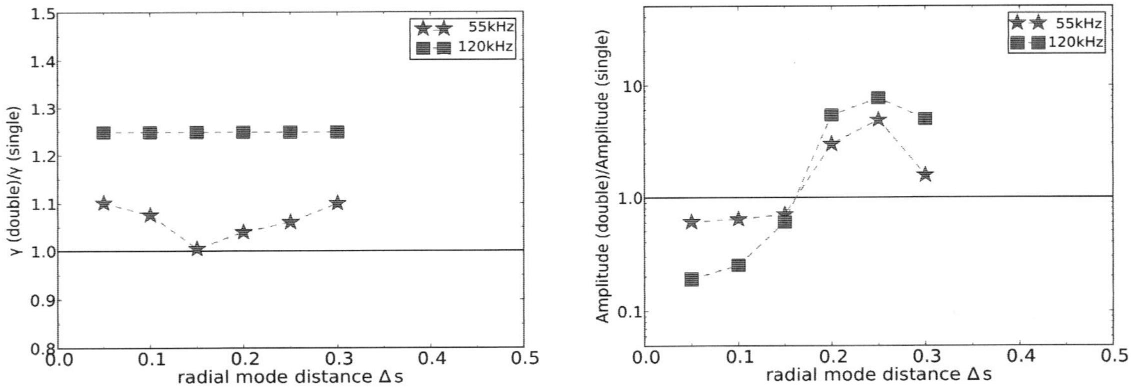

(a) *Growth rates enhancement* (b) *Amplitudes enhancement (at t ≈ 300 T)*

Figure 2: *Mode evolution scanned over the radial mode distance Δs in the inverted q profile case. Depicted is the Δs dependence of the ratios of growth rates (a) and amplitudes (b) in double mode simulations over those from single mode simulations. Red star: RSAE, blue square: TAE. One can clearly see that larger radial mode distances lead to higher amplitudes, whereas amplitudes are even lower than in the single mode case for $\Delta s \leq 0.15$. The linear growth rates, however, are higher than in the single mode simulation throughout the Δs range. The RSAE growth rate experiences a small drop at $\Delta s \approx 0.15$.*

It is important to note that the distance $\Delta s = 0.25$ giving maximum amplitude ratios depends strongly on the absolute mode positions with respect to the radial distribution function, and especially on the amplitude regime (stochastic or non-stochastic) of each mode. The same applies for the value $\Delta s = 0.15$ at which the transition towards double-resonant amplitude enhancement takes place.

Summary (I) Growth rates are generally enhanced by the presence of another mode, whereas for the amplitudes, this is only the case, if the modes have sufficiently large radial distance. For small distances, modes at radial positions, where the initial distribution function is already relatively flat, double-resonance leads to strong mode stabilization with respect to the single mode case. If the amplitudes are enhanced, their amplification level is, however, mainly determined by whether the mode reaches the stochastic regime. The stochastic threshold is reached much earlier or even only if a second mode is present.

4. Fast Particle Losses in Numerical Simulations vs. Experimental Measurements

Experimental Loss Observations In the AUG discharge #23824, during the inverted q profile equilibrium at $t = 1.16$ s, a high loss signal is observed at the Fast Ion Loss Detector (FILD), with the majority of losses characterized as *incoherent*. Later, at $t = 1.51$ s, during the monotonic q profile equilibrium, only very few losses occur, and all are identified as *coherent* [6]. In the following, simulations within this experimental reference frame are presented, that were performed to identify the origin of the different loss types.

Simulation Conditions The simulations presented in this section are carried out with an extended HAGIS version [5], including the vacuum region, and are based on more realistic plasma conditions: The radial particle loading follows a Fermi-like potential law $f(s) = (1 - s^2)^5$, that decays to zero at $s = 0.6$. It was chosen in a way that does not produce prompt losses in the simulations with monotonic q profile, as seen in the experiment. Furthermore, the markers are initialized in pitch and poloidal angle distribution according to an analytical function that models the distribution created by ICRH (main heating method in AUG #23824). β_f is set to 0.1%. As MHD perturbation, the analytic, gauss functions (figure 1) are used, without background damping. The mode frequencies, positions and widths are chosen to be consistent with experimental data: at $t = 1.16$ s, a 120 kHz TAE at $s = 0.45$ and a 55 kHz RSAE at $s = 0.35$, whereas at $t = 1.51$ s, there is a TAE of 180 kHz at $s = 0.6$ and a BAE of 80 kHz at $s = 0.35$ [13]. These data were obtained from Fourier spectrograms of the SXR signal, Mirnov Coils and the FILD signal [14].

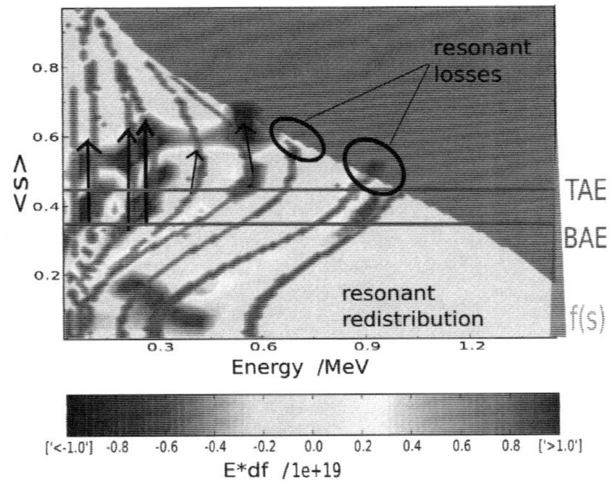

(a) *redistribution in phase space*

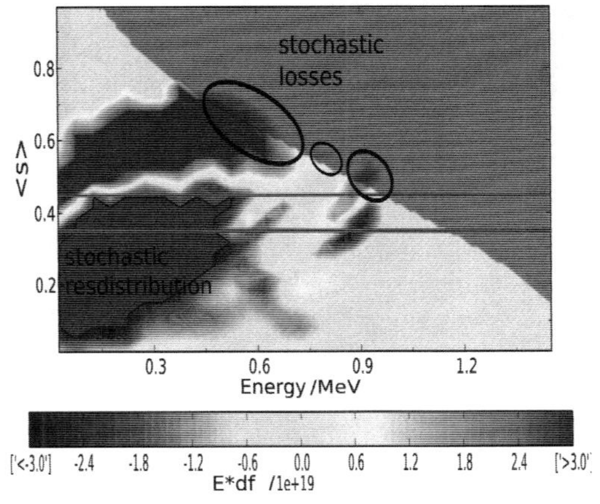

(b) *losses in phase space*

Figure 3: *Double mode simulation in the inverted q profile equilibrium: Redistribution in E-s space during (a): redistribution phase (t ≈ 0.6 ms), (b) stochastic phase (t ≈ 1.8 ms) Red: particles accumulate, blue: particles move away. One can see the good accordance of redistribution areas (black arrows) and the resonance lines (pink) in the modes' vicinity (horizontal lines). The redistribution in the areas, where the resonance lines meet the loss boundary (black circles) coincide with the losses (see figure 4a).*

Simulation Results Consistent with results from the numerical study, it is found that the mode amplitudes grow slower with the monotonic q profile and do not reach the stochastic threshold (at $\approx 2 \cdot 10^{-3}\delta B/B$). A superimposed oscillation frequency is visible in both scenarios, originating from the double-resonance mechanism of inter-mode energy exchange. Its frequency matches with the beat frequency of both modes. In the inverted q profile, the modes reach the stochastic regime. There, particle redistribution in phase space takes place broadly over the whole energy range (figure 3b) at the modes' radial position. Slightly before, during the resonant phase, however, redistribution in phase space occurs along the resonance lines (figure 3a). The energy of the losses obtained during this "redistribution phase" (see figure 4a) matches roughly with the energy, at which the higher resonance lines meet the loss boundary. This indicates, that losses in this time period are *resonant*, i.e. *coherent* losses. In the simulation with the monotonic q profile, there is no redistribution into the loss region. Thus, very few losses appear, consistent with experimental measurements.

To learn more about the identity of the *incoherent* losses, their **phase space pattern** is analyzed in the following. The distribution function of the fast particles is not known exactly, therefore, the markers are loaded according to an estimated distribution function. To learn how sensitive the loss pattern is with respect to the parameters defining the distribution function, a scan is performed over the radial, as well as over the poloidal distribution function. Combining the results of both the radial and the poloidal marker loading scan, it turns out that there is a phase space region that cannot be covered by prompt losses in neither loading scenario: the phase space region of energies below 600 keV (45 mm gyroradius) at negative pitch angles smaller than 63°. These losses are very likely to be stochastic losses. For the rest of the phase space, it is not possible to discern prompt from later losses yet.

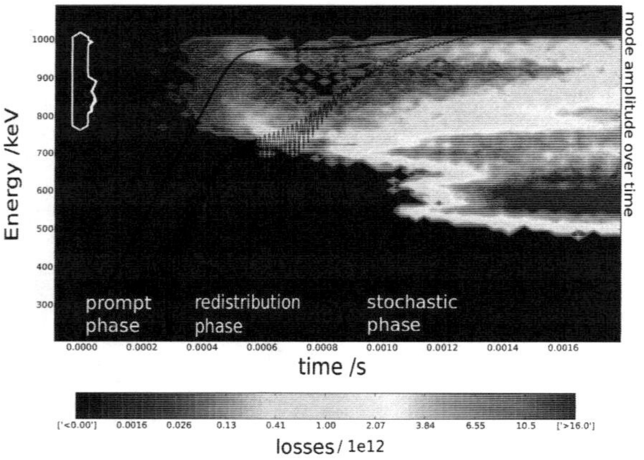

Figure 4: *Lost particles at the first wall over simulation time in the inverted q profile: in the stochastic phase, losses spread over a large energy range, even to lower energies around 500 keV. For comparison, the edge of the phase space region of the prompt losses in this simulation is shown (white lines). Black lines: amplitude evolution of RSAE (upper) and TAE (lower).*

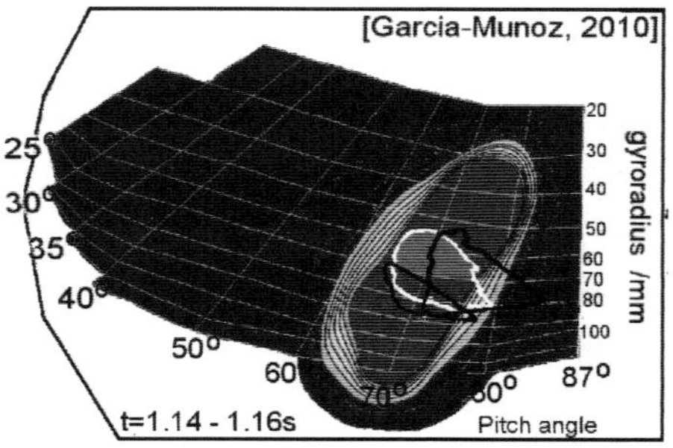

Figure 5: *Fast ion losses in phase space (AUG discharge #23824): colors give the FILD measured loss pattern [6], lines give the boundary of the numerical calculated loss region. Black lines: upper estimate for the (drift corrected) border of the prompt losses appearance resulting from the radial and poloidal marker loading scan. White lines: same for the losses during the stochastic phase of the MHD perturbations. The lighter shaded area indicates the peak region of stochastic losses.*

Figure 5 shows the good accordance of the numerically calculated losses and the experimentally measured ones: the whole loss area as well as the peak region of the numerical values lies almost entirely within the respective experimentally measured region of phase space. The experimental peak being slightly broader results mainly from over exposure of the FILD loss plate. The higher energetic losses (i.e. the pattern at gyroradii > 90 mm or energy > 1 MeV) do not appear in the simulation, as the fast particle energy was cut at 1 MeV. The lower energetic losses below 50 mm

Journal of Physics: Conference Series **401** (2012) 012022 doi:10.1088/1742-6596/401/1/012022

gyroradius are expected to appear when simulating with more than two modes at lower frequencies (work in progress).

When looking closely at the loss time traces, one can perceive an ejection modulation. A **Fourier analysis of the loss signal** gives a peak at 65 kHz, which is exactly the beat frequency. In the above simulation, the mode frequencies result in beat frequencies of 65 kHz and 175 kHz. Unfortunately, the RSAE frequency is very close to the lower beat frequency, and the TAE frequency almost coincides with the first harmonic of it. Therefore, another simulation was performed, using $\omega_{\text{RSAE}} = 70$ and $\omega_{\text{TAE}} = 100$ kHz. A Fourier analysis of the loss signal is performed for the resonant phase (figure 6a), and the stochastic phase (figure 6b). In the resonant phase, the RSAE frequency peak at around 70 keV is visible, whereas the TAE's amplitude is still too small to produce significant losses. Later on in the simulation, the TAE grows towards the same amplitude range as the RSAE, likewise, the beat frequency peak becomes visible in the Fourier spectrum. This is consistent with FILD spectra, showing the beat frequency as well.

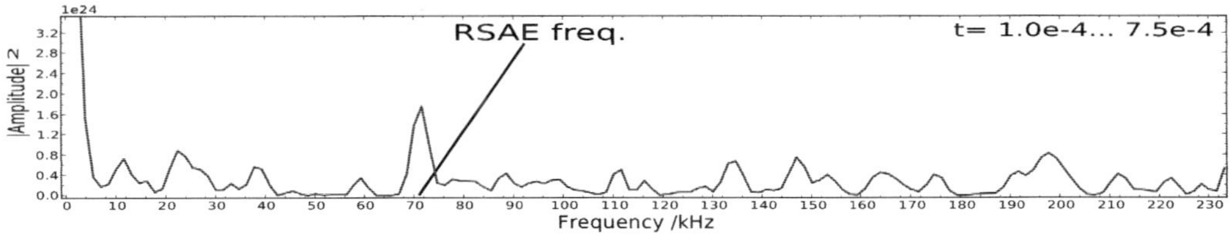

(a) *FFT of losses' time signal during resonant phase*

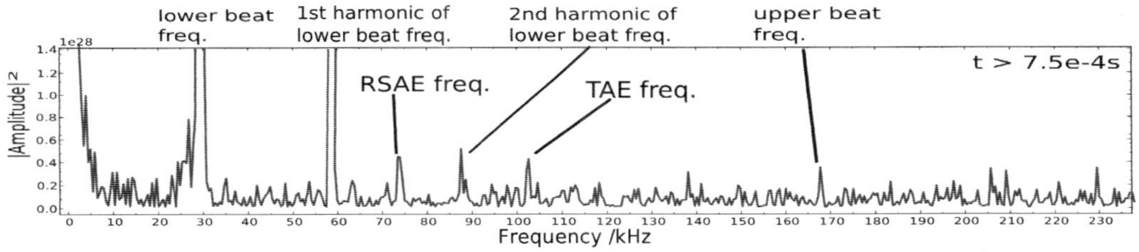

(b) *FFT of losses' time signal during stochastic phase*

Figure 6: *Fourier spectra of losses time signal. (a) in the resonant phase: a significant number of particles is ejected with the frequency of the dominating RSAE (70 kHz). (b) in the stochastic phase: the beat frequency (30 kHz) is by far dominating (and its higher harmonics). All frequencies experience a slight down-shift.*

Summary (II) With the extended version of HAGIS, losses were simulated and compared with experimentally observed losses at the Fast Ion Loss Detector. The simulated losses' phase space pattern coincides very well with the experimental one. Especially in multi-mode scenarios with different mode frequencies, stochastic redistribution sets in over a broad energy range, leading to lower energetic losses, that are incoherent. The resonant losses appear from the late linear phase on, mainly in the high energy regime, showing good coherence with the modes' frequencies and especially their beat frequencies. The higher energetic part of experimentally measured incoherent losses was identified as mainly prompt losses. The internal redistribution, as well as the losses can be understood very well as processes in phase space, when combining resonance lines, loss boundary and radial positions of the Alfvénic modes.

Journal of Physics: Conference Series **401** (2012) 012022 doi:10.1088/1742-6596/401/1/012022

References

[1] Cheng C Z, Chen L and Chance M 1985 *Ann. Phys.* **161** 21 – 47

[2] Berk H L, Borba D N, Breizman B N, Pinches S D and Sharapov S E 2001 *Phys. Rev. Lett.* **87**

[3] Heidbrink W W, Strait E J, Chu M S and Turnbull A D 1993 *Phys. Rev. Lett.* **71**(6) 855–858

[4] Pinches S, Appel L, Candy J, Sharapov S, Berk H, Borba D, Breizman B, Hender T, Hopcraft K, Huysmans G and Kerner W 1998 *Comput. Phys. Commun.* **111** 133 – 149

[5] Brüdgam M 2010 Ph.D. thesis Technische Universität München

[6] García-Muñoz M et al 2010 *Phys. Rev. Lett.* **104**(18)

[7] Mc Carthy P J and the ASDEX Upgrade Team 2012 *Plasma Phys. Control. Fusion* **54** 015010

[8] Huysmans G T A et al 1990 Conf. on Computational Physics Proc. World Scientific, Singapore (AIP)

[9] Berk H L, Breizman B and Ye H 1992 *Phys. Rev. Lett.* **68**(24) 3563–3566

[10] Fu G Y and Dam J W V 1989 *Phys. Fluids B* **1** 1949–1952

[11] Berk H L and Breizman B N 1990 *Phys. Fluids B* **2** 2226–2252

[12] Lauber Ph, Günter S, Könies A and Pinches S D 2007 *J. Comp. Phys.* **226** 447 – 465

[13] Lauber Ph et al 2009 *Plasma Phys. Control. Fusion* **51** 124009

[14] García-Muñoz M et al 2011 *Nuclear Fusion* **51**

Hamiltonian structure and stability analysis of a reduced four-field model for plasmas in the presence of a strong guide field

E Tassi[1], **T S Ratiu**[2], **and E Lazzaro**[3]

[1] Centre de Physique Théorique, CNRS – Aix-Marseille Universités, Campus de Luminy, case 907, F-13288 Marseille cedex 09, France
[2] Séction de Mathématiques and Bernoulli Center, École Polytechnique Fédérale de Lausanne, CH-1015, Lausanne, Switzerland
[3] Istituto di Fisica del Plasma CNR, Euratom-ENEA-CNR Association, 20125 Milan, Italy

E-mail: tassi@cpt.univ-mrs.fr

Abstract. We analyze the collisionless and inviscid limit of a reduced fluid model for tokamak plasma dynamics. We show that the model under consideration possesses a noncanonical Hamiltonian structure with four infinite families of Casimir invariants. Sufficient conditions for energy stability are derived and formulated in terms of requirements on the current, electron pressure, and parallel ion velocity gradients, as well as on the ion temperature. In particular, the stability condition requires a gradient of the parallel flow in the presence of ion temperature and a negative upper bound on the current gradient, in the presence of pressure gradients and/or parallel velocity gradients. Examples of stable configurations are derived.

1. Introduction

Reduced fluid models play an important role in the modeling of tokamak plasmas. Although they cannot describe important phenomena such as wave-particle interaction, their advantage, in terms of computational efficiency, when compared to kinetic models, make them a valuable tool for the investigation of phenomena such as turbulence, magnetic reconnection, or formation and propagation of coherent structures. Derivation of reduced fluid models from a parent model is typically carried out at the level of the equations of motion, operating with truncations and averages. Because the parent model ultimately has to go back to the Hamiltonian system describing the interaction between charged particles and electromagnetic fields, it is expected that the resulting reduced model possesses a non-trivial Hamiltonian core, plus dissipative terms, obtained as the result of the truncations and averages [1].

Identifying such a structure in the reduced model has several advantages. For instance, it gives access to an unambiguous definition of conserved energy (the Hamiltonian in the non-dissipative limit), which is necessary for correctly evaluating energy transfers in turbulent plasmas. Moreover, from the Hamiltonian structure, further non-trivial conservation laws can be systematically derived and stability methods can be applied [2].

In this spirit, we investigate the, initially supposed, dissipationless limit of a four-field fluid model for neoclassical tearing modes [3], which we obtain by omitting from the original model the terms proportional to the collisional and viscous coefficients. In such limit, of course, the

Journal of Physics: Conference Series **401** (2012) 012023 doi:10.1088/1742-6596/401/1/012023

model does not account for tearing modes and does not possess all the sink and forcing terms, due, for instance, to neoclassical mechanisms or externally driven current, and whose influence on the reconnection process was investigated in Refs. [3, 4]. Nevertheless, for the above reasons, it remains an interesting model to investigate. Making use of the Hamiltonian structure, we obtain also criteria for plasma stability in the dissipationless limit.

First, we show that the model possesses a noncanonical Hamiltonian structure, characterized by four infinite families of Casimir invariants. Subsequently, we apply the so-called Energy-Casimir method (see, e.g., Refs. [5, 6, 7, 8, 9, 10, 2]) to determine sufficient conditions for the energy stability of equilibria obtained from a variational principle and formulate the results in terms of conditions on pressure, parallel flow, current density gradients, and ion temperature. Finally, we provide examples of stable equilibria.

2. Model equations and Hamiltonian structure

Considering a Cartesian coordinate system (x, y, z), the dissipationless limit of the four-field model derived in Ref. [3], can be written as:

$$\frac{\partial \psi}{\partial t} + [\phi, \psi] = d_i[\exp p, \psi], \tag{1}$$

$$\frac{\partial U}{\partial t} + [\phi, U] = -[\nabla^2 \psi, \psi], \tag{2}$$

$$\frac{\partial v_{i\|}}{\partial t} + [\phi, v_{i\|}] = (1 + \tau_T)[\psi, \exp p], \tag{3}$$

$$\frac{\partial p}{\partial t} + [\phi, p] = \frac{5}{3}[\psi, v_{i\|}] + \frac{5}{3} d_i [\psi, \nabla^2 \psi], \tag{4}$$

where ψ is the poloidal flux function of a magnetic field $\mathbf{B} = \nabla\psi \times \hat{z} + \hat{z}$, ϕ is the electrostatic potential, $U = \nabla^2\phi$ the $\mathbf{E} \times \mathbf{B}$ vorticity, p the logarithm of the electron pressure, $v_{i\|}$ the ion parallel velocity, whereas the constants d_i and τ_T indicate the ion skin depth and the ratio between the ion and the electron temperature, respectively. The fields ψ, ϕ, p, and $v_{i\|}$ are assumed to be translationally invariant along z and the bracket $[\,,]$ is defined by $[f, g] \equiv \hat{z} \cdot (\nabla f \times \nabla g)$, for generic fields f and g.

In Eqs. (1)-(4), lengths and time are normalized with respect to a characteristic length scale and to the Alfvén time, respectively. Eq. (1) represents an ideal Ohm's law, (2) is the plasma vorticity equation, (3) determines the evolution of the parallel ion motion, whereas (4) is an adiabatic equation of state for the electron pressure.

In what follows, we shall assume that we work on a finite domain with periodic functions, or in function spaces with enough decay at infinity that ensures that all boundary terms appearing in integration by parts vanish. We shall also assume, that in these function spaces, for any smooth functions f, g, h, we have the identity $\int d^2x[f, g]h = \int d^2x f[g, h]$ (i.e., the boundary terms appearing in the integration by parts used to establish this formula, vanish).

We address now the question of whether the system (1)-(4) is Hamiltonian. First, we recall that a dynamical system describing the evolution of a, possibly infinite, set of dynamical variables $(\chi^1, \chi^2, \ldots)$, is Hamiltonian if it can be written in the form

$$\frac{\partial \chi^i}{\partial t} = \{\chi^i, H\}, \qquad i = 1, 2, \cdots, \tag{5}$$

w here $H(\chi^1, \chi^2, \cdots)$ is the Hamiltonian (the total energy of the system) and $\{\,,\}$ is a Poisson bracket, in general noncanonical (which is hence bilinear, antisymmetric, and satisfies the Leibniz and Jacobi identities).

In checking for the existence of a Hamiltonian structure of (1)-(4), great help comes from Ref. [11], where a Hamiltonian model (Eqs. (35)-(38) of Ref. [11]) very similar to (1)-(4) is considered. Indeed, the only differences between the two models consist in the presence of ion temperature and total pressure in (1)-(4), whereas [11], considers only pressure fluctuations but accounts also for magnetic field curvature and additional z-dependence of all variables. Thus, taking advantage of the results of Ref. [11], one sees by direct calculation that the functional

$$H = \int d^2x \left(\frac{|\nabla\psi|^2}{2} + \frac{|\nabla\phi|^2}{2} + \frac{v_{i\|}^2}{2(1+\tau_T)} + \frac{\exp p}{(5/3)} \right) \tag{6}$$

$$= \int d^2x \left(\frac{|\nabla\psi|^2}{2} + \frac{|\nabla\nabla^{-2}U|^2}{2} + \frac{D^2}{2d_i^2(1+\tau_T)} + \frac{(1+\tau_T)\psi^2}{2d_i^2} - \frac{D\psi}{d_i^2} + \frac{3}{5}\exp p \right), \tag{7}$$

where $D \equiv (1+\tau_T)\psi + d_i v_{i\|}$, is conserved by the dynamics generated by (1)-(4) and is, therefore, a good candidate for Hamiltonian (total energy) of the system. The first two terms of (6) can obviously be interpreted as poloidal magnetic and kinetic energy. The fourth term, is not of quadratic type, unlike in Ref. [11], which reflects the fact that the pressure has not been expanded about a constant equilibrium. This term represents internal energy, but only of the electron fluid. The third term, on the other hand, is analogous to the parallel kinetic energy term of the model of Ref. [12], with the same unfamiliar coefficient $1/(1+\tau_T)$.

Concerning the bracket, direct calculation shows that the operation

$$\begin{aligned}\{F,G\} = \int d^2x \Big(\psi \Big([F_\psi, G_U] + [F_U, G_\psi] + \frac{5}{3} d_i([F_\psi, G_p] + [F_p, G_\psi]) - \\ \frac{5}{3}(1+\tau_T)([F_{v_{i\|}}, G_p] + [F_p, G_{v_{i\|}}]) \Big) + U[F_U, G_U] + \\ p\Big([F_p, G_U] + [F_U, G_p] + \frac{5}{3} d_i [F_p, G_p] \Big) + v_{i\|}\Big([F_{v_{i\|}}, G_U] + [F_U, G_{v_{i\|}}] \Big)\Big),\end{aligned} \tag{8}$$

when combined with (6), according to (5) and using $(\psi, U, v_{i\|}, p)$ as dynamical variables, gives the equations (1)-(4). In (8) the subscripts indicate functional derivatives. The bracket (8) indeed satisfies the properties of a Poisson bracket and is of the same form as the Poisson bracket of Ref. [11]. Analogously to most of the brackets of reduced fluid models for tokamaks, this bracket is of Lie-Poisson type (see, e.g., Refs. [2, 5]).

As it is typical for noncanonical Poisson brackets, (8) possesses Casimir invariants, i.e., elements of the center of the Poisson bracket (8), which, in this specific case are the infinite families of functionals:

$$C_1 = \int d^2x\, \mathcal{S}(\psi), \quad C_2 = \int d^2x\, p\, \mathcal{C}(\psi), \quad C_3 = \int d^2x\, \mathcal{K}(D), \quad C_4 = \int d^2x \left(U - \frac{3}{5d_i} p \right) \mathcal{F}(D), \tag{9}$$

where $\mathcal{F}$, $\mathcal{C}$, $\mathcal{S}$, and $\mathcal{K}$ are arbitrary smooth functions of a real variable.

Casimir functions corresponding to the integral of an arbitrary function of field variables, such as C_1 and C_3, are associated with Lagrangian invariants. The conservation of the Casimir functions C_1 obviously reflects the fact that the magnetic field is frozen into the poloidal flow determined by the $\mathbf{E} \times \mathbf{B}$ and diamagnetic velocities. The conservation of C_3, on the other hand, is related to the parallel momentum conservation when electron inertia is neglected and ion pressure is assumed to be proportional to the electron pressure. The Casimir functions C_2 and C_4 are remnants of cross-helicities of the parent model from which the system (1)-(4) is derived [11]. We remark that the addition of electron inertia into the model changes the nature of the Casimir functions by replacing a family of Casimir functions of cross-helicity type, with one associated with a Lagrangian invariant [13].

Journal of Physics: Conference Series **401** (2012) 012023 doi:10.1088/1742-6596/401/1/012023

3. Equilibria and stability

One advantage of the Hamiltonian formulation, is that it makes it possible to apply the Energy-Casimir method in order to find formal stability conditions for equilibria of the system. In particular, the method can provide conditions for those equilibria which are extremals of the *free energy functional* F, obtained as a linear combination of the Hamiltonian with the Casimir functions. Then, according to the Energy- Casimir method, if $\delta^2 F$, evaluated at an equilibrium of interest, has a definite sign, the equilibrium is formally energy stable, which implies linear stability (for nonlinear stability in infinite dimensional systems, additional convexity arguments have to be provided [5, 9]). In the context of plasma equilibria, this method has been adopted in a number of cases (see for instance Refs. [14, 15, 16, 9, 17, 18]).

Equilibria. From (1)–(4), one obtains that equilibrium solutions have to satisfy

$$p = \ln\left(\frac{1}{d_i}(\mathcal{G}(D) - \mathcal{H}(\psi))\right), \tag{10}$$

$$\nabla^{-2} U = \mathcal{G}(D), \tag{11}$$

$$\left[\mathcal{G}(D), \nabla^2 \mathcal{G}(D)\right] = \left[\psi, \nabla^2 \psi\right], \tag{12}$$

$$\left[\mathcal{G}(D), \ln\left(\frac{1}{d_i}(\mathcal{G}(D) - \mathcal{H}(\psi))\right)\right] = \frac{5}{3d_i}\left[\psi, D\right] + \frac{5d_i}{3}\left[\psi, \nabla^2\psi\right], \tag{13}$$

with $\mathcal{G}, \mathcal{H}$ given arbitrary functions. Eq. (10) indicates that, at equilibrium, the pressure must be a function of the magnetic flux and of the modified ion canonical momentum D. Note that (10) implies also that we have to assume that $\mathcal{G}(D) - \mathcal{H}(\psi) > 0$. The relation (11), on the other hand, tells us that the equilibrium stream function must be a function of the modified ion canonical momentum D. Eqs. (12)-(13), finally, have to be considered as a system to be solved with respect to ψ and D, for given $\mathcal{G}(D)$ and $\mathcal{H}(\psi)$.

Two interesting examples of classes of equilibrium solutions are the following

Case 1: $\mathcal{G}(D) = \psi$. This case corresponds to the so called Alfvénic solutions, in which, at equilibrium, the $\mathbf{E} \times \mathbf{B}$ speed equals the Alfvén speed based on the poloidal magnetic field. Eq. (10) then implies that the equilibrium pressure is a function of ψ, i.e., it is a magnetic flux function. If $\mathcal{G}(D) = \psi$, a sufficient condition for (13) to be satisfied is $\nabla^2\psi + \frac{1}{d_i^2}D = \Psi(\psi)$ for some function Ψ. This corresponds to an equilibrium in which the electron parallel velocity is also a flux function.

Case 2: $\nabla^2\mathcal{G}(D) = \Xi(D)$ and $\nabla^2\psi = \Upsilon(\psi)$, for some functions Ξ and Υ. In this case, the equilibrium vorticity is a function of D, whereas the current density $J = -\nabla^2\psi$ is a magnetic flux function. Equation (13) then implies either

$$a.) \quad [\psi, D] = 0, \qquad \text{or} \qquad b.) \quad \frac{\mathcal{G}'(D)\mathcal{H}'(\psi)}{\mathcal{G}(D) - \mathcal{H}(\psi)} = \frac{5}{3d_i}.$$

Case *a.*) is solved if we assume that $D = \mathcal{D}(\psi)$ for some function $\mathcal{D}$. Case *b.*) can also be solved assuming the same condition, but, in addition to that, it also requires a further relation between $\mathcal{G}$ and $\mathcal{H}$.

A special case (not applicable in the situation *b*)) is obtained by requesting $\mathcal{G}'(D) = \Xi(D) = 0$, which refers to a case with no equilibrium $\mathbf{E} \times \mathbf{B}$ flow and with a current density being a magnetic flux function. The equilibrium magnetic field is then amenable to a number of solutions depending on the choice of Υ. A remarkable example is $\Upsilon(\psi) = \exp(2\psi)$, leading to the Liouville equation, which includes classical solutions such as the Harris sheet, the Kelvin-Stuart "cat's eye", and the Gold-Hoyle equilibrium (see, e.g., Ref. [19, 20]). The choice of $\Upsilon(\psi)$ linear in ψ, on the other hand, can give modon-like solutions [21, 22].

First variation. Following the Energy-Casimir method, we look for equilibria that are critical points of the free energy functional

$$F = H + \sum_i C_i,$$

where C_i are the Casimir functionals given in (9). In our case the free energy functional F is

$$F(\psi, U, D, p) = \int d^2x \left[\frac{1}{2} \left(|\nabla\psi|^2 - U\nabla^{-2}U + \frac{1}{d_i^2(1+\tau_T)} D^2 + \frac{1+\tau_T}{d_i^2}\psi^2 + \right.\right.$$
$$\left.\left. - \frac{2}{d_i^2} D\psi + \frac{6}{5}\exp p \right) + \mathcal{S}(\psi) + p\mathcal{C}(\psi) + \mathcal{K}(D) + \left(U - \frac{3}{5d_i} p \right) \mathcal{F}(D) \right]. \qquad (14)$$

Setting the first variation

$$\delta F = \int d^2x \left[\left(-\nabla^2\psi + \frac{1+\tau_T}{d_i^2}\psi - \frac{1}{d_i^2} D + \mathcal{S}'(\psi) + p\mathcal{C}'(\psi) \right) \delta\psi \right.$$
$$+ \left(\frac{1}{d_i^2(1+\tau_T)} D - \frac{1}{d_i^2}\psi + \mathcal{K}'(D) + \left(U - \frac{3}{5d_i} p \right) \mathcal{F}'(D) \right) \delta D$$
$$\left. + \left(\frac{3}{5}\exp p + \mathcal{C}(\psi) - \frac{3}{5d_i}\mathcal{F}(D) \right) \delta p + \left(-\nabla^{-2}U + \mathcal{F}(D) \right) \delta U \right]$$

of F equal to zero, leads to the following set of equations:

$$\mathcal{F}(D) - d_i \exp p = \frac{5}{3} d_i \mathcal{C}(\psi), \qquad (15)$$

$$\nabla^{-2}U = \mathcal{F}(D), \qquad (16)$$

$$-\nabla^2\psi + \frac{1+\tau_T}{d_i^2}\psi - \frac{D}{d_i^2} + \ln\left(\frac{\mathcal{F}(D)}{d_i} - \frac{5}{3}\mathcal{C}(\psi) \right) \mathcal{C}'(\psi) + \mathcal{S}'(\psi) = 0, \qquad (17)$$

$$\frac{1}{d_i^2(1+\tau_T)} D - \frac{\psi}{d_i^2} + \nabla^2\mathcal{F}(D)\mathcal{F}'(D) - \frac{3}{5d_i}\ln\left(\frac{\mathcal{F}(D)}{d_i} - \frac{5}{3}\mathcal{C}(\psi) \right) \mathcal{F}'(D) + \mathcal{K}'(D) = 0, \qquad (18)$$

where the prime denotes derivative with respect to the argument of the function.

Different choices of the free functions $\mathcal{S}$, $\mathcal{C}$, $\mathcal{F}$, and $\mathcal{K}$, of course lead to different equilibria.

Given equilibrium solutions of (10)-(13), on the other hand, are extremals of F, if the free functions $\mathcal{F}$, $\mathcal{C}$, $\mathcal{S}$, and $\mathcal{K}$ satisfy the appropriate constraints. Indeed, by comparing (11) with (16) and (10) with (15) one immediately sees that the functions $\mathcal{F}$ and $\mathcal{C}$ are determined by $\mathcal{F} = \mathcal{G}$ and $\mathcal{C} = (3/5d_i)\mathcal{H}$. The case with no $\mathbf{E} \times \mathbf{B}$ flow, then corresponds to $\mathcal{F} = 0$. The relations determining $\mathcal{S}$ and $\mathcal{K}$, on the other hand, are less straightforward. For the solutions of Case 2, for instance, they are determined by

$$\mathcal{S}'(\psi) = \Upsilon(\psi) - \frac{1+\tau_T}{d_i^2}\psi + \frac{\mathcal{D}(\psi)}{d_i^2} - \frac{3}{5d_i}\ln\left(\frac{\mathcal{G}(\mathcal{D}(\psi)) - \mathcal{H}(\psi)}{d_i} \right) \mathcal{H}'(\psi),$$

$$\mathcal{K}'(D) = \frac{\mathcal{D}^{-1}(D)}{d_i^2} - \frac{1}{d_i^2(1+\tau_T)} D - \Xi(D)\mathcal{G}'(D) + \frac{3}{5d_i}\ln\left(\frac{\mathcal{G}(D) - \mathcal{H}(\mathcal{D}^{-1}(D))}{d_i} \right) \mathcal{G}'(D),$$

where we assumed $\mathcal{D}$ to be invertible.

We notice also that, in the absence of $\mathbf{E} \times \mathbf{B}$ flow, the choices $\mathcal{K}(D) =$ constant and $\mathcal{C}(\psi) =$ constant, correspond to no parallel flow and no pressure gradient at equilibrium, respectively. Indeed, in that limit, one has $v_{i\parallel}(D) = -d_i(1+\tau_T)\mathcal{K}'(D)$ and $p(\psi) = \ln((5/3)|\mathcal{C}(\psi)|)$ at equilibrium.

Second variation The expression for the second variation of F for our model, after some manipulation, can be written as

$$\delta^2 F = \int d^2x \left[|\nabla\delta\psi|^2 + |\nabla\delta\phi|^2 + \left(\frac{\tau_T}{d_i^2} + p\mathcal{C}''(\psi) + \mathcal{S}''(\psi) - \mathcal{C}'(\psi)\right)|\delta\psi|^2 + \frac{|\delta\psi - \delta D|^2}{d_i^2} \right.$$
$$+ \left(-\frac{\tau_T}{d_i^2(1+\tau_T)} + U\mathcal{F}''(D) - \frac{3}{5d_i}p\mathcal{F}''(D) + \mathcal{K}''(D) - \left(1 + \frac{3}{5d_i}\right)\mathcal{F}'(D)\right)|\delta D|^2 - \mathcal{F}'(D)|\delta U|^2 +$$
$$\left. + \left(\frac{3}{5}\exp p - \frac{3}{5d_i}\mathcal{F}'(D) - \mathcal{C}'(\psi)\right)|\delta p|^2 + \mathcal{F}'(D)|\delta D + \delta U|^2 + \frac{3}{5d_i}\mathcal{F}'(D)|\delta p - \delta D|^2 + \mathcal{C}'(\psi)|\delta\psi + \delta p|^2 \right].$$

We consider now a generic equilibrium $(\psi, U, D, p) = (\psi_{eq}, U_{eq}, D_{eq}, p_{eq})$, solution of (15)-(18). A sufficient condition for $\delta^2 F$, evaluated at such equilibrium, to be positive, is given by:

$$\mathcal{F} = 0, \quad (19)$$

$$-\frac{\tau_T}{d_i^2(1+\tau_T)} + \mathcal{K}''(D_{eq}) \geq 0, \quad (20)$$

$$\frac{\tau_T}{d_i^2} + p_{eq}\mathcal{C}''(\psi_{eq}) + \mathcal{S}''(\psi_{eq}) - \mathcal{C}'(\psi_{eq}) \geq 0, \quad (21)$$

$$\frac{3}{5}\exp p_{eq} - \mathcal{C}'(\psi_{eq}) \geq 0, \quad (22)$$

$$\mathcal{C}'(\psi_{eq}) \geq 0. \quad (23)$$

Condition (19) means that we are restricting to the case of no $\mathbf{E}\times\mathbf{B}$ flow. The above conditions could then apply to equilibria of Case 2. We remark that it is well known (see, e.g. Refs. [23, 2]), that an equilibrium flow can lead to the presence of negative energy modes, which can imply indefiniteness of $\delta^2 F$. We observe also that our stability conditions are sufficient but not necessary. Therefore, there could exist, in general, spectrally stable equilibria of the system, that do not satisfy our stability conditions.

For $\mathcal{F} = 0$, from (15) one has $p_{eq} = \ln((5/3)|\mathcal{C}(\psi_{eq})|)$ and the condition $\mathcal{C}(\psi_{eq}) < 0$ is required in order for the equilibrium relation (15) to be valid. Conditions (22) and (23) can then be written as

$$-1 \leq p'_{eq}(\psi_{eq}) \leq 0, \quad (24)$$

which implies a restriction on the equilibrium pressure gradient.

Because $v_{i\|}(D_{eq}) = -d_i(1+\tau_T)\mathcal{K}'(D_{eq})$, the presence of a parallel equilibrium sheared flow is necessary in order to satisfy condition (20). This is not required for negligible ion temperature. On the other hand, this condition can be reformulated as $(dv_{i\|}/dD)(D_{eq}) \leq -\tau_T/d_i$, which directly shows the role of the parallel velocity shear. Finally, inequality (21) can be interpreted as a condition on the current density gradient. Indeed, because for the equilibria of Case 2 one has $D_{eq} = \mathcal{D}(\psi_{eq})$, it follows that the equilibrium current density can be written as

$$J_{eq}(\psi_{eq}) = -\Upsilon(\psi_{eq}) = -\frac{1+\tau_T}{d_i^2}\psi_{eq} + \frac{\mathcal{D}(\psi_{eq})}{d_i^2} + \frac{3}{5}p_{eq}(\psi_{eq})p'_{eq}(\psi_{eq})\exp p_{eq}(\psi_{eq}) - \mathcal{S}'(\psi_{eq}), \quad (25)$$

and the stability condition (21) can be expressed as

$$J'_{eq}(\psi_{eq}) \leq \frac{\mathcal{D}'(\psi_{eq}) - 1}{d_i^2} + \frac{3}{5}p'_{eq}(\psi_{eq})(p'_{eq}(\psi_{eq}) + 1)\exp p_{eq}(\psi_{eq}), \quad (26)$$

which relates the current with pressure and velocity gradients. In the absence of parallel ion flow and ion temperature (corresponding to $\mathcal{K}'(D) = 0$ and $\mathcal{D}(\psi) = \psi$) , this condition becomes

$$J'_{eq}(\psi_{eq}) \leq \frac{3}{5}p'_{eq}(\psi_{eq})(p'_{eq}(\psi_{eq}) + 1)\exp p_{eq}(\psi_{eq}). \quad (27)$$

Note that, because of (24), the right-hand side of (27) is negative (except for $p'_{eq} = 0$ or $p'_{eq} = -1$), so that formal stability is attained if the current gradient is negative and sufficiently steep. An analogous situation occurs for flat pressure equilibrium profiles, with ion parallel flow and temperature satisfying (20). Indeed, in this case the stability condition (26) reduces to

$$J'_{eq}(\psi_{eq}) \leq \frac{\mathcal{D}'(\psi_{eq}) - 1}{d_i^2}. \tag{28}$$

Condition (20), on the other hand, can be reformulated as

$$\mathcal{D}^{-1'}(D_{eq}) \geq 1. \tag{29}$$

Consequently, for $\psi_{eq} = \mathcal{D}^{-1}(D_{eq})$, one has $\mathcal{D}'(\psi_{eq}) \leq 1$, so, because of (28), also a parallel gradient of the flow imposes a negative upper bound to the current density gradient. Bounds on the current gradient have been shown, by means of dynamical accessible variations [2], to provide stability also in the presence of electron inertia and ion temperature for a collisionless reconnection model [24]. In that case, the bound on the current density gradient depended on the ion temperature and on the electron skin depth. We remark also that, in the limit of a flat pressure profile, with cold ions and no parallel flow, one recovers, from our conditions, the stability condition for reduced magnetohydrodynamics [16], $J'_{eq}(\psi_{eq}) \leq 0$, which reflects the fact that a monotone current density profile provides stability.
To construct a simple concrete example of stable equilibrium one can start from the case $\tau = 0$, and consider an equilibrium of Case 2 with $\mathcal{G} = 0$, $\mathcal{H}(\psi) = -5d_ic/3$, $\Upsilon(\psi) = \lambda^2\psi$, and $\mathcal{D}(\psi) = \psi$ (corresponding to $\mathcal{F} = 0$, $\mathcal{C}(\psi) = -c$, $\mathcal{S}'(\psi) = \lambda^2\psi$ and $\mathcal{K}'(D) = 0$), where $c > 0$ and λ are constant. An equilibrium solution is then given by:

$$U_{eq} = 0, \qquad D_{eq}(\psi_{eq}) = \psi_{eq}, \qquad p_{eq}(\psi_{eq}) = \ln\left(\frac{5}{3}c\right), \qquad \psi_{eq}(x) = \cosh(\lambda x), \tag{30}$$

which describes a one-dimensional magnetic field with a neutral line at $x = 0$. This static equilibrium, with a flat pressure profile and no parallel ion flow, has $J'_{eq} = -\lambda^2$, and thus turns out to be stable, for every λ, according to the criterion (19)-(23) (we observe that such equilibrium can be shown to be also tearing stable according to the Δ' criterion introduced in Ref. [25]).

The inclusion of pressure or parallel velocity profiles that still guarantee stability, can be obtained by choosing, for instance, $p_{eq}(\psi_{eq}) = 1 - \tanh\psi_{eq}$ and $v_{i\|_{eq}}(\psi_{eq}) = \gamma[(1 + \tau_T)/(1 - \gamma d_i)]\psi_{eq}$, with γ constant (corresponding to an equilibrium of Case 2 with $\mathcal{G} = 0$, $\mathcal{H}(\psi) = -d_i\exp(1 - \tanh\psi)$, $\mathcal{D}(\psi) = (1 + \tau_T)/(1 - \gamma d_i)\psi$ and $\Upsilon(\psi) = \lambda^2\psi$). In this way, conditions (22) and (23) are automatically satisfied. Condition (26) reads

$$-\lambda^2 \leq \frac{1}{d_i^2}\frac{\tau_T + d_i\gamma}{1 - d_i\gamma} - \frac{3}{5}\frac{1}{\cosh^2\psi_{eq}}\left(1 - \frac{1}{\cosh^2\psi_{eq}}\right)\exp(1 - \tanh\psi_{eq}). \tag{31}$$

Given that, for this equilibrium, $\psi_{eq}(x) \geq 1$ for all x, condition (31) can be satisfied, if $\lambda^2 \geq (1/d_i^2)((\tau_T + d_i\gamma)/(d_i\gamma - 1)) + (3/5)\mathrm{e}$, which is more restrictive than the condition for cold ions in the absence of pressure and parallel flow. Condition (20), on the other hand, is satisfied if $\gamma \leq -\tau_T/d_i$. In particular, in this stable configuration, the parallel flow turns out to be in the direction opposite to that of the toroidal magnetic field.

4. Summary and outlook

We presented the Hamiltonian structure of a reduced model for plasma dynamics in the presence of a strong guide field, and obtained sufficient stability criteria for its equilibria, making use of the Energy-Casimir method. The stability conditions apply to equilibria with no poloidal flow and can be interpreted as restrictions on the pressure, parallel velocity and current density gradients, thus generalizing previous stability criteria for reduced magnetohydrodynamics. Examples of equilibria with magnetic shear, satisfying the stability conditions, are presented.

As a further remark, we observe that our stability criteria hold also for two-dimensional equilibria, for which traditional spectral stability methods based on the Fourier transform become more difficult to apply. In this respect, we point out that an identification of the relations between the stability conditions derived for this model, with well established stability conditions obtained, for instance, from the classical energy principle, is under progress, as well as an extension of the stability conditions to include poloidal flow, which will be the subject of a forthcoming paper.

A further direction of development would consist of extending the model in such a way that the analysis could be more directly applicable to tokamaks. In particular, we believe that adapting the model to tokamak geometry and reintroducing, in a Hamiltonian framework, the neoclassical effects decribed in the original version of the model, could lead to promising developments.

Acknowledgments

The authors acknowledge useful discussions with Fulvio Militello, Cesare Tronci, Phil Morrison, Luca Comisso, and with the Nonlinear Dynamics group at the Centre de Physique Théorique, Luminy. ET and EL were partially supported by the European Community under the contracts of Association between EURATOM, CEA, and the French Research Federation for fusion studies and the Agence Nationale de la Recherche (ANR GYPSI). TSR was partially supported by Swiss NSF grant 200020-126630 and by the government grant of the Russian Federation for support of research projects implemented by leading scientists, Lomonosov Moscow State University under the agreement No. 11.G34.31.0054.

The views and opinions expressed herein do not necessarily reflect those of the European Commission or the Governments of the Swiss and Russian Federations.

[1] Morrison P J 2005 *Phys. Plasmas* **12**, 058102-1
[2] Morrison P J 1998 *Rev. Mod. Phys.* **70** 467
[3] Lazzaro E, Comisso L and Valdettaro L 2010 *Phys. Plasmas* **17** 052509
[4] Comisso L and Lazzaro E 2010 *Nucl. Fusion* **50** 125002
[5] Marsden J E and Ratiu T S 1999 *Introduction to Mechanics and Symmetry* (New York: Springer)
[6] Arnold V I 1964 *Amer. Math. Soc. Transl.* **79** 267
[7] Holm D D, Marsden J E, Ratiu T S and Weinstein A 1983 *Phys. Lett.*, **98A** 15
[8] Abarbanel H D I, Holm D D, Marsden J E and Ratiu T S 1984 *Phys. Rev. Lett.* **52** 2352
[9] Holm D D, Marsden J E, Ratiu T S and Weinstein A 1985 *Physics Reports*, **123** 1
[10] Abarbanel H D I, Holm D D, Marsden J E and Ratiu T S 1986 *Phil. Trans. Royal Soc. London A* **318** 349
[11] Hazeltine R D, Hsu C T and Morrison P J 1987 *Phys. Fluids* **30** 3204
[12] Hazeltine R D, Kotschenreuther M and Morrison P J 1985 *Phys. Fluids* **28** 2466
[13] Tassi E, Morrison P J, Waelbroeck F L and Grasso D 2008 *Plasma Phys. and Controlled Fusion* **50** 085014
[14] Morrison P J and Eliezer S 1986 *Phys. Rev. A* **33** 4205
[15] Kruskal M D and Oberman C 1958 *Phys. Fluids* **1** 275
[16] Hazeltine R D, Holm D D, Marsden J E and Morrison P J 1984 *Proc. Int. Conf. on Plasma Physics (Lausanne)* vol 1 ed Tran M Q and Sawley M L (Lausanne: Ecole Polytechnique Federale de Lausanne) p 203.
[17] Khater A H, Moawad S M and Callebaut D K 2005 *Phys. Plasmas* **12** 012316
[18] Tassi E, Grasso D, Pegoraro F and Morrison P J 2009 *J. Plasma Fusion Res. Series* **8** 159
[19] Ceccherini F, Montagna C, Pegoraro F and Cicogna G 2005 *Phys. Plasmas* **12** 052506
[20] Tassi E, Pegoraro F and Cicogna G 2008 *Phys. Plasmas* **15** 092113
[21] Larichev V D and Reznik G M 1976 *Dokl. Akad. Nauk SSSR* **231** 1077
[22] Meiss J and Horton W 1983 *Phys. Fluids* **26** 990

[23] Morrison P J and Kotschenreuther M 1989 *Nonlinear World* **2** 910
[24] Kuvshinov B N, Pegoraro F and Schep T J 1994 *Phys. Lett. A* **191** 296
[25] Furth H P, Killeen J and Rosenbluth M N 1963 *Phys. Fluids* **6** 459

Joint Varenna-Lausanne International Workshop 2012 IOP Publishing
Journal of Physics: Conference Series **401** (2012) 012024 doi:10.1088/1742-6596/401/1/012024

Monte-Carlo model for nonlinear interactions of Alfvén eigenmodes with energetic ions

S Tholerus and T Hellsten

Division of Fusion Plasma Physics, KTH School of Electrical Engineering, SE-100 44 Stockholm, Sweden

E-mail: simon.tholerus@ee.kth.se

Abstract. A Monte-Carlo model for interactions between a single Alfvén eigenmode and energetic ions in a tokamak is presented. A phenomenological decorrelation of the wave-particle phase is introduced to mimic decorrelation by collisions or by other waves. Analysis is dedicated to how the strength of the phase decorrelation affects the nonlinear wave-particle interactions. Several of the phenomena that have been observed in some earlier models describing the nonlinear dynamics of Alfvén eigenmodes have been verified, such as the growth and saturation of the wave mode amplitude giving rise to a localized flattening of the distribution function, as well as the generation of coherent structures in the distribution function. The degree of phase decorrelation is shown to have a strong effect on the dynamics of the Alfvén eigenmode excitation.

1. Introduction

Populations of energetic ions are routinely produced in present day experiments on magnetic confinement fusion, e.g. by fusion nuclear reactions, ion cyclotron resonant heating (ICRH) and heating by neutral beam injection (NBI). Shear Alfvén waves, such as the toroidal Alfvén eigenmodes (TAEs), can be excited [1, 2] by these energetic and resonant ion populations, to such an extent that they eject a large amount of resonant ions from the plasma, resulting in a significant reduction of heating efficiency [3–5]. Theoretical understanding of the nonlinear dynamics of such waves is of importance for the development of viable fusion power plants.

Models successfully describing the excitation of TAEs have been developed earlier. Major contributions to the field of TAE dynamics have been based on the model by Berk and Breizman [6–8], with contributions also made by Pekker [9]. Their model is constructed such that it takes into account the lowest order contributions to the nonlinear wave growth rate. It has been shown that systems with marginal linear instability of the wave growth rate give a variety of wave amplitude dynamics, including regimes of stable saturation, periodic and chaotic modulation, and explosive amplitude evolution [9–12]. In particular, it was found that when dynamical friction (drag) is much larger than velocity space diffusion of particles the mode grows explosively [11, 12]. Other important results include the creation of phase space hole and clump pairs around the wave resonance [12–14].

A hybrid magnetohydrodynamic-gyrokinetic code (HMGC) [15–17] was developed by Vlad *et al* to simulate linear and nonlinear dynamics of Alfvén eigenmodes. It was shown that the growth rate of energetic particle continuum modes (EPMs) [18], with frequencies dependent on characteristic energetic particle motion, increases rapidly with increasing energetic particle pressure gradient above a certain threshold of the pressure gradient. It was also shown that the saturation of EPMs was caused by an outward displacement of energetic ions, which could have significant consequences on the energetic ion confinement.

The HAGIS (HAmiltonian GuIding centre System) code [19], developed by Pinches *et al*, is a gyro-kinetic code using a δf method based on Hamiltonian formalism to simulate nonlinear interactions of an arbitrary distribution of energetic ions and a set of discrete Alfvén eigenmodes, which are allowed to vary both in amplitude and phase. In the limit of marginally unstable TAEs the code has shown the existence of frequency sweeping events with a sweeping rate in quantitative agreement with theory [13, 20].

A Monte-Carlo model in the quasilinear regime, developed by Bergkvist *et al* [21], has described the excitation of TAEs by superthermal ions and renewal of the distribution function by ICRH. The model gave the correct magnitude of the eigenmode frequency side bands [22] by ICRH and the time scale of the rapid damping of modes as the ICRH was instantly turned off.

In this paper, a one dimensional Monte-Carlo model for the nonlinear interaction between an oscillating wave field and an ensemble of charged particles is developed. The model attempts to mimic the scenario of a fusion plasma confined in a tokamak, in which a set of energetic ions interacts with a discrete Alfvén eigenmode, such as a TAE. The model is constructed *ad hoc* to include basic dynamics necessary for the nonlinear excitation of a wave mode, describing phenomena such as the growth and saturation of a wave mode, and the generation of coherent structures in the distribution function. One of the major issues to be treated in this paper is the regime of partial phase decorrelation, i.e., regimes in between those of complete correlation and quasilinear theory. The presented one dimensional model will be used as a basis for the construction of a more complete model, not presented in this paper.

2. Wave-particle interaction model

The interaction of energetic particles with a single discrete wave mode is considered, with particle dynamics treated along one dimension. The component of the wave mode along this dimension has the form

$$E(\phi, t) = \hat{E}(t)\cos(n\phi - \omega_w t - \phi_0(t)) = E_1(t)\cos(n\phi - \omega_w t) + E_2(t)\sin(n\phi - \omega_w t). \quad (1)$$

The amplitudes $E_1(t)$ and $E_2(t)$ are reparameterizations of the two quantities $\hat{E}(t)$ and $\phi_0(t)$. The Lorentz force equation for particle interaction with this mode is

$$\frac{dv_i}{dt} = \frac{q}{m}\Big[E_1(t)\cos(n\phi_i - \omega_w t) + E_2(t)\sin(n\phi_i - \omega_w t)\Big], \quad (2)$$

where v_i and ϕ_i are the velocity and phase of particle i.

Interactions with magnetically trapped particles[1] are considered in this paper. Note that passing particles in resonance with the wave field may as well excite TAEs. The restriction of the study to trapped particles is motivated partly by simplicity and partly by the fact that trapped particles in the MeV range are more likely to be resonant with TAEs. For this reason the conceptual study of trapped particle interactions with TAEs may be of interest.

The velocity and phase (v_i and ϕ_i) are the parallel velocity and toroidal phase, respectively, at some particular angle of the trapped orbit. A good approximation for magnetically trapped particles is that the precession frequency is linearly dependent on the particle kinetic energy. A first order expansion around the wave mode resonance would be

$$\frac{d\phi_i}{dt} = \omega_i = \frac{\omega_w}{n} + \frac{\partial\omega}{\partial W}\left(\frac{mv_i^2}{2} - W_{\text{res}}\right), \quad (3)$$

[1] Trapped particles can refer to two different phenomena in this paper. Either the particles are toroidally trapped by the equilibrium magnetic field (referred to as magnetically trapped particles) or they may be trapped by the field of the wave mode (wave field trapped particles).

where $\partial\omega/\partial W$ is constant, and $W_{\text{res}} = mv_{\text{res}}^2/2$ is the resonant energy. Note that $W_{\text{res}}\partial\omega/\partial W \neq \omega_w/n$ can be a most valid assumption locally around the resonance $mv_i^2/2 = W_{\text{res}}$. This assumption becomes nonphysical close to zero kinetic energy, though, since $v_i = 0$ then results in $d\phi_i/dt \neq 0$. This fact should be kept in mind by ensuring that the wave mode only interacts with particles close to the resonance.

Using the coordinate substitution $\phi_i \to \phi_i + \omega_w t/n$ and $\omega_i \to \omega_i + \omega_w/n$ simplifies the problem. Equations (2) and (3) then become

$$\frac{dv_i}{dt} = \frac{q}{m}\Big[E_1(t)\cos(n\phi_i) + E_2(t)\sin(n\phi_i)\Big], \tag{4}$$

$$\frac{d\phi_i}{dt} = \omega_i = \frac{\partial\omega}{\partial W}\left(\frac{mv_i^2}{2} - W_{\text{res}}\right) \tag{5}$$

It should be noted that the actual value of ω_w does not influence this particular problem, since the dynamics of the particles only depend on their relative phase velocity compared to the wave. The coefficient $\partial\omega/\partial W$ depends on the characteristics of the magnetically trapped orbit. Considering that the resonant interacting particles are localized in the same region of phase space, the assumption that $\partial\omega/\partial W$ is constant and the same for all particles is valid.

The total wave mode energy is assumed to be of the form

$$W_w = \epsilon\langle E^2\rangle_\phi = \frac{\epsilon(E_1^2 + E_2^2)}{2}, \tag{6}$$

with ϵ being a proportionality factor depending on the characteristics of the wave mode and on plasma parameters. Energy balance for wave-particle interaction yields

$$\frac{dE_1}{dt} = -\frac{q}{\epsilon}\sum_i^{N_0} v_i\cos(n\phi_i) - \gamma_d E_1 \approx -\frac{qs}{\epsilon}\sum_i^{N_0/s} v_i\cos(n\phi_i) - \gamma_d E_1, \tag{7}$$

$$\frac{dE_2}{dt} = -\frac{q}{\epsilon}\sum_i^{N_0} v_i\sin(n\phi_i) - \gamma_d E_2 \approx -\frac{qs}{\epsilon}\sum_i^{N_0/s} v_i\sin(n\phi_i) - \gamma_d E_2, \tag{8}$$

where the contributions to the orthogonal components of the wave mode have been separated. The parameter γ_d is the wave field net damping rate, which can be used e.g. to model interactions with a thermal background of particles. The parameter N_0 can be considered as the physical number of particles in the system, whereas s (typically $\gg 1$) is a resolution parameter that can be used to simulate $N_p \equiv N_0/s$ *macroparticles*. The ratio ϵ/s is a parameter quantifying the interaction strength between the wave mode and the macroparticles.

An important property of the model is the inclusion of a phenomenological decorrelation of the wave-particle phase, which mimics decorrelation by collisions between particles or by interactions with other waves. Mathematically, decorrelation is modeled using a Monte-Carlo approach. Specifically, it is introduced as a Wiener process in the toroidal angular position of the particles according to

$$\Delta\phi_i = \frac{\partial\omega}{\partial W}\left(\frac{mv_i^2}{2} - W_{\text{res}}\right)\Delta t + \pi\zeta_i\sqrt{\frac{\Delta t}{t_c}}, \tag{9}$$

where ζ_i is a random variable of unit variance, and t_c is the phase decorrelation time. In numerical simulations, the differentials $\Delta\phi$ and Δt are small compared to macroscopic phase and time scales, but finite. Partial wave-particle phase decorrelation is defined by the condition

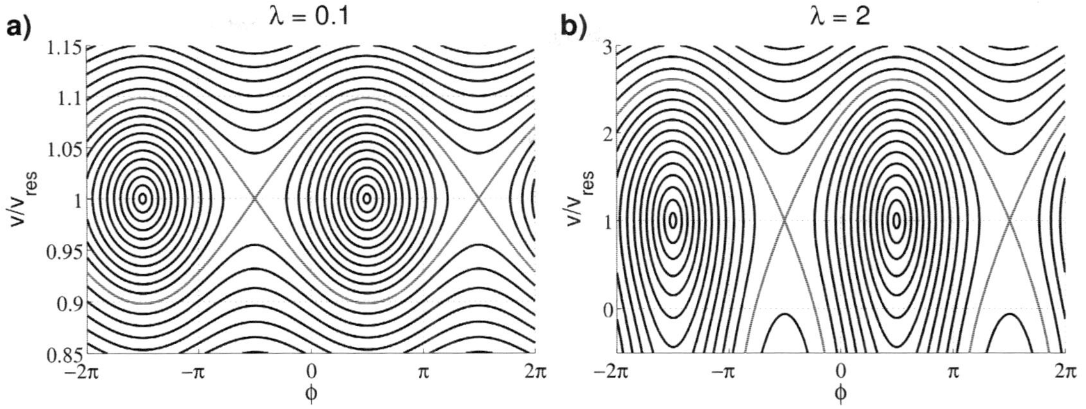

Figure 1. Phase space trajectories for trapped particles, with $n = 1$, $\phi_0 = 0$, and $q > 0$. The separatrices are marked with red curves.

$t \lesssim t_c$ for a finite t_c. The stochastic phase variation will in turn give rise to velocity space diffusion via interactions with the wave mode. Quasilinear theory is valid in the limit of t_c much less than any other macroscopic time scale of the system. Variation of t_c allows complete study of the transition between the correlated and the quasilinear regimes.

In the regime of a quasi-stationary wave, i.e., in the limit where the wave amplitude and phase are in principle unaffected by the transfer of energy from the particles, the dynamics of the particle interactions with the wave mode are similar to that of a pendulum interacting with the gravitational field. The parameter

$$\lambda \equiv \frac{2}{m}\sqrt{\frac{|q|\hat{E}}{nv_{\text{res}}^3}\left(\frac{\partial\omega}{\partial W}\right)^{-1}}, \tag{10}$$

quantifies the width of the potential well where particles are trapped by the wave field, and $E_{\text{norm}} \equiv \lambda^2$, referred to as the normalized field strength, is a dimensionless quantification of the wave field amplitude. For $\lambda \ll 1$ the parameter is approximately the separatrix width in $\hat{v}$ space (see figure 1.a), where $\hat{v} \equiv v/v_{\text{res}}$ is the normalized velocity. If, on the other hand, $\lambda \gtrsim 1$ the phase space trajectories become asymmetric around the resonance, as can be seen in figure 1.b. In this regime, almost all closed phase space orbits pass through a region of $v \approx 0$. Hence, the model becomes invalid in this limit. The parameters have to be set such that the wave never grows to such large amplitudes. Another characteristic quantity is

$$\omega_B = \sqrt{nv_{\text{res}}|q|\hat{E}\frac{\partial\omega}{\partial W}}, \tag{11}$$

which is identified as the bounce frequency for particles deeply trapped by the wave field.

The values of λ and ω_B change as $\hat{E}$ varies with time. A useful parameter to derive would be the time scale at which $\hat{E}$ varies initially. Consider starting from a distribution function

$$f(\phi, v, t = 0) = \frac{1}{2\pi}F(v), \tag{12}$$

where $F(v)$ is the initial distribution of macroparticles in velocity space. Since the distribution function is initially flat and uncorrelated in phase ϕ, $d\hat{E}/dt = 0$ to first order in time. Specifically,

$$\Delta\hat{E} = -\frac{q^2 N_0}{\epsilon m}\frac{\hat{E}\Delta t^2}{2} + \mathrm{O}(\Delta t^3), \tag{13}$$

where $N_0 = s\int dvF(v)$ is the total number of energetic particles. The time scale for field amplitude variation can be identified as

$$t_E \equiv \frac{1}{|q|}\sqrt{\frac{\epsilon m}{N_0}}. \tag{14}$$

The inclusion of a collision operator would modify the second order time scale, and a phase dependent distribution function would give a nonzero first order variation of $\hat{E}$. Equation (13) implies an initial decrease of the amplitude. However, if a considerably non-flat phase distribution is built up during time scales $\lesssim t_E$, the amplitude may instead increase, initially.

3. Dynamics of the wave-particle interactions

3.1. Amplitude saturation

Assuming an initial distribution function $f_{\text{init}}(\phi, v)$ inverted in velocity space ($\text{sgn}(v)df_{\text{init}}/dv > 0$) the wave mode is expected to saturate at a level corresponding to the energy gain by a local flattening of the distribution function within the separatrices of the wave. In the limit of low amplitude ($\lambda \ll 1$), the theoretical saturation level is

$$E_{\text{norm}}^{(\text{sat})} = \left(\frac{16}{3}\frac{q}{nv_{\text{res}}}\left[\frac{\partial\omega}{\partial W}\right]^{-1}\sqrt{\frac{2}{\epsilon m^3}\frac{\partial f_{\text{init}}}{\partial v}}\right)^4. \tag{15}$$

This expression assumes that $\partial f_{\text{init}}/\partial v$ is constant within the trapped region of the saturated wave mode, and that $\partial f_{\text{init}}/\partial\phi = 0$. A simulation of the nonlinear growth and saturation of the wave mode amplitude using our model compared with the theoretical saturation level is shown in figure 2.a, indicating a good agreement between the numerical and theoretical saturation levels, given by (15).

From the assumption of a local flattening of the distribution function the final velocity distribution function will be on the form

$$F_{\text{final}}(v) = \begin{cases} F_0 + \dfrac{dF}{dv}(v - v_{\text{res}}) & \text{for } |v - v_{\text{res}}| > \lambda v_{\text{res}} \\ F_0 + \dfrac{dF}{dv}(v - v_{\text{res}}) \times \dfrac{2}{\pi}\sin^{-1}\left(\dfrac{|v - v_{\text{res}}|}{\lambda v_{\text{res}}}\right) & \text{for } |v - v_{\text{res}}| \leq \lambda v_{\text{res}} \end{cases}, \tag{16}$$

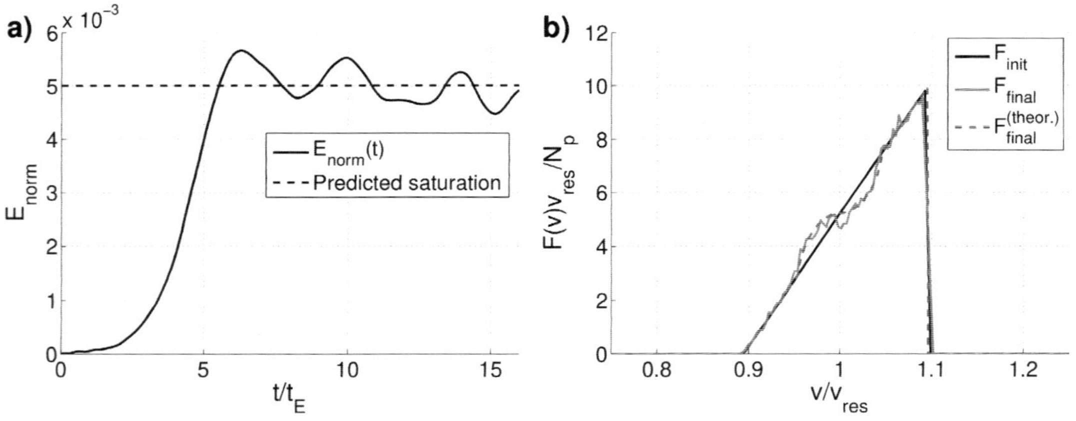

Figure 2. a) Saturation of a wave mode with the corresponding theoretical prediction ($\gamma_d = 0$). b) A comparison between the theoretical and the numerical distribution function after wave saturation ($t = 16t_E$).

Journal of Physics: Conference Series **401** (2012) 012024 doi:10.1088/1742-6596/401/1/012024

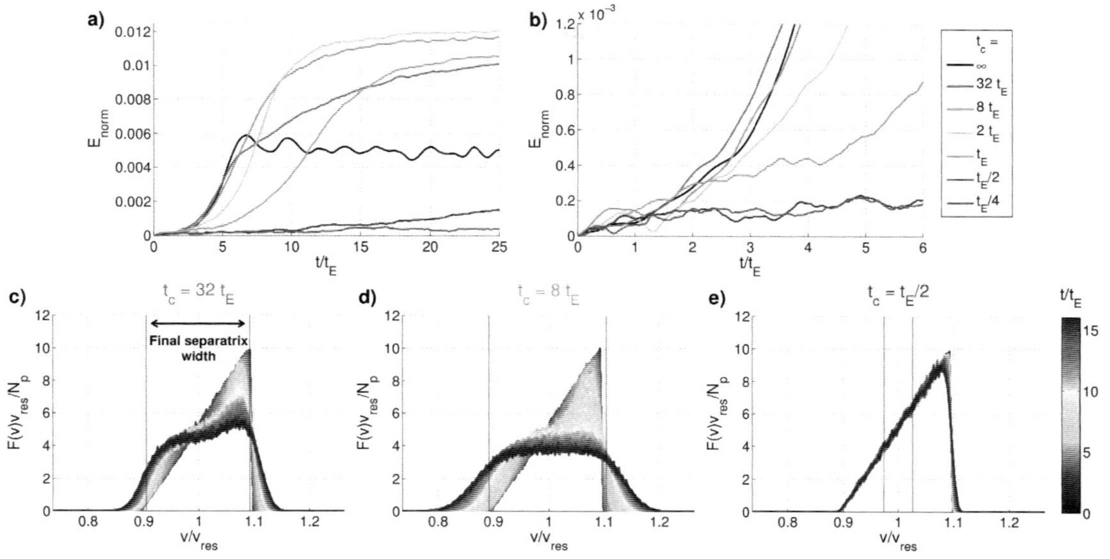

Figure 3. a and b) The time evolution of E_{norm} for different values of t_c. c – e) The time evolution of the distribution function in velocity space for c) $t_c = 32t_E$, d) $t_c = 8t_E$, e) $t_c = t_E/2$. The solid vertical lines show the separatrix width at $t = 16t_E$.

where $F(v)$ is the integral of $f(\phi, v)$ over phase ϕ. A comparison of the model equation (16) and the numerical distribution function after saturation was reached is shown in figure 2.b. Again, a good agreement with the model can be observed. Deviations from the predicted distribution function may come from asymmetries of the phase space trajectories (see figure 1), as well as from fluctuations due to discrete particle effects.

When including decorrelation of the wave-particle phase the saturation level and saturation time changes as can be seen in figure 3.a. Three additional regimes can be observed for different degrees of decorrelation, referred to as the weakly, intermediately and strongly decorrelated regime. In figure 3, $t_E\omega_B = 40\sqrt{E_{\text{norm}}} \sim 1$, implying that time scales for evolution of particles in phase space and for wave dynamics are similar.

In the weakly decorrelated regime, characterized by finite $t_c \gg t_E$, the amplitude grows until saturation at time scales much longer than in the correlated regime ($t_c = \infty$), and to a higher level of saturation. Amplitude oscillations around the saturated level are also damped by decorrelation. In the intermediately decorrelated regime, approximately identified as $2t_E \leq t_c \leq 8t_E$, both the saturation level and saturation time are rather insensitive to variations of t_c, and the saturation time is similar to that in the correlated regime. It can be viewed as a transition regime between the weakly and strongly decorrelated regime. The strongly decorrelated regime can be found below a relatively sharp limit of $t_c \approx t_E$. It can be identified as the quasilinear regime if t_c is much less than any macroscopic time scale. Here, the growth of the wave mode amplitude becomes strongly damped on investigated time scales as compared to less decorrelated regimes. The mode will eventually grow and saturate even in the strongly decorrelated regime, but on much longer time scales ($t \sim 50t_E$ for $t_c = t_E/2$ and $t \sim 500t_E$ for $t_c = t_E/4$). However, there is an upper time limit for the validity of the model due to the neglect of e.g. dissipative mechanisms.

More can be said about initial growth rates from figure 3.b. They are similar for $t_c \geq 8t_E$, and then decrease for increasing decorrelation until $t_c \leq t_E/2$, where they become more or less independent of t_c. Simulations with the SELFO code in quasilinear regime have demonstrated

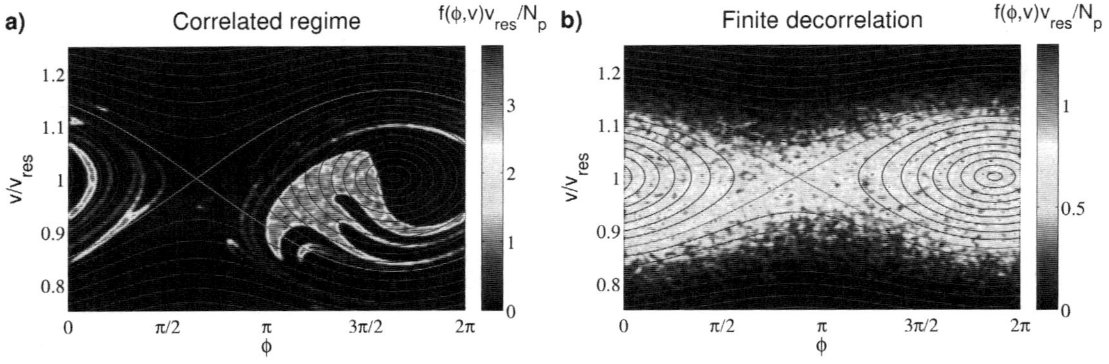

Figure 4. The distribution function $f(\phi, v)$ at $t = 8t_E$, given an initially flat and localized distribution in velocity space (the full width of the initial distribution $\Delta v/v_{\rm res} = 0.05$). The lines show phase space trajectories for a momentarily static wave field. a) $t_c = \infty$, b) $t_c = 2t_E$.

an initially exponential growth of the mode in time [21]. However, statistical fluctuations are too large to tell whether the initial growth is exponential or not in figure 3.b for $t_c \leq t_E/2$.

Via interactions with the wave the wave-particle phase decorrelation translates into diffusion in velocity space, as seen in figure 3.c and d. The velocity diffusion explains the higher saturation levels in the weakly and intermediately decorrelated regimes, since it allows a wider range in the velocity distribution to become flattened. The final saturation level depends on the initial distribution function of the entire population of energetic particles, rather than on the distribution locally around the resonance. The velocity space diffusion increases for increasing wave-particle phase decorrelation, until a certain limit. In figure 3.e one can see that the diffusion almost vanishes at the investigated time scales for $t_c = t_E/2$. As the mode grows on longer time scales velocity diffusion will eventually kick in even in the strongly decorrelated regime.

3.2. Generation of coherent structures

When wave-particle phase decorrelations are negligible, coherent structures in phase space may appear. Formation of such structures strongly affects the dynamics of the wave mode amplitude. The irregular structures in figure 4.a are possible when the time scales of particle evolution in phase space (approximately the inverse bounce frequency of the particles deeply trapped by the wave field) are similar to the time scales of wave amplitude and phase evolution.

The growth of the mode is a nonlinear effect, requiring an initial build-up of phase correlation. In the strongly decorrelated regime the necessary build-up of correlation cannot occur. Velocity space diffusion is also a process that can transfer energy to the wave mode, since it results in a flattening of the velocity distribution. At low mode amplitudes the wave-particle phase decorrelation and the velocity diffusion are decoupled. Thus, velocity space diffusion is also damped in the strongly decorrelated regime.

In the intermediately decorrelated regime, where not all phase correlations are immediately smeared out, the wave interactions with the particles also give rise to diffusion in velocity space on time scales $t \gtrsim t_c$. This can be seen in figure 4.b. Even though the modes grow to almost the same level in figures 4.a and b, there are different physics behind the growth.

4. Conclusions and discussion

The presented model successfully describes several phenomena regarding the nonlinear interactions of Alfvén eigenmodes with energetic ions in a tokamak, such as the nonlinear growth and saturation of the wave mode, due to a local flattening of an inverted distribution function in the region of phase space where particles are trapped by the wave field. The discrete particle picture that has been used allows the study of nonlinear structures in the distribution function.

The dynamics of the system are fundamentally different for different degrees of wave-particle phase decorrelation. In addition to the fully correlated regime, three regimes have been observed: the weakly, the intermediately, and the strongly decorrelated regime. In the weakly decorrelated regime the mode amplitude grows until saturation at time scales larger than in the fully correlated regime, although initial growth rates are similar. The intermediately decorrelated regime is a transition regime where velocity space diffusion is the largest. In this regime the initial growth rate decreases for increasing degree of decorrelation, although other dynamics of the wave mode are fairly unchanged. Velocity space diffusion may increase the saturation level of the mode depending on the initial distribution function. In the strongly decorrelated regime, both the generation of coherent structures and velocity space diffusion are suppressed relative to the more weakly decorrelated regimes, which heavily damps the nonlinear growth of the wave mode.

The purpose of the presented model was to investigate the basic principles needed to successfully describe the nonlinear dynamics of TAEs that include decorrelation of the wave-particle phase. The investigation was a first step in the development of a new model with a more complete description of the wave-particle interactions, e.g. by considering ICRF- or NBI-driven scenarios and dissipative mechanisms, such as wave damping and particle drag. Other possible generalizations are to consider a 3D toroidal plasma and more advanced, time dependent gyro center orbit geometries. Collisions can then be modeled e.g. as a decorrelation in the pitch angle.

Acknowledgments

We would like to thank Dr. Thomas Johnson, co-supervisor for Simon Tholerus's current PhD studies, for his support throughout the project regarding both theoretical and technical issues.

References

[1] Rosenbluth M N and Rutherford P H 1975 *Phys. Rev. Lett.* **34** 1428
[2] Fu G Y and van Dam J W 1989 *Phys. Fluids B* **1** 1949
[3] Wong K L *et al* 1991 *Phys. Rev. Lett.* **66** 1874
[4] Heidbrink W W, Strait E J, Doyle E, Sager G and Snider R T 1991 *Nucl. Fusion* **31** 1635
[5] Hawryluk R J *et al* 1994 *Phys. Plasmas* **1** 1560
[6] Berk H L and Breizman B N 1990 *Phys. Fluids B* **2** 2226
[7] Berk H L and Breizman B N 1990 *Phys. Fluids B* **2** 2235
[8] Berk H L and Breizman B N 1990 *Phys. Fluids B* **2** 2246
[9] Berk H L, Breizman B N and Pekker M S 1996 *Phys. Rev. Lett.* **76** 1256
[10] Breizman B N, Berk H L, Pekker M S, Porcelli F, Stupakov G V and Wong K L 1997 *Phys. Plasmas* **4** 1559
[11] Lilley M K, Breizman B N and Sharapov S E 2009 *Phys. Rev. Lett.* **102** 195003
[12] Lilley M K, Breizman B N and Sharapov S E 2010 *Phys. Plasmas* **17** 092305
[13] Berk H L, Breizman B N and Petviashvili N V 1997 *Phys. Lett. A* **234** 213
[14] Berk H L, Breizman B N, Candy J, Pekker M S and Petviashvili N V 1999 *Phys. Plasmas* **6** 3102
[15] Vlad G, Kar C, Zonca F and Romanelli F 1995 *Phys. Plasmas* **2** 418
[16] Briguglio S, Vlad G, Zonca F and Kar C 1995 *Phys. Plasmas* **2** 3711
[17] Briguglio S, Zonca F and Vlad G 1998 *Phys. Plasmas* **5** 3287
[18] Chen L 1994 *Phys. Plasmas* **1** 1519
[19] Pinches S D *et al* 1998 *Comput. Phys. Commun.* **111** 133
[20] Vernon Wong H and Berk H L 1998 *Phys. Plasmas* **5** 2781
[21] Bergkvist T, Hellsten T, Johnson T and Laxåback M 2005 *Nucl. Fusion* **45** 485
[22] Fasoli A, Breizman B N, Dorba D, Heeter R F, Pekker M S and Sharapov S E 1998 *Phys. Rev. Lett.* **81** 5564

Joint Varenna-Lausanne International Workshop 2012 IOP Publishing
Journal of Physics: Conference Series **401** (2012) 012025 doi:10.1088/1742-6596/401/1/012025

Fluid moments and spectral diagnostics in global particle-in-cell simulations

C Wersal[1], A Bottino[1], P Angelino[2], B D Scott[1]

[1] Max-Planck-Institut für Plasmaphysik, Boltzmannstrasse 2, EURATOM Association, D-85748 Garching, Germany

[2]Ecole Polytechnique Fédérale de Lausanne (EPFL), Centre de Recherches en Physique des Plasmas, Association Euratom-Confédération Suisse, CH-1015 Lausanne, Switzerland

E-mail: `christoph.wersal@ipp.mpg.de`

Abstract. The convergence properties of global electrostatic particle-in-cell gyrokinetic simulations have been studied by means of a set of new 3D diagnostics recently implemented in the code ORB5, allowing for measurements of electromagnetic potentials and relevant fluid quantities (density, temperature, vorticity) as well as turbulence spectral analysis. For all the quantities considered, non converged simulations show flatter spectra. However, not all the physically relevant quantities converge with the same rate: density fluctuation spectra can still show an unphysical flattening of the spectrum at high $k_\theta \rho_s$, even when electrostatic potential and heat fluxes appear to be converged. A remarkable results is that resolution in terms of number of markers, required to correctly describe the turbulence density fluctuations, is found to be size-dependent: larger systems (ITER like) require less numerical particles per active mode than smaller systems.

1. Introduction

Global linear and nonlinear PIC codes are routinely used for solving the gyrokinetic equations, often in the limit of the electrostatic approximation. Most of the existing gyrokinetic PIC codes are based on the δf method [1, 2, 3]. In the δf method the distribution function f of each plasma species is split into a time independent background distribution function f_0 and a time dependent perturbation δf, $f = f_0 + \delta f$. In the δf method, the perturbed part only (δf) is discretized using numerical particles called markers. As long as the perturbation δf remains small as compared to f_0, the δf method reduces the statistical noise. The δf method can be interpreted as a "control variate" algorithm [4, 5], a variance reduction technique widely used for Monte Carlo methods. In this paper we present a systematic study of the convergence properties of global electrostatic δf PIC simulations when the number of markers is varied.

2. Simulation model

All the simulations presented in this work have been performed using the code ORB5 [6]. The ORB5 gyrokinetic model is based on the gyrokinetic Vlasov-Maxwell system of equations of Hahm and Brizard [7, 8, 9, 10]. The latter consists of a set of self-consistent and energy conserving nonlinear gyrokinetic equations for particles and fields. The code ORB5 solves the full-f Vlasov equation using a particle-in-cell δf method: the fluctuating part of the distribution function, δf is discretized using a population of numerical particles, called markers. Each marker

Journal of Physics: Conference Series **401** (2012) 012025 doi:10.1088/1742-6596/401/1/012025

is characterized by a position in the 5D phase-space and by a weight, carrying the information of the average value of δf in a small portion of the phase-space. The linearized field equations, the polarization equation and the parallel Ampère's law [11], are discretized using finite elements. The code is based on straight-field-line coordinates and includes collision operators [12, 13]. In collisionless simulations, like the one performed in this paper, the dissipation necessary to assure entropy saturation and to reach a steady state is provided either by a zonal flow residual preserving Krook-operator [14] or by a coarse graining algorithm [15, 16]. The equations of the model, in the electrostatic limit, are:

$$\frac{\mathrm{D}f}{\mathrm{D}t} \equiv \frac{\partial f}{\partial \mathbf{R}} \cdot \dot{\mathbf{R}} + \frac{\partial f}{\partial p_z}\dot{p_z} = C(f) + S \tag{1}$$

$$\dot{\mathbf{R}} = \frac{p_z}{m}\frac{\mathbf{B}^*}{B_{\parallel}^*} + \frac{1}{B_{\parallel}^*}\mathbf{b} \times \left[\mu\frac{m}{e}\nabla B + \nabla J_0 \Phi\right]$$

$$\dot{p_z} = -\frac{m\mathbf{B}^*}{B_{\parallel}^*} \cdot \left[\mu \nabla B + \frac{e}{m}\nabla J_0 \Phi\right]; \qquad \dot{\mu} = 0 \tag{2}$$

$$\sum_{\text{species}} \delta n = -\nabla_{\perp}\left(\sum_{\text{species}} \frac{m n_0}{eB^2}\right)\nabla_{\perp}\Phi \tag{3}$$

where Φ is the electrostatic potential, $(\mathbf{R}, p_z, \mu)$ are the gyrocenter coordinates with $\mu \equiv v_{\perp}^2/2B$, magnetic moment per unit mass, and $p_z \equiv m v_{\parallel}$ is the canonical parallel momentum, J_0 is the gyroaverage operator; $\mathbf{b}$ is the unit vector for $\mathbf{B}$, m and e are the species mass and charge; δn is the perturbed gyrocenter density $\delta n = \int \mathrm{d}W\ \delta f \delta(\mathbf{R} + \boldsymbol{\rho} - \mathbf{x})$, $\boldsymbol{\rho}$ being the particle Larmor radius and $\mathrm{d}W = 2\pi B_{\parallel}^* m^{-2} \mathrm{d}\mu \mathrm{d}p_z$, $B_{\parallel}^*$ is the variable part of the velocity space volume element. For electromagnetic simulations, not considered in this paper, the system of equation is completed by the linearized gyrokinetic parallel Ampère's law, which determines the evolution of the magnetic potential $A_{\parallel}$. In this case, the canonical parallel momentum becomes $p_z \equiv m v_{\parallel} + e J_0 A_{\parallel}$.

For $C(f) = 0$ and $S = 0$, this set of gyrokinetic equations is energy and momentum conserving. An elegant proof is given in Ref. [17].

In the absence of sources, $S = 0$, transport processes tend to relax the temperature profile. The heat source has the form of a Krook operator $S_H \propto -\gamma_H \tilde{\delta f}(\psi, v^2, t)$, modified to be particle conserving. The symbol $\tilde{}$ stands for the operator which projects a general function $A(\mathbf{R}, p_z, \mu, t)$ to a reduced space $\tilde{A}(\psi, \epsilon, t)$, where ψ is the poloidal magnetic flux and ϵ is the kinetic energy. The complete definitions of the projection operator and of the heat source S_H are given in Ref. [13]. This source term tends to readjust the temperature profile toward the background profile. Note that small profile variations are still allowed during the simulation. In this work, electrons have been modelled with a simple adiabatic fluid model $\delta n_e \propto (\Phi - \bar{\Phi})$, $\bar{\Phi}$ being the axisymmetric component of the electrostatic potential Φ. In the following, we will make use of the following definitions:

$$\text{radial heat flux} \quad Q_{\mathrm{h}} \equiv \langle \frac{1}{|\nabla \psi|} \int \mathrm{d}W \frac{1}{2} m v^2 \delta f \frac{\mathrm{d}\psi}{\mathrm{d}t}\Big|_{\mathrm{ExB}} \rangle$$

$$\text{effective heat diffusivity} \quad \chi \equiv Q_{\mathrm{h}}/n\nabla T$$

$$\text{generalized vorticity,} \quad \Omega \equiv -e\nabla_{\perp}\left(\sum_{\text{species}} \frac{m n_0}{eB^2}\right)\nabla_{\perp}\Phi = \sum_{\text{species}} e\delta n$$

$$\text{density fluctuation,} \quad \delta n \equiv \int \mathrm{d}W\ \delta f$$

$$\text{temperature fluctuations,} \quad \delta T \equiv \frac{1}{3}m\left[\int \mu B \mathrm{d}W\, \delta f + \int v_\parallel^2 \mathrm{d}W\, \delta f - \left(\int v_\parallel \mathrm{d}W\, \delta f\right)^2\right]$$

where ψ is the poloidal magnetic flux. Moments are defined according to the prescriptions of the GEM gyrofluid model [18]. The generalized vorticity, expressed as a frequency, is $\Omega \simeq eB/m_D(n_e - n_i)/n_0$ further details can be found in Ref.[19]. The work presented in this paper is based on a new set of 3D diagnostics implemented in ORB5, allowing for instantaneous measurements of electromagnetic potentials and relevant fluid quantities (density, temperature, vorticity) on a full resolution 3D spatial mesh, making a massive use of parallel I/O. The same set of diagnostics can be used for 3D data visualization.

3. Results

The nonlinear simulations described in this Section are based on parameters and profiles of the ITM Cyclone base case [20]. This is a standard benchmark case based on a circular low-β equilibrium, deuterium plasma, $\rho* \simeq 1/185$ (mid radius), $T_e = T_D$ and flat initial R/L_T profiles between $0.2 < s < 0.8$, $s \propto \sqrt{\psi}$. The q profile is parabolic, matching the local value of the local Cyclone case ($q = 1.4$) at mid-radius. A detailed description of the physical parameters and profiles can be found in [20]. Figure 1 shows the results of a scan on the initial temperature gradient, with all the other parameters kept fixed. Those simulations are equivalent in terms of physics and numerical parameters, to the standard ITM global case of Ref. [20], except for the presence of the heat source preventing profile relaxation. The ion heat diffusivity in gyro-Bohm units ($\chi_{GB} \equiv \rho_s^2 c_s/a$, with $c_s^2 \equiv T_e/m_\mathrm{D}$ and $\rho_s^2 \equiv T_e m_\mathrm{D}/(eB_0)^2$) is plotted versus R/L_T, both the quantities correspond to radial averages between $0.4 < r/a < 0.6$. Here it is evident that the heat source prevents profile relaxation but still allows for local modifications of the gradient profile. The solid line represents the original Dimits fit for this curve, derived as a fit to the results of the LLNL gyrokinetic flux-tube PIC electrostatic turbulence code [21]. The ORB5 results follow the Dimits curve, although with lower diffusion coefficients, as expected by global simulations. They also correctly reproduce the linear $R/L_T|_\mathrm{lin} \simeq 4$ and nonlinear critical thresholds $R/L_T|_\mathrm{nonlin} \simeq 6$. This is particularly evident when considering the simulation with initial $R/L_T \simeq 5$ (dark green in Figure 1): after the initial linear phase, the heat flux drops rapidly to zero, showing that the mode was linearly unstable, but nonlinearly fully stabilized by zonal flow dynamics. The density, temperature, vorticity and potential spectra for a simulation

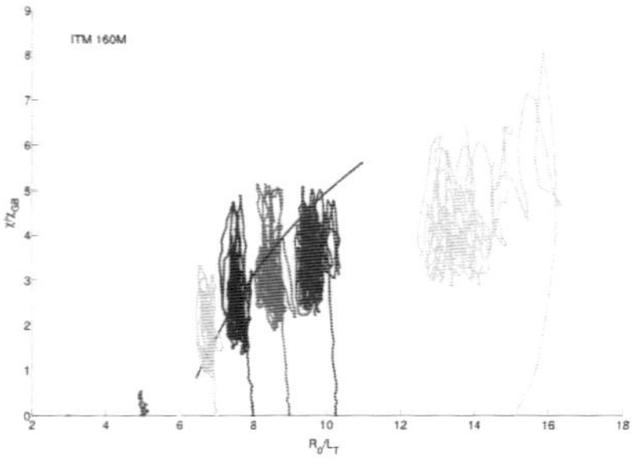

Figure 1. Radial averaged ion heat diffusivities in gyro-Bohm units versus R/L_T for the Cyclone base case with sources. During the saturation phase, χ/χ_{GB} lies close to the Dimits fit curve.

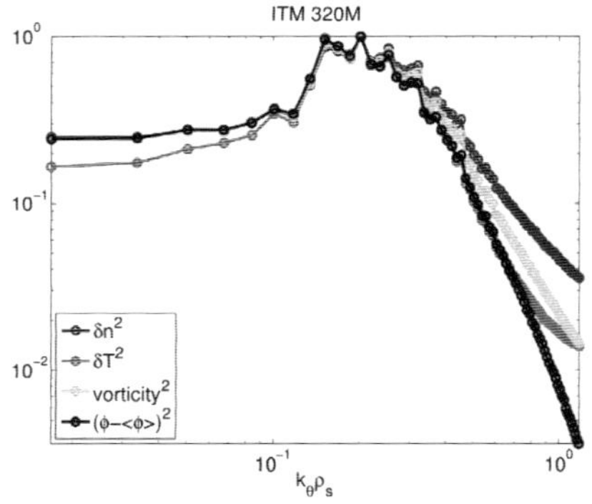

Figure 2. Different time averaged spectra for the initial $R/L_T \simeq 10.3$ simulations for the Cyclone base case with sources, calculated from 3D data.

with initial $R/L_T \simeq 10.3$ are presented in Figure 2. Time averaged spectra show evidence of the non-linear cascades to lower $k_\theta \rho_s$. Figure 3 shows, for the same simulation, poloidal cross-sections reconstructed from the 3D diagnostics data.

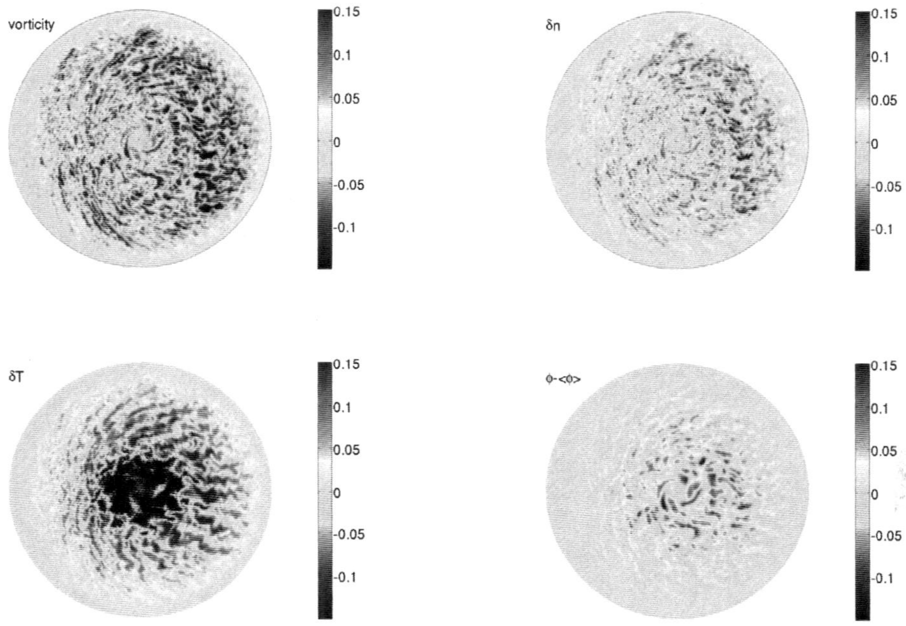

Figure 3. Poloidal cross-sections of different quantities for the initial $R/L_T \simeq 10.3$ simulations, reconstructed from 3D data.

3.1. Convergence in number of markers

The simulations with initial $R/L_T \simeq 10.3$ has been chosen for a systematic study of the convergence in number of markers for real space time traces and for the corresponding spectra. The quantities considered are: density fluctuations, electrostatic potential, temperature fluctuations and vorticity. Spectra have been obtained by performing a time average during the saturation phase of the simulation, while real space data correspond to radial averages between $0.5 < r/a < 0.7$. The convergence results are illustrated in Fig.4. For all the quantities considered, non converged time traces correspond to flat spectra at high $k_\theta \rho_s$. On the other hand, converged spectra always exhibit a clear power law down to the highest $k_\theta \rho_s$ values kept in the simulation. The most remarkable result is that not all the physically relevant quantities converge with the same rate. Figure 5 summarizes the results of Fig.4 by normalizing the time averaged quantities to the results of the simulations with the highest number of markers, plotted versus the number of markers per mode kept in the simulation. The number of markers per active mode has been already shown to be the relevant convergence quantity for the ORB5 code [22]. The electrostatic potential converges much faster than all the other quantities. The slowest converging field is the density fluctuation, δn. Remarkably, the ion heat flux diffusivities (and the heat fluxes) converge as fast as the potential, as it is illustrated in Fig.6. Although the heat flux, important for predicting turbulence induced transport in experiments, requires relatively few markers to converge (20M), comparisons with experimental measurements (for example, with

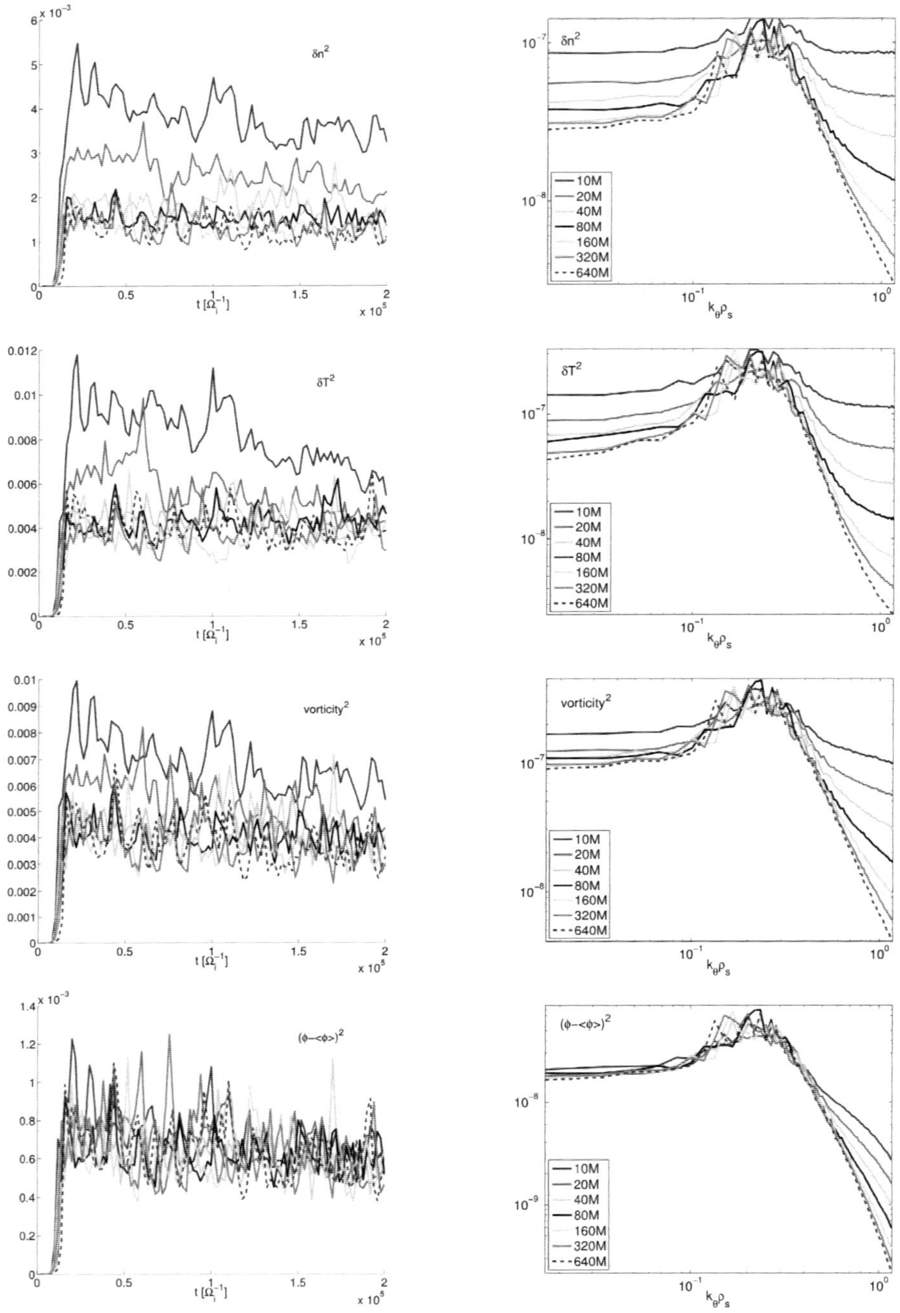

Figure 4. Time evolution of the radial averaged ($0.5 < r/a < 0.7$) signals (left) and time averaged spectra (right) for density, temperature, vorticity and nonzonal electrostatic potential for the initial $R/L_T \simeq 10.3$ simulations. Non converged time traces correspond to flatter spectra.

reflectometry data) rely on accurate description of the density fluctuation spectra. According to our results, the latter requires at least a factor of 10 more markers then what convergence studies based on potentials or fluxes suggest.

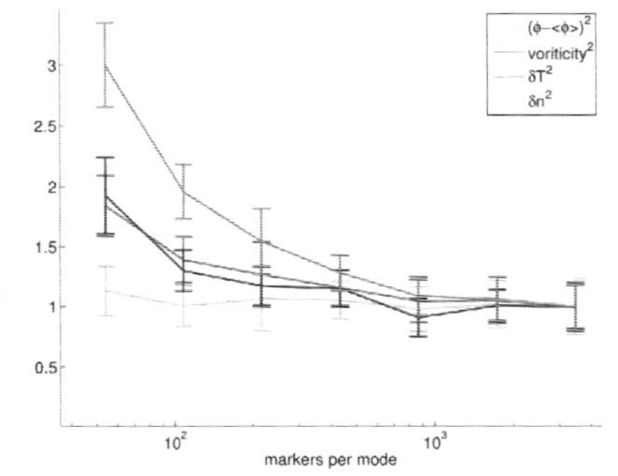

Figure 5. Radial and time averaged quantities of Fig.4, normalized to the 640M marker results, as a function of the number of markers per active mode.

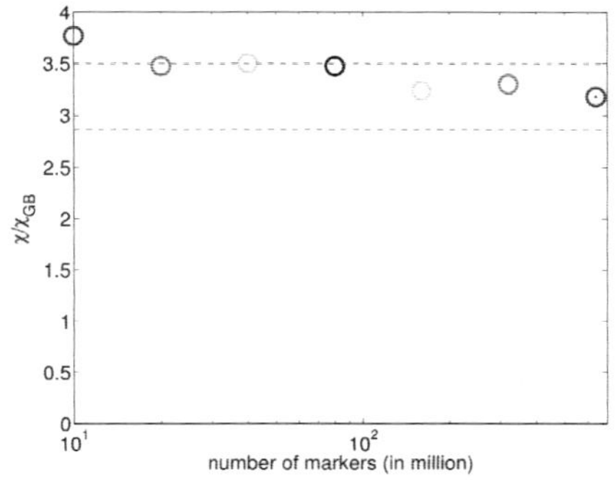

Figure 6. Radial and time averaged ion heat diffusivity for the initial $R/L_T \simeq 10.3$ simulations. Dashed lines indicate values of $\pm 10\%$ of the 640M markers simulation.

As a final test for the convergence in number of markers, we have reconstructed the total density, n, as the sum over markers of the contribution coming from the background Maxwellian, calculated at the marker position, plus the δf contribution given by the marker weights. By binning the contributions of all the markers, it is possible to construct a projection of the total density on a 3D grid. The spectrum of this quantity is shown in red in Fig.7.

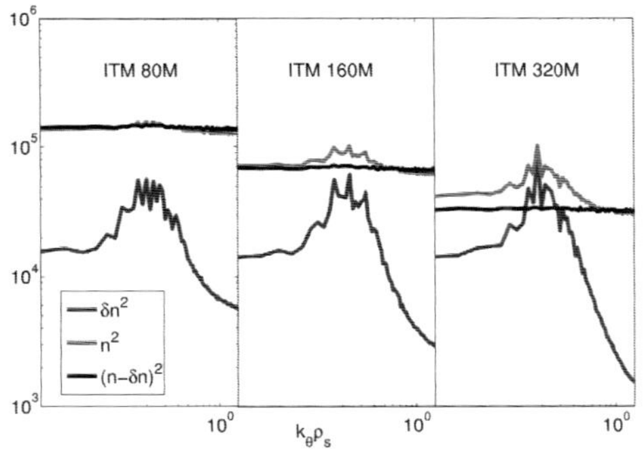

Figure 7. Spectrum of the total density n reconstructed by binning the marker weights (δn) and the background distribution calculated at the marker position (n_0), compared with the spectrum of the density fluctuation.

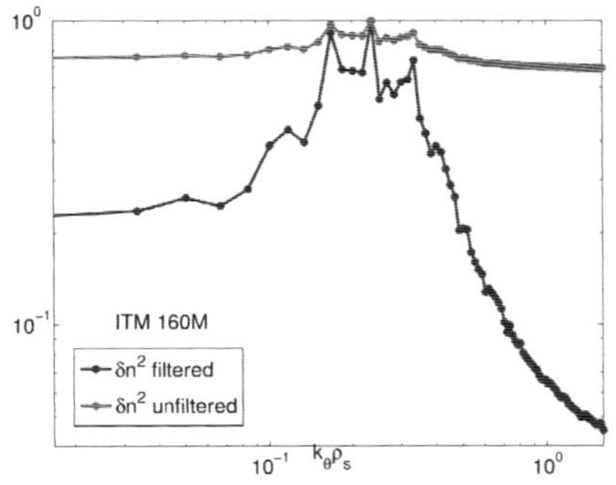

Figure 8. Spectrum of δn before and after having filtered the physically irrelevant modes (noise).

The spectrum of the total density is flat, with an amplitude 4 orders of magnitude smaller than the axisymmetric part, (not included in the figure). This level of "noise" is comparable with the amplitude of the density fluctuations described by the markers (in blue on Fig.7), even when 320M markers are used. Although this way of computing n is generally not used by a δf PIC code for computation, it is often used for diagnostics purposes. A better way of reconstructing the total density (and temperature) consists in using the analytical expression of the background distribution (not necessarily Maxwellian) in each bin and adding it to the δf contribution coming from the marker weights.

3.2. $\rho*$ scaling

The same convergence study has been performed for a set of simulations in which only the machine size has been varied, from small devices $\rho*^{-1} \simeq 46$ to ITER like machines $\rho*^{-1} \simeq 740$. Previous simulations [22] showed that the convergence in number of markers depends on the ratio of number of markers to the number of active modes included in the field equations. Note that in ORB5 the number of active modes does not correspond to the number of grid points, since in ORB5 field equations are solved on a non field aligned grid. All modes having $k_{\|}$ too large to contribute to the instability, are filter out from the simulation by applying a straight field line Fourier filtering on the right hand side of the field equations [6]. The same filtering procedure is applied to the 3D diagnostics. The effect of the Fourier filtering is shown in Fig.8. Filtering is required to get converged spectra since it reduces the amplitude of the noise (flat part of the unfiltered spectrum) of several orders of magnitude.

The $\rho*$ scaling shows that the convergence in number of particles per mode is verified for all the values of $\rho*$. However, the total number of markers also plays a role, due to the intrinsic Monte-Carlo nature of the PIC discretization. Since the number of active modes scale with $(1/\rho*)^2$, smaller $\rho*$ require more markers to keep the ratio markers over number of modes constant. On

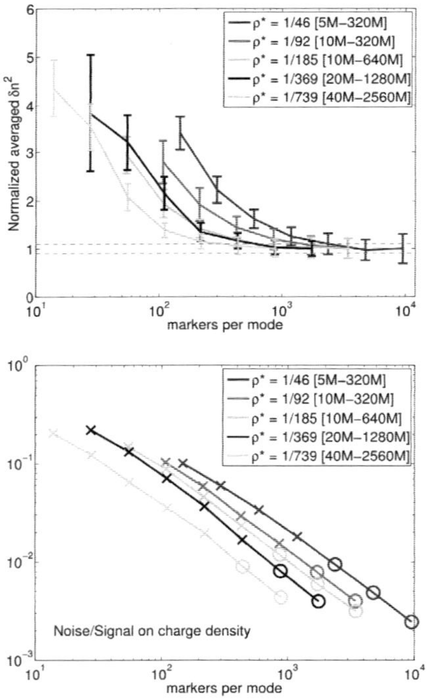

Figure 9. Top: convergence of δn in number of markers per mode for different $\rho*$ values. Bottom: noise/signal ratio, defined as the ratio of the average background noise to the average signal contained in the active modes.

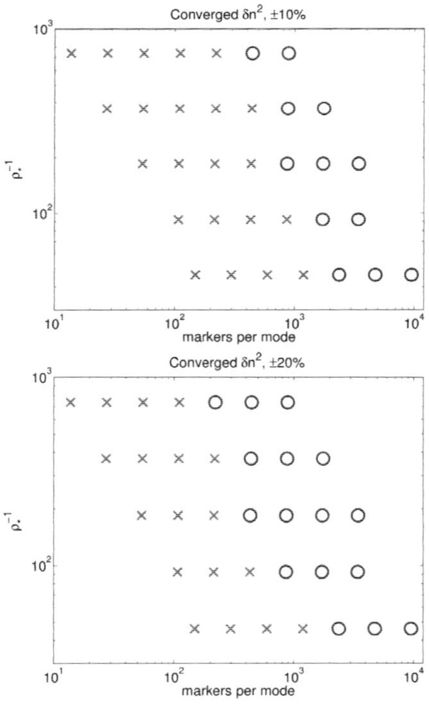

Figure 10. Different convergence criteria: results are considered converged when δn is within $\pm 10\%$ (top) or $\pm 20\%$ (bottom) of the value of the best resolved simulation.

the other hand, more markers also imply a better discretization of the entire phase-space. In other words, the scaling in number of particles per mode holds, but the convergence depends on the machine size: the larger is the system the less particles per mode are needed for convergence. This is shown in Fig.9, top, where δn, averaged over time and radius, is plotted for different values of $\rho*$. For each $\rho*$ a number of simulations with different number of markers has been performed. As a result, ITER size plasma ($\rho* \simeq 740$) require at least a factor of 2 less markers per mode than present day tokamaks ($\rho* \simeq 185$). Therefore, the number of markers, and consequently the computational time does not scale with $(1/\rho*)^2$, as it was predicted by dimensional arguments in the original analysis [23], but because of the statistical process of the PIC method it scales even better than this [24], making ITER size PIC simulations computationally affordable. The two graphs of Fig.9 show that this result holds even when different criteria for defining convergence are applied. Another remarkable result is that all the simulations seem to converge when the signal to noise ratio is around 1%, independent of $\rho*$ (Fig.9 bottom). Here the signal to noise ratio (s2n) is defined as the inverse of the ratio between the average signal contained in the modes actively contributing to the instability and the average amplitude of the modes having $k_{||}$ too large to contribute to the instability.

Figure 11 shows radial (left) and time (right) averages of the ion heat diffusion coefficients. The heat diffusivities χ/χ_{GB} are relatively small at large $\rho*$ while they converge, as expected, toward similar values and profiles (at least for $s > 0.4$) for sufficiently small $\rho*$. Figure 11 (right) also shows the time averaged R/L_T profile. The heating operator prevented profile relaxation at small $\rho*$. However, a significant relaxation is still present for simulations with larger $\rho*$, this suggest that the chosen value for the relaxation coefficient, γ_H, was probably not adequate for those cases.

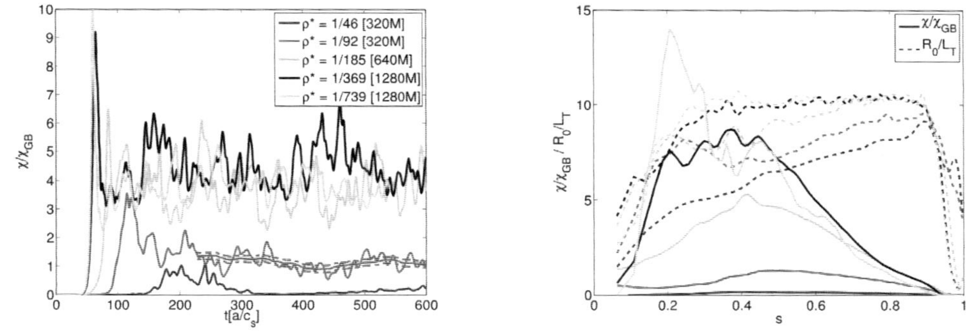

Figure 11. Spatial (left) and time (right) averages of the heat diffusion coefficients for different values of $\rho*$. The heat source prevented temperature gradient relaxation only for small $\rho*$, steady state values are influenced by the different R/L_T.

4. Conclusions

The convergence properties of global electrostatic particle-in-cell gyrokinetic simulations have been studied for the Cyclone base case by means of a set of 3D diagnostics originally introduced for visualization purposes. The new diagnostics allowed for high resolutions 3D measurements of electromagnetic potentials, density, temperature and vorticity as well as turbulence spectral analysis. For all the quantities considered, non converged simulations show flat spectra at high $k_\theta \rho_s$. However, not all the physically relevant quantities converge with the same rate: density fluctuation spectra can still show an unphysical flattening, even when electrostatic potential and heat fluxes are converged. The main results of the $\rho*$ scaling is that the resolution in terms of number of markers, required to correctly describe the turbulence density fluctuations, is found to be size-dependent: larger systems (ITER like) require significantly less numerical particles per active mode than smaller systems. Therefore, the number of markers, and consequently the computational time does not scale with $(1/\rho*)^2$, as it was predicted by dimensional arguments, but even better, making ITER size PIC simulations computationally affordable.

Acknowledgements: simulations were performed on the EFDA HPC-FF (project EMCONS) and IFERC-CSC (project EMFAST) supercomputers

[1] Lee W W 1983 *Physics of Fluids* **26** 556
[2] Parker S E and Lee W W 1993 *Physics of Fluids B* **5** 77
[3] Dimits A M and Lee W W 1993 *Journal of Computational Physics* **107** 309
[4] Aydemir A Y 1994 *Physics of Plasmas* **1** 822
[5] Hatzky R, Könies A and Mishchenko A 2007 *Journal of Computational Physics* **225** 568
[6] Jolliet S, Bottino A, Angelino P *et al.* 2007 *Computer Physics Communications* **177** 409
[7] Hahm T S, Lee W W and Brizard A 1988 *Physics of Fluids* **31** 1940
[8] Hahm T S 1988 *Physics of Fluids* **31** 2670
[9] Brizard A 1989 *Journal of Plasma Physics* **41** 541
[10] Sugama H 2000 *Physics of Plasmas* **7** 466
[11] Bottino A, Vernay T, Scott B *et al.* 2011 *Plasma Physics and Controlled Fusion* **53** 124027
[12] Vernay T, Brunner S, Villard L, McMillan B F, Jolliet S, Tran T M, Bottino A and Graves J P 2010 *Physics of Plasmas* **17** 122301
[13] Vernay T, Brunner S, Villard L, McMillan B F, Jolliet S, Tran T M and Bottino A 2012 *Physics of Plasmas* **19** 042301
[14] McMillan B F, Jolliet S, Tran T M, Villard L, Bottino A and Angelino P 2008 *Physics of Plasmas* **15** 052308
[15] Brunner S *et al.* 1999 *Physics of Plasmas* **6** 4504
[16] Chen Y and Parker S E 2007 *Physics of Plasmas* **14** 082301
[17] Scott B and Smirnov J 2010 *Physics of Plasmas* **17** 112302
[18] Scott B 2010 *Physics of Plasmas* **17** 102306 (*Preprint* `arXiv:0710.4899`)
[19] Scott B, Kendl A and Ribeiro T 2010 *Contributions to Plasma Physics* **50** 228
[20] Falchetto G L, Scott B D, Angelino P, Bottino A, Dannert T, Grandgirard V, Janhunen S J, Jenko F, Jolliet S, Kendl A, McMillan B F, Naulin V, Nielsen A H, Ottaviani M, Peeters A G, Pueschel M J, Reiser D, Ribeiro T T and Romanelli M 2008 *Plasma Physics and Controlled Fusion* **50** 124015
[21] Dimits A M and et al 2000 *Physics of Plasmas* **7** 969
[22] Bottino A, Peeters A G, Hatzky R, Jolliet S, McMillan B F, Tran T M and Villard L 2007 *Physics of Plasmas* **14** 010701
[23] Lee W W 1987 *Journal of Computational Physics* **72** 243
[24] Lee W W 1992 Private communication with Scott B at the 14th IAEA Conference on Plasma Physics and Controlled Fusion, Wuerzburg, Germany

Journal of Physics: Conference Series **401** (2012) 012026 doi:10.1088/1742-6596/401/1/012026

Gyrokinetic simulations of collisionless tearing modes

O Zacharias, R Kleiber
Max Planck Institute for Plasma Physics, EURATOM Association, Wendelsteinstr. 1, 17491 Greifswald, Germany

E-mail: oliver.zacharias@ipp.mpg.de

R Hatzky
Max Planck Institute for Plasma Physics, EURATOM Association, Boltzmannstr. 2, 85748 Garching, Germany

Abstract. We investigate the effects of temperature and density gradients on collisionless magnetic reconnection by means of PIC-δf simulations with standard gyrokinetic equations using gyrokinetic ions and drift kinetic electrons. In a slab model ion and electron temperature gradients act stabilising on the tearing mode instability in the cases that have been considered. An additional ion and electron density gradient completely stabilises the mode below a critical η (ratio of temperature gradient to density gradient). The critical η becomes larger with decreasing equilibrium current.
Nonlinear single-mode simulations show that the island width saturates at the level predicted by [1], if the linear instability growth rate is sufficiently small. Including ion and electron temperature and density gradients in nonlinear calculations the saturated island width vanishes below a critical η.

1. Introduction

Magnetic reconnection is an important subject in astrophysics and magnetic confinement physics [2]. For instance solar flares are thought to be caused by reconnection events strongly accelerating charged particles [3]. The complicated structure of the Sun's surface can be simplified with locally antiparallel magnetic field line structures, which are unstable with respect to tearing modes. In these high-temperature plasmas resistivity can play a significant role, as described e. g. by the resistive MHD [2, 4]. However magnetic reconnection in Earth's magnetotail could not be explained this way since the conductivity in Ohms law almost vanishes completely in this collisionless plasma. Additional inertia terms could solve the problem of triggering reconnection [4, 5]. It was also suggested that nongyrotropic effects could influence reconnection [6].
Here we investigate laboratory aspects of magnetic reconnection, where a strong magnetic guiding field and a small sheared magnetic field in the reconnection plane are present. We can employ standard gyrokinetics assuming that the Larmor radius is much smaller than the equilibrium and gradient scales and the growth rates are much smaller than the gyrofrequencies. Up to date there are only very few particle-in-cell (PIC) simulations of tearing modes which include density and temperature gradients [7]. Although previous gyrofluid approaches show that a complete stabilisation must occur [8], a comparison with PIC results was not available.

Another aspect are nonlinear simulations of tearing modes. There are numerical results of saturated island widths resulting from kinetic theories [9], but they have not been compared with other PIC or gyrofluid codes in an extended parameter space. Kinetic nonlinear tearing theories [1] predict saturated one-mode behaviour and saturated island widths depending on the shear length l_s and the perpendicular wave vector k. Here we compare our PIC simulations with this theory for saturated island widths, which could be important for toroidal systems in future . Moreover, there are only very few simulations using nonlinear gyrokinetics that include temperature and density gradients [7]. It has been suggested that gradients should influence the saturated island width due to diamagnetic effects [8]. This is generally of importance since for most confined plasmas a sheared magnetic field is accompanied by gradients due to the equilibrium pressure condition.

2. The model

We use the PIC-δf code EUTERPE, where the distribution function f_s for each species s (i: ions, e: electrons) is split into a background distribution function $f_{0,s}$ and a perturbed part δf_s. Here the background is assumed to be a Maxwellian with a bulk velocity $u_{0,e}$ for the electrons in the rest frame of the ions ($u_{0,i} = 0$). EUTERPE solves the gyrokinetic electromagnetic Vlasov-Maxwell-system globally in toroidal 3D equilibria. Here the tearing mode has been simulated in a slab geometry. The guiding field $B_{z,0}$ is along the periodic z-direction. The y-direction is also treated periodically, whereas the equilibrium quantities only depend on the spatial x-coordinate ranging from $0 \ldots L_x$. In the x-direction Dirichlet boundary conditions are employed for the vector potential $A_\|$ and the electrostatic field Φ. We restrict our studies to two spatial dimensions by setting $\partial/\partial z = 0$. The local approach for the gradients of the temperature and densities has been used, but not for the current i. e. , the gradients are constant over the whole simulation domain as well as the temperature and density itself. We work in normalised units, i. e. $r_* = \sqrt{T_* m_p}/e B_*$, $v_* = \sqrt{T_*/m_p}$, $\Omega_* = v_*/r_*$ with B_* the guiding field $B_{z,0}$ and T_* the normalising electron temperature $T_{0,e}$. Charge and mass are normalised to the proton charge $q_i = e$ and mass m_p. Under these assumptions the gyrokinetic equations in the $p_\|$-formulation simplify to:

- Trajectories: ($\Psi = \Phi - v_\| A_\|$)

$$\dot{\vec{R}}_s = v_\|\vec{b} - \frac{q_s}{m_s}\vec{b}\langle A_\|\rangle + \frac{1}{B}\vec{b}\times\nabla\langle\Psi\rangle = v_\|\vec{b} + \dot{\vec{R}}^1 \quad (1)$$

$$\dot{v}_{\|,s} = -\frac{q_s}{m_s}\vec{b}\cdot\nabla\langle\Psi\rangle \quad (2)$$

- δf equation: $\dot{\delta f}_s = -f_{0,s}S_s$, $\kappa_s = \kappa_{n,s} - \left(\frac{3}{2} - \frac{v^2}{2v_{th,s}^2}\right)\kappa_{T,s} + \kappa_{u0,s}$

$$S_s = \kappa_s \dot{\vec{R}}^1\cdot\nabla x + \frac{q_s}{m_s}\frac{1}{v_{\mathrm{th},s}^2}\left(-\left(v_\| - u_{0,s}\right)\vec{b}\cdot\nabla\langle\Psi\rangle\right) \quad (3)$$

$$(4)$$

- Quasineutrality (long wave length limit):

$$\sum_{s=i,e} q_s n_s = 0 \quad (5)$$

for ions and drift kinetic electrons

$$n_i = \langle n_i\rangle + \frac{m_i}{q_i}\nabla\cdot\left(\frac{n_{0,i}}{B^2}\nabla_\perp\Phi\right) \quad (6)$$

$$(7)$$

- Ampère's law:

$$-\frac{1}{\beta}\nabla_\perp^2 A_\parallel + \sum_{s=i,e} n_{0,s}\frac{q_s^2}{m_s}A_\parallel = \sum_{s=i,e}\langle j_{\parallel,s}\rangle \tag{8}$$

with the gyroaveraging procedure $\langle\Phi\rangle = \frac{1}{2\pi}\int\Phi(\vec{R}+\vec{\rho})\,d\alpha$ and the gradients for temperature, density and current

$$\kappa_{T,s} = -\frac{T'_{0,s}}{T_{0,s}}, \quad \kappa_{n,s} = -\frac{n'_{0,s}}{n_{0,s}}, \quad \kappa_{u0,s} = \left(v_\parallel - u_{0,s}\right)\frac{u'_{0,s}}{v_{th,s}}. \tag{9}$$

B is the modulus of the equilibrium magnetic field, $v_{th,s}$ thermal speed of each species, v the modulus of the velocity and a prime denotes d/dx.
Here we only are interested in configurations with $\beta \ll 1$; thus only the parallel vector potential $A_\parallel$ in Ampères law is used. Because of the parallel momentum formulation, a skin term in Ampère's law appears. This formulation has the advantage that no time derivative of the vector potential enters the equations but necessitates the use of an adaptive control variate scheme [10].

3. Simulating the tearing mode

The magnetic field in the y-direction depends only on the x-coordinate:

$$B_{y,0}(x) = \frac{\beta}{2}\sqrt{\mu}\,C_1\,a\,\sqrt{\pi}\,\mathrm{erf}\left(\frac{x-\frac{L_x}{2}}{a}\right) \tag{10}$$

with the resonant surface at $L_x/2$. a is the half width of $B_{y,0}(x)$, C_1 is a free parameter which controls the strength of $B_{y,0}(x)$, and $\mu = 1836$ is the proton to electron mass ratio.
First EUTERPE has to be benchmarked for the tearing mode. Therefore we focus first on the simplest model of tearing, where we neglet electrostatic potential fluctuations Φ, and the ions only serve as a neutralising fixed background. The electrons are the only kinetic species (electron tearing mode). Later we include gyrokinetic ions as well together with the electrostatic potential Φ. If not otherwise stated, this will be the standard case (tearing mode).
To get the linear dispersion relation of the electron tearing mode, one applies the Fourier transform $A_\parallel(x,y,t) = e^{i(ky-\omega t)}A_\parallel(x)$. Then the problem of finding the dispersion relation $\omega(k)$ with k as the wave vector in y-direction, reduces to an eigenvalue problem:

$$\frac{d^2 A_\parallel}{dx^2} = -q(x,\omega)\,A_\parallel \tag{11}$$

$$\begin{aligned} q(x,\omega) &= -k^2 + \frac{\beta}{k_\parallel}\left(u_{0,e}\left[\omega + \frac{k}{\mu}u'_{0,e}u_{0,e}\right]\langle V^1\rangle\right. \\ &+ \left.\left[\omega + 2\frac{k}{\mu}u'_{0,e}u_{0,e}\right]\langle V^2\rangle + \frac{k}{\mu}u'_{0,e}\langle V^3\rangle\right). \end{aligned} \tag{12}$$

Here $k_\parallel(x) = k\,B_{y,0}/B_{z,0}$ and $\langle V^n\rangle(x,\omega)$ are the n-th moments of the plasma dispersion function :

$$\langle V^n\rangle(x,\omega) = \frac{1}{\sqrt{2\pi\mu}}\int_{-\infty}^{\infty} ds\frac{s^n e^{-\frac{s^2}{2\mu}}}{s + u_{0,e}(x) - \frac{\omega}{k_\parallel}}. \tag{13}$$

This eigenvalue problem is solved numerically using a shooting method. The shooting method reformulates the problem as a Riccati problem and integrates from both sides to a certain matching point. If the electrostatic potential is included together with drift kinetic ions the eigenvalue problem is of fourth order and can also be solved with the Riccati method.
The densities $n_{0,i} = n_{0,e}$ and temperatures $T_{0,i} = T_{0,e}$ are first assumed to be constant for the case of the electron tearing mode. The simulation domain in the perpendicular plane is set to $L_x \times L_y = 10 \times 10$ throughout the paper if not otherwise stated. When using noise as the initial condition for the fields, the instability evolves within about $t = 10$. The mode develops a very thin current layer of width smaller than the ion Larmor radius at the resonant surface $L_x/2$ in which magnetic energy is converted into thermal energy of the particles. Therefore the mode structure is localised around $L_x/2$. The equilibrium scale L_x is much larger than the current channel and therefore the boundary conditions in the x-direction do not play an important role. Figure (1) (left, green) shows the solution of the eigenvalue problem for the electron tearing mode with $\beta = 10^{-3}$ and γ the imaginary part of the eigenvalue ω. Varying the slab parameter a and fixing the wave vector k, the range of instability of the mode is depicted. The agreement between the two methods is better than two percent.
The more general case including gyrokinetic ions and the electrostatic potential Φ (the tearing mode case) is shown in red in the left panel of Fig. (1). The dispersion relations $\omega(k)$ show a

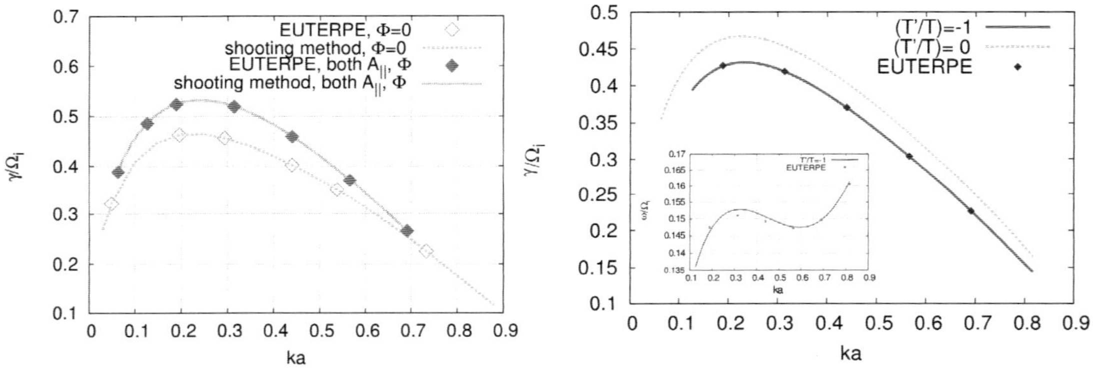

Figure 1. Benchmarks of EUTERPE against shooting method match very well. Left: electron tearing and tearing mode including the electrostatic potential. Right: electron tearing mode with temperature gradient. The insert shows the real frequency of the mode.

pure growing mode without real frequency. It turns out that here finite Larmor radius (FLR) effects have no influence, but the additional electrostatic potential slightly destabilises the mode. In our units, the Larmor radius is $\rho_i = 1$, while the electron skin depth $k_{0,e}^{-1} = 1/\sqrt{\beta\mu} \sim 0.7$. So the skin depth is somewhat smaller than the gyroradius ρ_i and therefore the influence of the electrostatic potential Φ is expected to be small [9]. It also turns out that this small difference of the growth rate between including and neglecting the electrostatic potential Φ vanishes as C_1 goes to zero. Since $C_1 = -1$ is a rather large value so that the bulk velocity $u_{0,e}$ is in the range of $v_{th,e}$, this parameter set gives rather benchmarks than realistic scenarios. Furthermore because the growth rate of the instability is proportional to $v_{th,e}$, the runs do not need much computing time. For these benchmarks we have used about 5×10^6 markers for each kinetic species. Using 8 grid points in the y-direction, 256 points in the x-direction and the time step $\Delta t = 0.1$ these runs need about 20 CPUh.

4. Influence of density and temperature gradients

A natural generalisation, which is more realistic, includes gradients of temperature and density for each kinetic species, i. e. diamagnetic effects. In this case the electron tearing instability acquires a drift frequency, which also has been benchmarked together with the growth rate and is shown in Fig. (1) (right) for $\kappa_{T,e} = 1$, $\kappa_{n,e} = 0$ and $\beta = 10^{-3}$. The eigenvalue problem eq. (11) can also be extended to include a temperature gradient $\kappa_{T,e}$ in the electron tearing mode case. The benchmark shows that the mode becomes slightly more stable if a temperature gradient is applied; the inserted figure shows the real part of the eigenvalue. Again the solution of the eigenvalue problem and the EUTERPE results agree very well.
If both temperature and density gradients are present in the tearing mode case, the mode can be completely stabilised. Figure (2) (left) shows the growth rate obtained with EUTERPE changing the density gradient $\kappa_{n,s}$ while fixing the temperature gradient at $\kappa_{T,s} = -0.05$, $\beta = 10^{-3}$ and $a = 0.5$. Here $\eta_s = \kappa_{T,s}/\kappa_{n,s}$ and $\eta_i = \eta_e = \eta$.

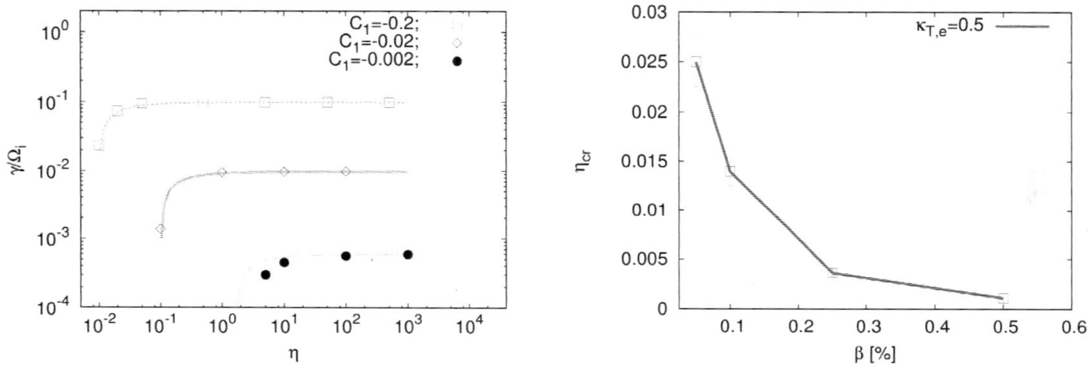

Figure 2. Critical behaviour of the tearing mode. Left: complete stabilisation below a critical η (points EUTERPE results, curves from eq. (14)) Right: Critical η depending on β for electron tearing mode.

For various C_1 in Fig. (2) (left) one oberves the same critical behaviour: below a certain η_{cr} the mode is completely stabilised. If C_1 is descreased η reduces to somewhat more realistic values in the range of $\eta \sim 1$. In this η-range the simulation time increases rapidly since the growth rates are very small. Especially for $C_1 = -0.002$ one needs at least 2×10^4 CPUh for each point. This is the first time such calculations have been performed within the PIC framework using standard gyrokinetics.
Previous works have pointed out that diamagnetic effects change the growth rates completely [8, 11]. There the linear gyrofluid growth rate has been estimated as

$$\gamma^2 = \gamma_0{}^2 - \beta \left[\frac{k\,\kappa_{T,e}}{2\eta} (1+\tau) \right]^2 \tag{14}$$

This result has been applied for the runs in Fig. (2) (left) with γ_0 as the growth rate with $\kappa_{T,s} = 0$, $\kappa_{n,s} = 0$ and $\tau = T_{0,i}/T_{0,e}$. The simulation results aggree very well with the predictions. As mentioned in [8], the analytical results are valid when $\rho_i \gg k_{0,e}^{-1}$. Here only $\rho_i > k_{0,e}^{-1}$ holds, but since the growth rate only depends weakly on τ one could increase τ and thereby ρ_i without changing γ_0 very much. Then this case here corresponds approximately to the large ρ_i-limit. As can be seen in the left panel of Fig. (2), C_1 acts as a free parameter, which measures just the strength of the $B_{y,0}$-field. It does not introduce new physical effects and only shifts η_{cr}. In Fig. (2) (left) η has low values down to $\sim 10^{-2}$, but the characteristic behaviour is the same

for all C_1.
Another interesting point concerns the β-dependence of the critical η. A scan is shown in Fig. (2) (right) for the electron tearing mode for $C_1 = -1$ and $a = 0.5$. Figure (2) (right) shows that η_{cr} decreases with higher β-values. Since the thermal velocity and density stay the same, increasing β means here that the guiding field changes its values towards lower strengths. So the ratio between $B_{y,0}$ and guiding field strength becomes larger. This corresponds to the case when fixing β and increasing C_1, which was shown in Fig. (2) (left). This means that for realistic β values and temperature and density gradients the tearing mode in a slab is always unstable.
Our results show that the kinetic tearing theory in [12] for linear electron tearing mode cannot adequately account for the influence of the density and temperature gradients on the growth rate and drift frequency. It predicts, for instance, that the growth rate is not affected by diamagnetic effects $\gamma \approx \gamma_0$.

5. Nonlinear simulations of the tearing mode

This section describes for nonlinear one-mode simulations of the collisionless tearing mode. The mode of interest is selected by applying a Fourier filter. We simulate one-mode runs to compare the saturated island widths w with nonlinear kinetic theories for collisionless electron tearing modes [1].
After the linear stage of the simulation the mode enters the nonlinear regime and saturates when the island width $w(t)$ approaches the width of the perturbed current layer. In the fluid picture third-order force terms become important, which stop the plasma inflow towards the current layer and reconnection is inhibited [13]. Then the field energy of the mode oscillates about a mean value with a characteristic frequency ω_B. According to kinetic theory electrons bounce within the islands with this bounce frequency ω_B and exchange energy with the wave and vice versa.
The saturated island width w has been predicted for the electron tearing mode for a single k [1]. In these calculations the constant Ψ-approximation has been used. To compare with this theory, we have used the perturbed vector potential $A_{\|}$ and the equilibrium vector potential $A_{\|,0}$ producing $B_{y,0}$, to calculate the island width geometrically at each time solving the equation:

$$A_{\|}(x = 0, y = 0, t) = A_{\|,0}(x = w) + A_{\|}(x = w, y = \frac{\pi}{k}, t) \tag{15}$$

and in the constant Ψ-approximation:

$$w = 2\sqrt{A_{\|}\, l_s} \tag{16}$$

which only needs the time-averaged perturbed vector potential at $L_x/2$ for $t \gg \gamma^{-1}$. l_s is the shear length defined by $l_s = B_{z,0}/B'_{y,0}$ at $L_x/2$. It is important to note, that the energy conservation of the runs is fulfilled better than a few percent. It turns out that the constant Ψ-approximation is valid since the relative deviation from the results of eq. (15) and (16) is less than two percent.
In Fig. (3) (left) w is shown as a function of the growth rate γ of the electron tearing mode with given wave vector k. The growth rate was increased by decreasing a starting with $a = 0.7$. This is similar to Fig. (1) (left) when increasing the small γ around $ka \sim 0.8$ to the maximum of the dispersion relation by decreasing a. The point at $\gamma = 0.5$ corresponds to a very small value ka left to the maximum of the dispersion relation. The simulation domain is $L_x \times L_y = 2.5 \times 2\pi$.

The points represent the saturated mean island width w and show a shows linear dependence on the growth rate if γ is small enough. These values correspond to large a: the $B_{y,0}$ field approaches a slab configuration in which the bulk velocity $u_{0,e}$ is a constant over the whole simulation domain. In this case $B_{y,0} \sim x/l_s$ resulting in the prediction:

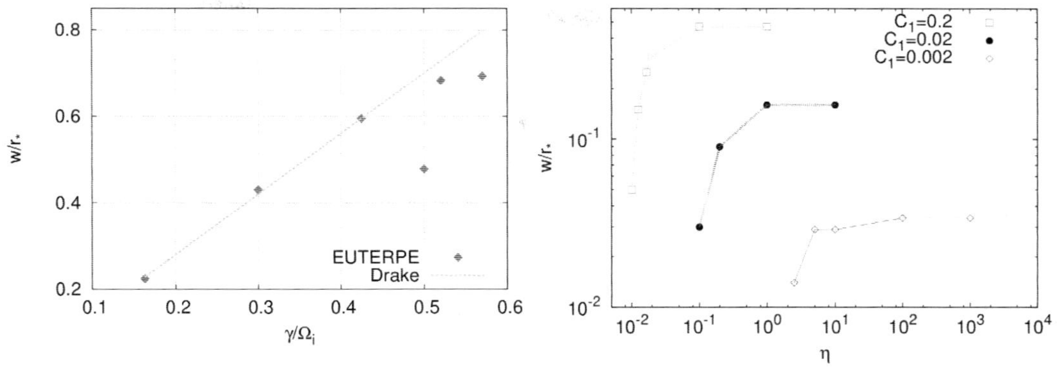

Figure 3. Left: theoretical prediction of saturated island width is confirmed for low growth rates. Comparison in between kinetic theory [9, 1] and EUTERPE. Right: saturated island widths w depending on η. Below a finite threshold w vanishes.

$$w = \frac{\gamma\, l_s}{\sqrt{\mu}\, k\, G} \tag{17}$$

with $G \sim 0.41$ (straight line in Fig. (3) (left)) [1]. This analytical result contains only information about the mode structure and mode dynamics around the resonant surface $L_x/2$ and not explicitly the boundary conditions. A variation of the simulation domain L_x would influence the growth rate and therefore the saturated island width. The growth rate already contains the outer behaviour and boundary conditions of $A_\parallel$. So the boundary conditions do not appear explicitly in the analytical expression of the island width. For larger growth rates γ the magnetic configuration deviates from the $B_{y,0} \sim x\,/\,l_s$ behaviour and the saturated island widths deviate from this theory. Additionally the point $w = 0.597$ was benchmarked with the one result given in [9] as well as the scaled field energy of the mode.
The next step of generalisation is to include gradients in the full tearing mode framework. Both electrons and gyrokinetic ions enter the simulation and the temperature gradient $\kappa_{T,e} = \kappa_{T,i} = -0.05$ was fixed. Then, varying the density gradient $\kappa_{n,e} = \kappa_{n,i}$ the saturated island widths are shown in Fig. (3) (right). The island width of the most unstable mode with $a = 0.5$ and $\beta = 10^{-3}$ has been calculated using the constant-Ψ approximation according to eq. (16). Since the linear growth rate drops to zero when $\eta \to 0$ the island width has to vanish. Also the qualitative behaviour of the growth rate at finite $\eta > \eta_{cr}$ is the same as for the island width.

6. Conclusions

First EUTERPE was benchmarked against a shooting method for electromagnetic tearing mode simulations. In case of the electron tearing mode the results of EUTERPE agree better then two percent with results from the shooting method. Calculations with and without temperature gradient also in the tearing mode case with drift kinetic ions the shooting method deliver the same results as EUTERPE.
Introducing both density and temperature gradients in linear simulations a critical behaviour regarding η has been observed. Analytical predictions of the growth rate have been shown to be in a good aggreement [8]. The kinetic theory of [12] is not able to predict this behaviour.
Nonlinear one-mode simulations have been carried out and have been benchmarked to the only published result from a PIC-δf code given in [9]. These simulations were extended to give a comparison between the predicted saturated island width [1] and our numerical results. For low

growth rates the predictions can be recovered.
Temperature and density gradients have also been applied to the nonlinear evolution of the magnetic island. It was shown that reconnection is also inhibited in the nonlinear regime when η is small enough. A parameter scan of C_1 has shown the dependence on the strength of the equilibrium current. If the bulk velocity $u_{0,e}$ is much smaller than the thermal speed the critical η is in the range of unity. It would be interesting to find out analytically whether the linear connection $w \sim \gamma$ also holds in the case when gradients are present.
A rich physics needs to be explored in the nonlinear regime, e. g. study the behaviour of higher k modes and increasing the filter size. Up to now, all nonlinear simulations assume that there is almost no coupling between the Fourier modes with different k. This issue of mode coupling has not been simulated yet and further simulations are currently under progress.

[1] Drake J F and Lee Y C 1977 *Phys. Rev. Lett.* **39** 453
[2] Priest E and Forbes T 2000 *Magnetic Reconnection* (Cambridge University Press)
[3] Yamada M, Kulsrud R and Ji H 2010 *Review of Modern Physics* **82** 603
[4] Vasyliunas V M 1975 *Review Geophysics* **13** 303
[5] Fitzpatrick R 2010 *Phys. Plasmas* **17** 042101
[6] Hoshino M, Mukai T, Nishida A, Kokubun S and Yamamoto T 2000 *Advanced Space Research* **26** 425
[7] Sydora R D 2001 *Phys. Plasmas* **8** 1929
[8] Tassi E, Waelbroeck F L and Grasso D 2010 *Journal of Physics Conference Series* **260** 012010
[9] Wan W, Chen Y and Parker S E 2005 *Phys. Plasmas* **12** 012311
[10] Hatzky R, Koenies A and Mishchenko A 2007 *J. Comp. Phys.* **225** 568
[11] Porcelli F 1991 *Phys. Rev. Lett.* **66** 425
[12] Drake J F and Lee Y C 1977 *Phys. Fluids* **20** 1341
[13] Rutherford P H 1973 *Phys. Fluids* **16** 1903

AUTHOR INDEX

Institute of Physics
Dirac House, Temple Back
Bristol BS1 6BE UK

CURRAN ASSOCIATES INC
proceedings.com

AIChE Spring Meeting 2005

DOE/ITP Funding of Novel Technologies 2005

April 10-14, 2005

Atlanta, Georgia, USA